A DIRTY HISTORY OF PHOTOGRAPHY

A DIRTY HISTORY *of* PHOTOGRAPHY

CHEMISTRY, FOG, AND EMPIRE

Michelle Henning

THE UNIVERSITY OF CHICAGO PRESS
CHICAGO AND LONDON

The University of Chicago Press, Chicago 60637
The University of Chicago Press, Ltd., London

Published 2025
Printed in the United States of America

34 33 32 31 30 29 28 27 26 25   1 2 3 4 5

ISBN-13: 978-0-226-84067-3 (cloth)
ISBN-13: 978-0-226-84068-0 (ebook)
DOI: https://doi.org/10.7208/chicago/9780226840680.001.0001

Library of Congress Cataloging-in-Publication Data

Names: Henning, Michelle, author
Title: A dirty history of photography : chemistry, fog, and empire / Michelle Henning.
Description: Chicago : The University of Chicago Press, 2025. | Includes bibliographical references and index.
Identifiers: LCCN 2025023085 | ISBN 9780226840673 (cloth) | ISBN 9780226840680 (ebook)
Subjects: LCSH: Photographic chemicals industry—Technological innovations. | Photographic chemicals industry—Environmental aspects. | Photographic industry—Technological innovations. | Photographic chemicals—Environmental aspects.
Classification: LCC HD9708.5.C442 H46 2025 | DDC 338.7/61771—dc23/eng/20250804
LC record available at https://lccn.loc.gov/2025023085

♾ This paper meets the requirements of ANSI/NISO Z39.48-1992 (Permanence of Paper).

Authorized Representative for EU General Product Safety Regulation (GPSR) queries: **Easy Access System Europe**—Mustamäe tee 50, 10621 Tallinn, Estonia, gpsr.requests@easproject.com
Any other queries: https://press.uchicago.edu/press/contact.html

FOR JOHN

It was conceived in a storm and in darkness, born within sight of coal mines, its sibling an internal combustion engine. It was made from the exports of the Levant, from the chemicals and metals sold at London's Apothecaries' Hall, in the paper mills that crushed metal buttons with the cotton. Long before its skin was formed of the distilled by-products of coal gas and the remains of cattle, it knew explosions and steam, shipyards and tanneries. Its companions were consumptives, diabetics, and addicts, drawn to water cures, high on laughing gas. It passed through the hands of entrepreneurial aristocrats, scientists and romantics, radicals and dissenters, before it was tamed and put to work in a factory.

CONTENTS

COLOR PLATES FOLLOW PAGE 246.

01 THE CREATURE

► There is a cave, not far from here, in which a giant creature lurks. It takes refuge in the absolute darkness there, unable to survive the tiniest chink of light. It can breathe only the perfect, purified air of the cave. It is wet and only just beginning to form a skin, and so it cannot bear to be touched: no human hands have ever stroked it. It shrinks from most metals, from liquids, from dust. The unvarying temperature of the cave suits it, for it cannot bear to be too hot or too cold. Deep in this cave, the creature is undergoing a metamorphosis. Soon it will be more resilient and will even be able to emerge into the world outside, though only at night. It will remain sensitive, until one day when it turns toward the light and is changed forever. One single exposure is all it takes. And the change will not be evident at once. It will be imperceptible until the creature has been plunged into a series of baths. Only after bathing three times will the metamorphosis be complete and the creature take on its new form.[1]

Every photograph was once a consequence of a metamorphosis that goes something like this, not creaturely but chemical. This is no longer the case, since most of our photographs are electronic, but it was true for about a century and a half. Even now, the industrial manufacture of photosensitive films and papers continues, though on a much smaller scale. It still requires huge buildings to house the coating beds in darkness, and the latter are made of specific metals and plastics that will not react with the emulsion. These enormous rooms are air-conditioned to ensure the perfect temperature and the purest atmosphere. The freshly coated photographic films and papers are today overseen using digital cameras and computer screens. During this stage, when they are still wet, and afterward to a lesser extent, photographic materials are sensitive not only to light but to atmosphere, humidity, dust, temperature, and the oils in people's hands.

In 2018, I visited the factory of Harman Technology Ltd., in the village of Mobberley, Cheshire, just outside Manchester in England. Harman Technology is the current manufacturer of Ilford brand film (trading as Ilford Photo), and its factory is one of the original factories of Ilford Limited, which took it over from the smaller Rajar company in the interwar period. I arrived there after several months researching in the archives of Ilford Limited, which was the largest British photographic materials manufacturer in the 1920s and '30s. I was studying this period between the two World Wars, when the company expanded, to learn how new photographic films and materials were shaping photographic practice. I had begun this work because I noticed that very little research has been done on Ilford Limited, even though it was Kodak's largest British competitor at that time. The brand continued to be strongly identified with monochrome photography around the world in the second half of the twentieth century, and, even now, where darkroom photography is still practiced, Ilford films and chemicals are widely used.

The factory visit changed the direction of my research, giving me a stronger awareness of how precise and controlled photographic manufacturing processes must be, and with what care and delicacy the materials are treated. Standing in a small control room with computer screens monitoring the enormous coating bed, I imagined this creature in the cave beyond and became curious about what it would mean to approach chemical photography as atmospherically and environmentally sensitive. I began to wonder what a history of photography that took atmosphere into account would look like, and how it might contribute to a rethinking of the chemical photography that once dominated how we saw our world. How might we understand photographic chemistry and photographic aesthetics in relation to the changing atmosphere? In the context of anthropogenic climate change resulting from the emission of greenhouse gases, it seems necessary to consider photography as a phenomenon of atmosphere, dependent on weather, pollution, and climate.

Envisaging the photographic surface that I could not see laid out in the huge room beyond, I was struck by how the reactivity of photochemistry made it seem alive to me. It seemed that the sensitivity of emulsions and films transgressed the boundaries we erect between the inanimate and the animate. Chemical photographic materiality appeared as a large and quasi-mythical beast, one which once dominated the world and is now endangered and rarely sighted. It is so easy to underestimate the size of this beast, to forget how much the world once depended on it, and to lose sight of its many qualities and characteristics.

To say that photographic materials are atmospherically sensitive is also to suggest that they can register the changes in the atmosphere introduced by industrialization and that have led to the current environmental crisis. While climate change and extinction events threaten us as societies and even as a species, it is photography and its related technological media (film and video) that have been decisive in shaping how we see our planet and what is happening to it. Photographs can document environmental degradation and they are atmospheric too in the way they can attach feelings to places. Meteorology and mood are heavily associated in daily experience as well as in visual and literary imagery. People talk of forgetting in terms of "brain fog," of moods as "black clouds," and associate rain with sadness and sunsets with beautiful melancholy. In photographs, phenomena such as fog, mist and rain convey feeling in ways that undercut any straightforward reading of the photograph as a document or record. But to speak of photography's atmospheric sensitivity is also to go beyond a consideration of what photographs picture or communicate, to pay attention to the vulnerability of the technology to pollution and environmental change. The visit to the factory taught me that photographic emulsions and chemicals can sense and respond to chemical changes in their surroundings that are invisible to eyes and to digital cameras.

The period in which the photographic industry started to fully take control of these other sensitivities, meaning both to exploit and suppress them, was the one I was researching, the interwar moment of the 1920s and '30s. This was a pivotal time in the transformation of photographic materials, with the introduction of new panchromatic emulsions, the advance of nuclear emulsions, infrared plates, and new color films. In the 1920s and '30s, a few international corporations dominated the photographic industries, and their trade in materials and hardware was growing rapidly with the popularization of snapshot photography across the British Empire.

Photography had become a global medium very quickly, extending across Europe, Africa, Asia, and the Americas from the 1840s.[2] The globalized trade in sensitized photographic materials, cameras, and accessories resulted from another trade headed in the reverse direction: a trade in papers, collodion, gelatin, nitrate, silver, and other minerals and chemicals. En route to Britain, this trade passed through European networks, through Belgium, Germany, and France, but much of it originated further afield, in the landscapes of colonized places and the labor of people subjugated by the various European empires.

The British photographic industry's trade was fueled by coal depots established across the empire, and heavily dependent on coal, petrochemicals, and the wider chemical industries. In Ilford, the town on the edge of London that was home to Ilford Limited's headquarters, the proximity of photographic to chemical industries was reflected in the physical closeness of the photography factory to the chemical works of Howard and Sons. The relationship of such industries to the changing skies is strangely prefigured in the fact that this firm was founded by Luke Howard, a manufacturing chemist famous for his taxonomy of the clouds. He named cirrus, cumulus, stratus, and nimbus, and is sometimes described as "the father of meteorology."[3]

Photographic manufacturing also had ties to the military and was one of the trades that flourished in wartime. The close connections between empire, the military, and industry in the interwar photographic industry were embodied by men like Major-General Sir Ivor Philipps, a veteran of colonial military campaigns in India, one-time managing director and chairman of the board at Ilford Limited, who oversaw the company's massive interwar expansion.[4] They were also

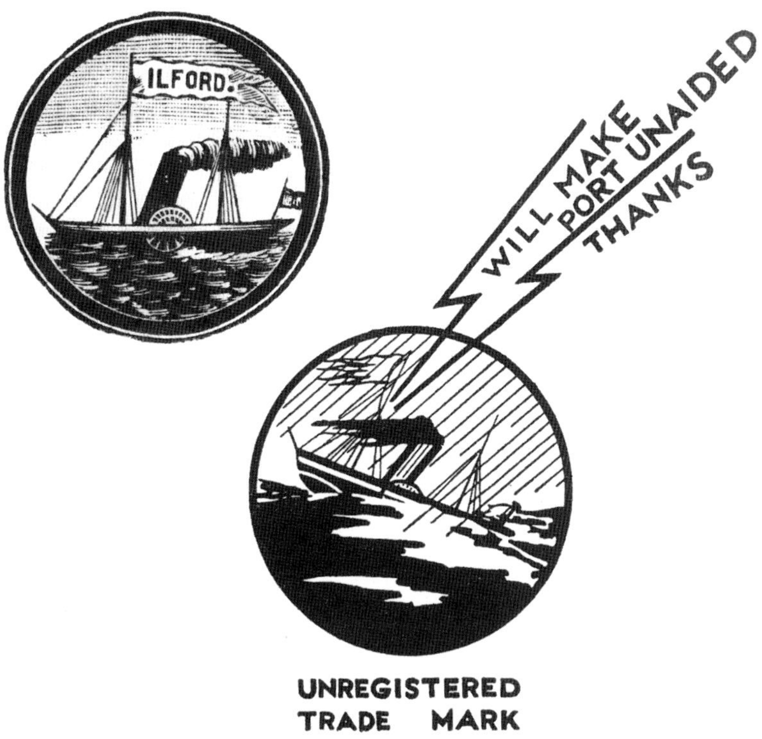

FIGURE 1. Ilford Limited's Trademark (left) first introduced in 1886, and parody (right) produced for the Ilford Advertising Department Report for the Year ended October 1932. Science and Industry Museum, Manchester, MS0232/49/8. Reproduction © Board of Trustees of the Science Museum.

suggested by Ilford's trademark in this period—a paddle steamer with a black cloud pouring from its sloping smokestack—and its Selo roll-film mascot, a toy soldier constructed from roll-film boxes and spindles.

In this period, despite great technical advances in the sensitivity of photographic emulsions and in the design of cameras, photographers continued to struggle with low light and difficult atmospheric conditions, including the dense, polluted fogs of London and the industrial north. By the 1920s, photography seemed most at home in the sun, and sun exposure and tanned skin, previously signifiers of manual labor, now became associated with the health of both indi-

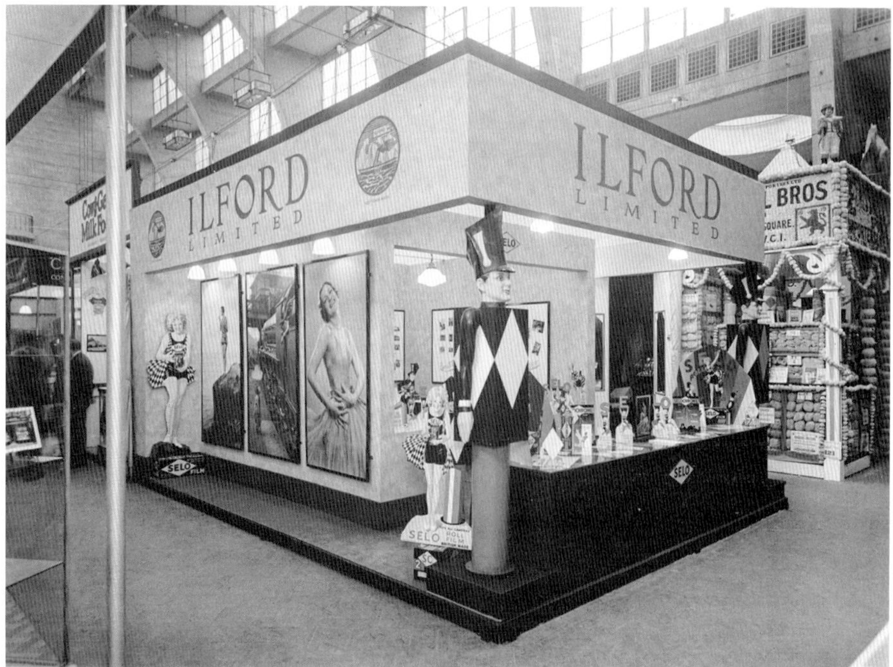

FIGURE 2. Ilford Limited display stand with a model of the Selo soldier mascot at the forty-second annual Chemists Exhibition, at the Royal Horticultural Hall, London, 25–29 September 1933. Ilford Limited collections, Redbridge Museum and Heritage Centre. Courtesy of Redbridge Museum and Heritage Centre 2025.

vidual and empire.[5] Seaside holidays became more common in this period and, encouraged by the photographic firms, this was when most people would take photographs (see plates 1 and 2). Having emerged from a highly protective and purified darkness, my creature now found itself on the beach, its light sensitivity increasingly refined and calibrated.[6]

As Harman Technology proves through its continuing operation, chemical photography is not extinct, but it has become rare and rarefied, no longer part of everyday life for most people.[7] Why write about a medium and a practice now largely usurped? One answer is that it was part of a world that continues to shape our present. As photography curators Boaz Levin and Esther Ruelfs argue, to represent climate change "in its full historical specificity" as a product of fossil-fueled industrial capitalism requires recognizing how the photographic "means of representation" are part of this.[8] As the precursor to digital photography and video, chemical photography shaped the forms they take and the modes of photographic seeing they reproduce and recirculate. It has also left its traces as chemical residues in the air, soil, and water.

How to tell the story of this strange creature? Reading about the early photography pioneers Thomas Wedgwood and Humphry Davy, in the context of Davy's other experiments with laughing gas (nitrous oxide), I came across a quotation from one of the people who imbibed the gas with Davy (known as a "bibber"): "We must either invent new terms to express these new and peculiar sensations, or attach new ideas to old ones, before we can communicate intelligibly with each other on the operation of this extraordinary gas."[9] Commenting on this, and also the "stuttering and stumbling" in Davy's notebooks of these experiments, philosopher of science Birgit Griesecke suggests that it lays down a gauntlet to historical accounts more generally.[10] How does the historian preserve the unnarratability of experience, or the possibility that things do not always come together to make a neat kind of sense? Linear narrative leans toward continuities and joins fragments sequentially, concealing their jagged edges. Resolving those fragments into a coherent story might be to flatten and deaden them. My solution has been to organize this book into thirty-six short but interlinked essays, thirty-six being the standard number of exposures on the 35 mm roll film that dominated twentieth-century photography. Treating the chapters as snapshots enables me to shift between different narrative scales and registers. I want to keep a sense of the strange and alien qualities of the media of chemical photography. I am hoping for something a little hallucinatory, as if nitrous oxide or toxic darkroom chemicals had left their faint traces on these pages.

02 ARCHIVE

Writing about it now, I feel myself charged with responsibility toward a dead (or dying) chemical photography, wanting to protect my monstrous creature. Not from extinction, which is out of my hands. Perhaps from being forgotten—though that forgetting has already begun. At Harman Technology, new, young employees do not have the same knowledge of film as earlier employees had, because film is not part of their everyday experience. In my desire to protect and to prevent obliteration, I have something in common with generations of archival researchers and historians, like the French historian Jules Michelet, who saw his task as "the care and protection of the forgotten poor and the forgotten dead."[1] Or Walter Benjamin, who wrote of the historian's task as making good a promise to the dead.[2] Except their dead were people.

It is easy to empathize strongly with a person whose private letters you are reading. Or, as Robert Darnton found out when reading the archives of the French revolutionary Jacques Pierre Brissot, to start to dislike them. Letters reveal connections between people that would have been impossible to know otherwise and tempt you to form and re-form assumptions about the kinds of people they were, so that from a tone of writing that is dismissive, you infer that the person was gruff and rude, from a considerate and sensitive letter, that they were sweet and kind. As Darnton says, the temptation is to pass judgment. He writes that during his archival research on both Brissot and Jean-Paul

Marat, "Without meaning to, I had declared Brissot guilty, or at least probably guilty, of spying and Marat innocent of theft."[3] As a young researcher "propelled by the sensation of belonging to a minority of one and by the eternal temptation of revisionism," Darnton pronounced Brissot a liar and (probably) a spy; later he reflected, "I cannot help but ask what Brissot, poor bastard, ever did to me?"[4]

The various Ilford archives and collections are more impersonal than archives drawn from individuals' estates, and than the national or state archives that concern themselves with people and life events—births, marriages and deaths, criminal proceedings, police and court records. Ilford's archives are visually interesting—there are labels and packaging, documentary photographs of the factories and the labs—but they don't have the emotional resonance of those other kinds of archives. Even so, they are still capable of evoking both empathy and antagonism.

The shape of the Ilford Limited collections, especially the one at Redbridge Museum in the suburb of Ilford, was formed by researchers who came before me, especially by R. J. Hercock and G. A. Jones, authors of the sole existing full history of the company, *Silver by the Ton: A History of Ilford Limited, 1879–1979*. This is a corporate history written by insiders: Hercock worked for the company from 1933, and Jones from 1955 (after a stint at Kodak). The foreword is written by Ilford's then chairman. The firm's centenary commemoration was presumably the impetus for the writing of *Silver by the Ton*, but the book is also a kind of eulogy to a company that had already been absorbed by a foreign conglomerate.[5] Hercock and Jones do not express it this way, but Ilford Limited had struggled to compete in color photography from the 1950s, and by the time they wrote their book the company had been subject to a series of takeovers, by ICI (Imperial Chemical Industries) and the Swiss firm CIBA in the 1950s and '60s, which later merged with Geigy to become Ciba-Geigy. Ciba is a name known to photographers and artists because of the extraordinarily vivid Cibachrome process, coproduced with Ilford Limited from 1963. Geigy may have darker connotations, at least for environmentalists, since this was the company that invented the toxic pesticide DDT. In 1976, under ownership of the newly merged Ciba-Geigy, the original Ilford Limited factory in Ilford closed, after ninety-seven years of operation, although other British factories remained.[6] The brand was split across two companies, with the Swiss manufacturers (no longer owned by Ciba-Geigy) taking the color print process Cibachrome and the digital printing papers, and the newly formed British firm Harman Technology Ltd. continuing to produce black-and-white films, photographic papers, and darkroom chemicals under the brand name Ilford, though at a much smaller scale than the original Ilford Limited company.

Silver by the Ton is a chronicle of Ilford Limited and a memorial to its hero, the company founder, Alfred Hugh Harman.[7] Drawing together many of the materials that have ended up in the archive, Hercock and Jones set out to assemble an accurate and detailed account of the firm, with a strong emphasis on the company leadership and key chemists. Despite its title, *Silver by the Ton* is a story of people, though some people, the vast majority of the workers at Ilford Limited, barely appear. Perhaps a dedicated social historian could reconstruct something of the daily lives of the factory workers through the Ilford Limited archives, despite their omissions.[8] There are people here, pictured at work in photographs depicting the different operations and processes in the factories and labs, or sitting stiffly at company banquets, or among the large groups arriving and leaving the factories (not employees, it turns out, but members of the Professional Photographers Association who visited the Illingworth Park Royal factory on May 18, 1927). Workers' names are listed in the wage books and ledgers, some are discussed in letters, and we get a glimpse of personality in the slant of a handwritten formula, or a news clipping about a car crash stuck inside the cover of a chemist's experiment book. Human stories are implied in the surveys done to understand changes in sales, in the debates over whether the company should allow itself to be bought by Kodak, and in the advertising department's grumbles about budgets.

Yet when I open the archival boxes, lifting leather-bound experiment books out of their tissue paper, what strikes me are the materials. Rather than an archive of people, I find one populated by chemicals, glass, silver, paper, and celluloid; by formulas, experiments, samples, and product lists; and by frequent reminders of the difficulty and recalcitrance of materials, which do not always behave

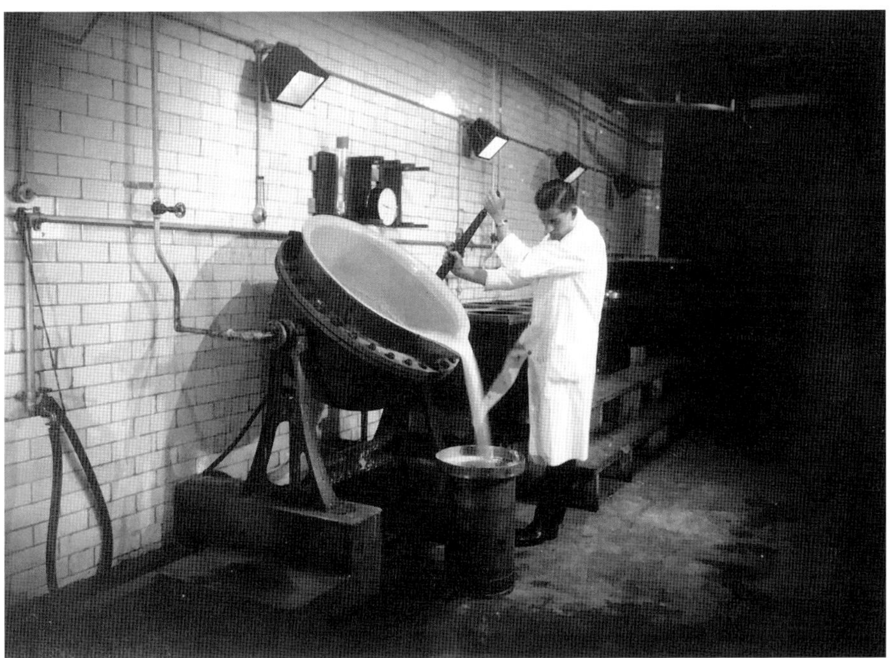

FIGURE 3. Emulsion manufacture at the Selo Works, Brentwood, 1938. The emulsion pans had warm water plumbed into them, to keep the emulsion at a constant temperature. The room is lit only by colored safety lights. Ilford Limited collections, Redbridge Museum and Heritage Centre. Courtesy of Redbridge Museum and Heritage Centre 2025.

as expected. It is not unusual to come across a negative on a piece of nitrate film—unstable, fragile, potentially explosive—slipped between the pages of a handwritten experiment book. Nor is it rare to find descriptions of the struggles with materials, for instance, of how, when flat drawn glass instead of blown glass was introduced for photographic plates, hundreds of plates would stick together "like a block of ice."[9] In a speech given at the end of the Second World War, the Ilford plate factory manager Cecil N. Potter described glass as "not an inert substance but a material which has many qualities. It can be hard or soft, flat or curved, having a thick skin or a thin skin, flexible or rigid, and in fact appearing at times to be almost temperamental."[10]

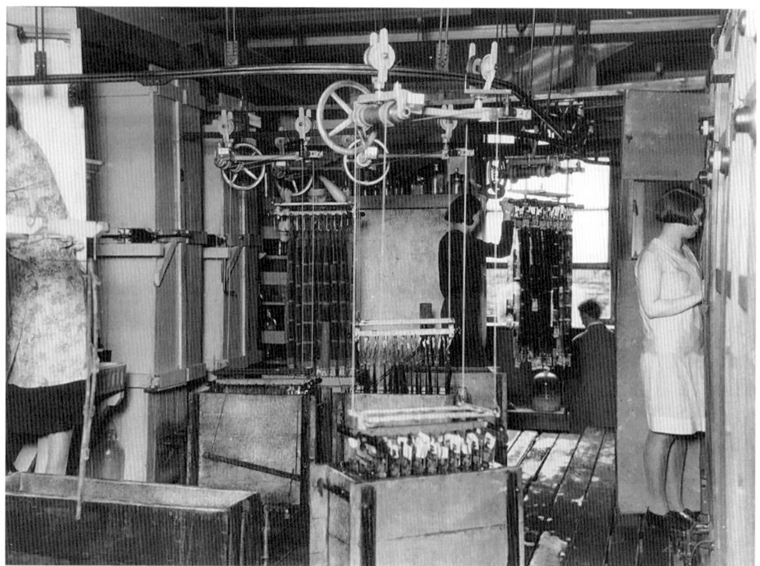

FIGURE 4. Tank Processing of Roll Films at the Selo Works, Brentwood, 1938. Ilford Limited collections, Redbridge Museum and Heritage Centre. Courtesy of Redbridge Museum and Heritage Centre 2025.

The preoccupation with questions of atmosphere came late in my research, and I have had to read back through my archival notes, piecing together a quite different pattern than I had initially imagined. Archival documents rarely directly discuss atmospheric and meteorological conditions in any depth, but these are suggested in the records of photographic experiments and formulas, as well as in marketing materials, labels, and packaging. For example, there is reference to materials like infrared film being deployed to cut through the bluish haze in aerial photography, and to Ilford's "Tropical Hardener," invented to deal with high humidity and heat in the darkroom. Reading the archives for atmosphere entails seeking out mentions of weather, the tropics, fog, and air-conditioning, as well as paying close attention to atmospheric traces in photographs. But it also involves attending to the range of sensitivities of photographic materials, and especially to emulsions—highly sensitive concoctions that are largely, though not entirely, ignored by historians and theorists more concerned with tangible materiality of photographs and cameras.[11] I am interested in emulsions not only for how they successfully sense different wavelengths of visible light, but also for how they fail to sense them, or how they sense other things entirely—for example, other sources of radiation, contamination in water, and the gases and particles in the atmosphere. The Ilford Limited archive contains books full of handwritten emulsion formulas (as well as formulas for developers, toners, and other photographic concoctions). They remind me of handwritten recipe books passed down in my family, except I do not have the skills to interpret them. Not being a chemist, I have no way of judging if the combination of ingredients and the procedures by which the emulsions are made are conventional or counterintuitive.

The changing meaning of atmospheric effects, how fog, for example, could signify differently at different moments, may be gleaned through photographs from the period. But photographs from the interwar years are not always the best for thinking with and through these ideas. Sometimes I am attracted to specific images that help make sense of how photographs can be atmospheric and how they can articulate the materiality of light, atmosphere, and the photographic apparatus itself. For that reason, this book will often stray from interwar Britain and from the British Empire, seeking other ways of thinking in different kinds of photographs.

In the archive and at the factory, I form attachments. Partly this is an attachment to a brand, as Ilford has been present in my life since before I can remember. The familiarity I felt looking through the postwar Ilford materials, with their distinctive blocky black logo, was only outstripped by my experience of looking through the archives of Boots the Chemist, which once sold and processed films, and which is still the dominant pharmacy chain in the UK. But unlike the official historians of Ilford Limited, who were also its employees, I have no direct connection to the company. Ilford Limited provides me with an archive, but this is not really its story. My greatest attachment and empathy lies not with Ilford as a company or a brand, but with something else: my imaginary, metaphorical creature. Mine is not a fantasy of rescue or redemption. In one respect, I am as treacherous as Darnton was with Brissot: I protect my monster in order to put it on trial. I want to reveal its place in the fossil economy, situate it as part of that choking legacy. At the same time, I show it to have been constrained, held back from its full potential by an industry that wanted to suppress and control its full range of sensitivities. And I endow it with great power: most of us may have forgotten it, but it lives on in the ways in which we see and represent the world.

03 PHOTOSENSITIVITY

Photographic materials are sensitive, to light and to much more than light. They show how sensing and sensitivity extend beyond humans and animals to our technologies, and how such technologies might facilitate new kinds of human experience and help shape our own sensory capacities. Technological sensing is entangled with our own, as new technologies produce new aesthetic experiences, and—as Walter Benjamin argued—provide a kind of training for the human sensorium. Via the research laboratories of companies like Agfa, Eastman Kodak, and Ilford, industrial photochemistry participated in the reconfiguration of the human senses that Benjamin described as characteristic of modernity.[1] Photography and film are central to the modern transformation of experience, not least because they mediate the human body itself, through people pictured on screens and in photographs. We experience and imagine our bodies differently, too, because of the ability of photographic technologies to turn us inside out and allow us to view ourselves at a microscopic level. If photography transforms human perception, it follows that in "tuning" photographic sensitivities, the photographic industry also "tuned" what it was possible for people to perceive, and to ignore.

Like our own senses, photographic sensitivity is subject to a "sensory economy" and to a "distribution of the sensible." Sensory economy refers to the management, circulation and distribution of sensory experiences and organization of the relationships between them. It brings together the broad sense of economy as administration and distribution with the everyday sense of it as thrift and budgeting. It can be used to describe how photographic materials are designed to prioritize certain kinds of sensitivity: to be able to sense radiation while not reacting to other substances (metals, oils, sulfur compounds in the air, and so on), and to be more receptive toward one side of the electromagnetic spectrum than the other, so that they are more blue or red sensitive, for example, or able to register X-ray, ultraviolet, or infrared radiation.

The "distribution of the sensible," a phrase from the philosopher Jacques Rancière, refers to the social distribution of competencies and how "a social destination is anticipated by the evidence of a perceptive universe, of a way of being, saying and seeing."[2] It describes the unequal availability of varieties of sensory and aesthetic experience across social groups. Transferred to the photographic context, it suggests that sensitive materials are organized according to certain functions and specific destinations: X-ray materials head for the hospitals and dental practices, infrared to the police and the air force, other specialized emulsions to nuclear research, and so on. Ilford Limited's records reveal the diversity of sensitized films and glass plates they manufactured during the 1930s, including aerial films and infrared but also clydonograph film, which was used to measure changes in electrical voltage, and cardiographic film, used to diagnose heart and circulation problems.

One might reasonably object that, in applying the concept of the distribution of the sensible to photosensitive materials, I am stripping it of its political meaning in relation to society and specifically to capitalism, since Rancière intends it to show how aesthetic experience is connected to social class. However, writers such as Jane Bennett, Donna Haraway, and Bruno Latour have argued for the importance of a politics that engages nonhumans as active coproducers of our world.[3] These theories of distributed, nonhuman agency emerge from a critique of modern science and political philosophy, and from ecological, environmentalist, and feminist perspectives. In rejecting the view that "nature" is mere background or environment for human activity, they acknowledge what many Indigenous societies have

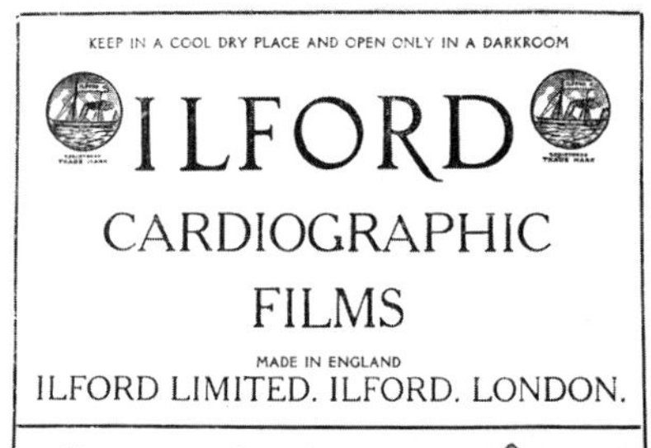

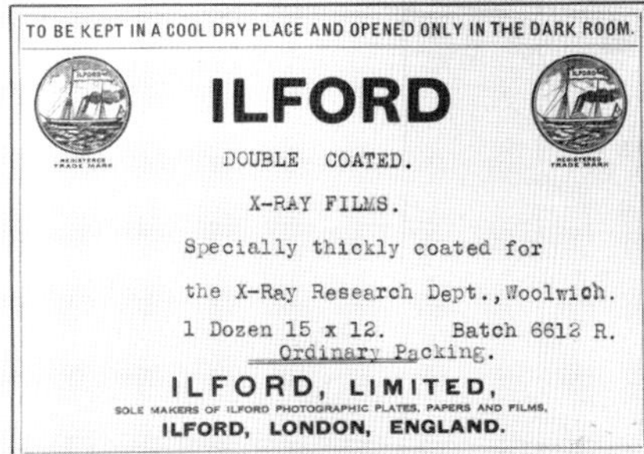

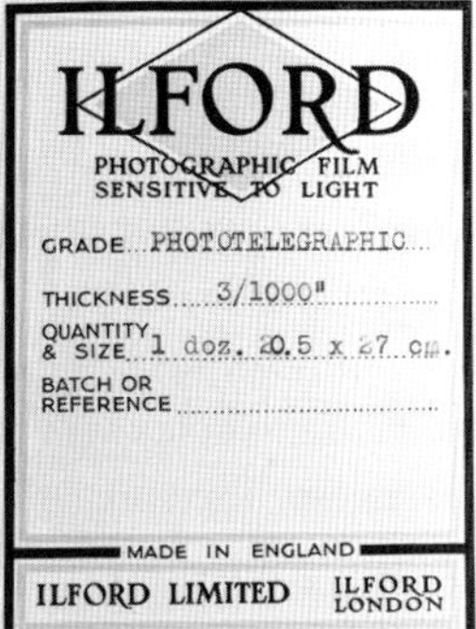

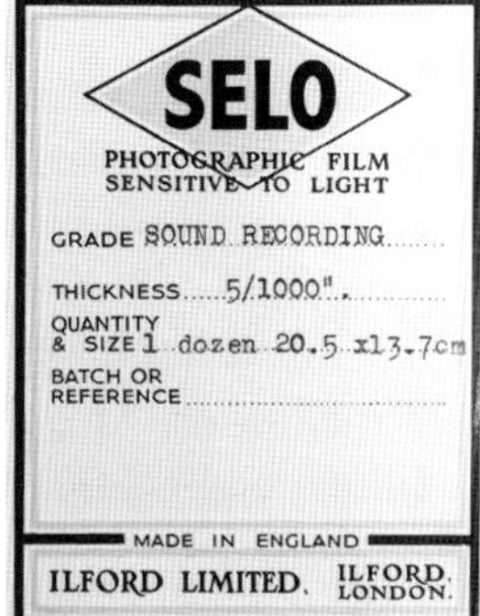

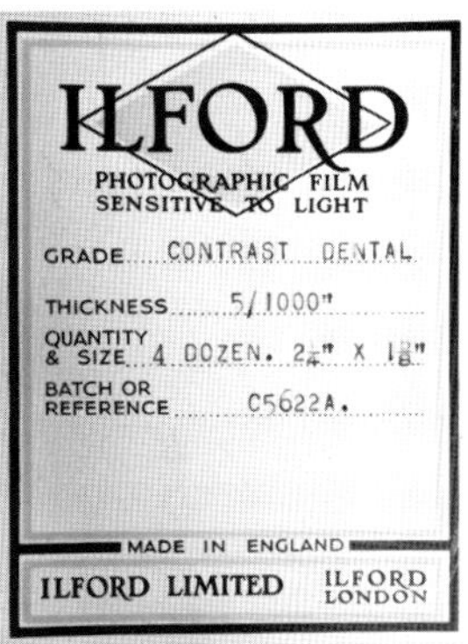

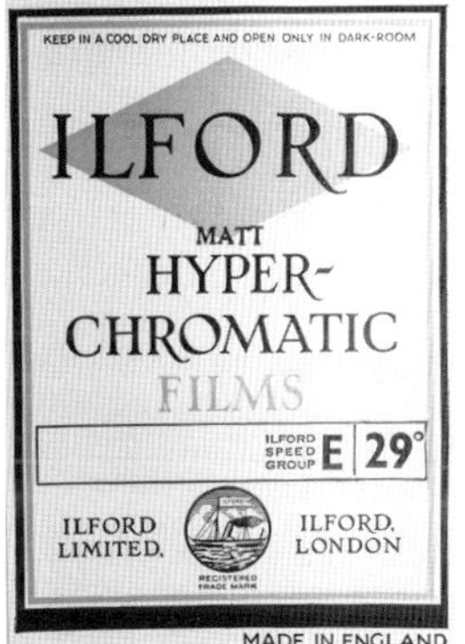

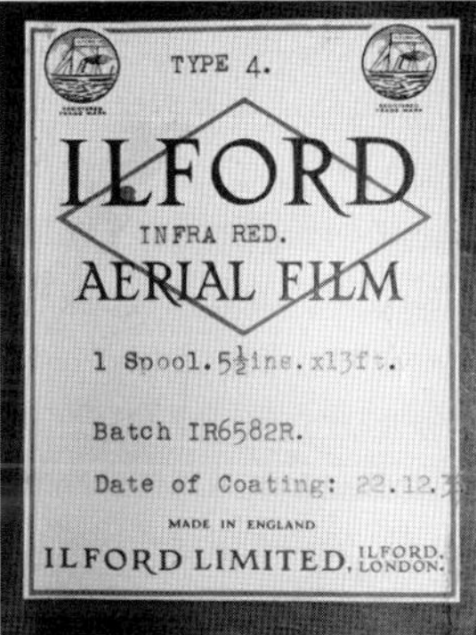

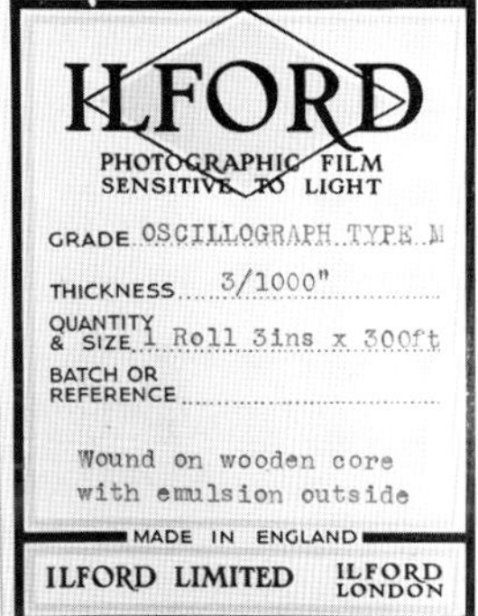

FIGURE 5. Ilford film labels, showing a diversity of uses. Ilford Limited collections, Redbridge Museum and Heritage Centre. Courtesy of Redbridge Museum and Heritage Centre 2025.

long known. The repression of animist and vitalist philosophies was a central part of the European colonial project and core to the development of early capitalism.[4] As Amitav Ghosh explains, the modern view of nature as matter to be exploited emerges out of colonial relations: "It was the rendering of humans into mute resources that enabled the metaphysical leap whereby the Earth and everything in it could also be reduced to inertness."[5]

Like the sensing capacities of living animals and plants, the sensitivities of technologies do not exist outside of and separately from human sensual and aesthetic experience but instead help to determine it. Since human sensibilities and competencies are so entangled with technological ones, their social destinations are also entangled; indeed, the introduction of new kinds of photographic films and plates required new kinds of trained sensitivities on the part of medical staff, scientists, military personnel, and others. Moreover, if sensitivity is, as Latour suggests, about "detecting and responding rapidly to small changes," then we need to recognize the sensitivity of the complex, interconnected systems of the Earth—or what he (after James Lovelock) calls Gaia—and the impact of our technologies upon it.[6] It is in this context that photographic sensitivity becomes politicized, since it determines what we are capable of detecting and responding to, and since these sensitive and reactive substances have effects on ecosystems, too.

The labs of the manufacturers shaped photographic materials according to the demands of the culture and the corporation: tuning and attuning their sensitivities, heightening them in some ways and dulling them in others. By the interwar period, light-sensitive films, plates, and papers were manufactured through complex and highly regulated chemical processes, adjusted and calibrated. In 1938, Ilford opened its Selo sensitometric testing department (fig. 6). Sensitometry is the study and measurement of the light sensitivity of given materials, standardizing their response to different intensities of light and to different parts of the spectrum. As the 1930s marketing of "Selo" brand film put it, this is "the power behind the lens" (fig. 7). It is the culmination of processes of diversification and specialization of photographic materials, a new sensory economy that started

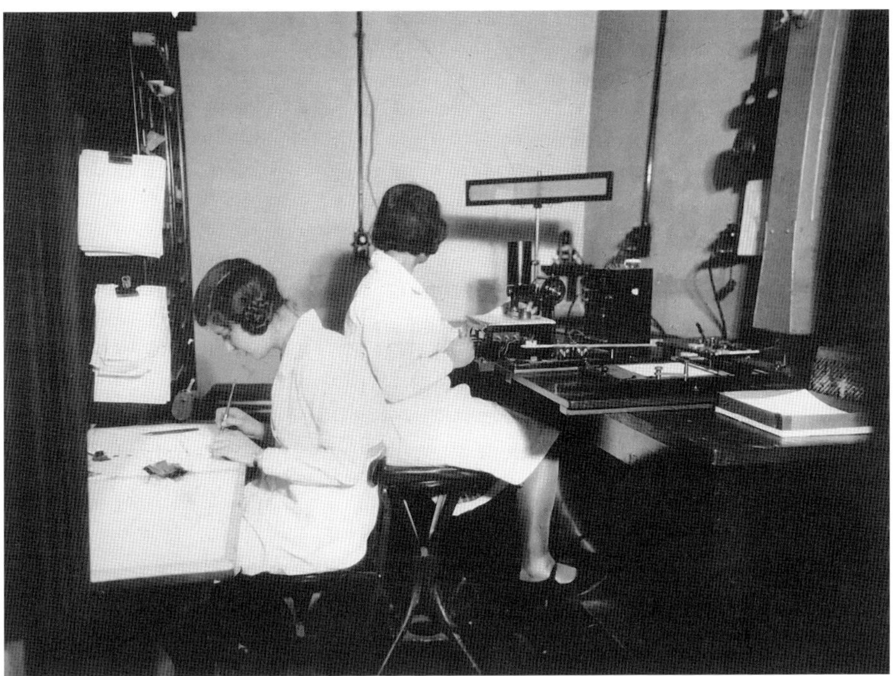

FIGURE 6. The Selo Sensitometric Laboratory, Selo Works, Brentwood, 1938. Ilford Limited collections, Redbridge Museum and Heritage Centre. Courtesy of Redbridge Museum and Heritage Centre 2025.

in the late nineteenth century and expanded rapidly during and after the First World War.

In the 1920s and '30s one of the major research tasks in the photography industry was to explore the various ways in which silver halide emulsions (containing silver bromide, silver chloride, and silver iodide) could be refined through the addition of specific kinds of dyes. These dyes were derived from coal tar, and brand names later associated with photography, such as Ciba-Geigy and Agfa, began as dyestuffs companies. The coal tar dyes were discovered in experiments first driven by the moral and economic imperative to minimize waste (in this case, waste from coal-fired gasworks), and later by the demands of empire, national competition, and profit. Some of

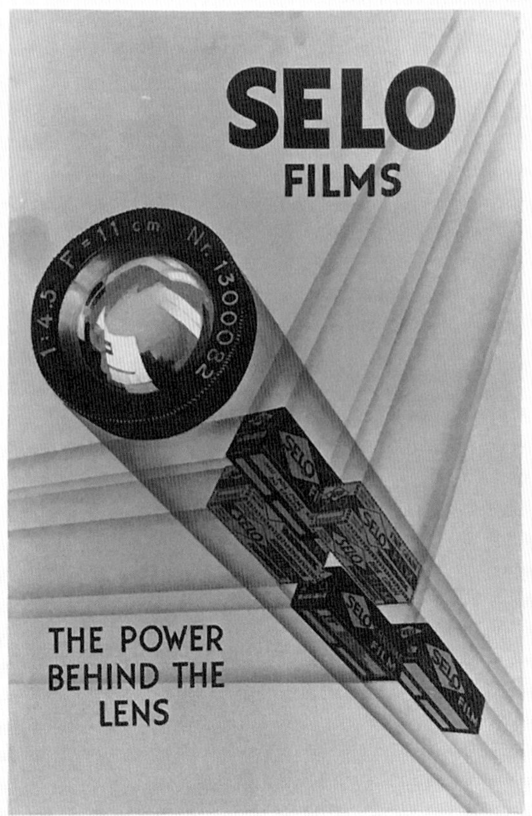

FIGURE 7. "The Power Behind the Lens," 1937 advertisement for Selo film. Ilford Limited collections, Redbridge Museum and Heritage Centre. Courtesy of Redbridge Museum and Heritage Centre 2025.

them were used to tailor the sensitivity of photographic emulsions, extending or narrowing what part of the spectrum the photographic film or plate could register. Known as photographic sensitizers, they made it possible to control photosensitivity to a greater degree, including making visible parts of the spectrum that were otherwise invisible to the human eye. For nonspecialist markets, they were used in the manufacture of films whose spectral sensitivity mimicked that of human vision.

At the same time, the factories and labs were hard at work trying to standardize their products, eliminating flaws and inconsistencies.

This work, too, was about trying to control photographic sensitivity, not to open but to close its ability to sense certain things, especially substances in air and water. The difficult climates of the British Isles and parts of the British Empire galvanized attempts by the photographic materials industries to control the vagaries of atmosphere.

When photographic emulsions sense things they are not designed or intended to sense, these are designated flaws and accidents. In his book *Inadvertent Images*, Peter Geimer makes the point that excess information or "noise" is not a flaw or a "deficit" but "a specific potential of photography."[7] And if the interwar photography industry devoted itself to the expansion of photography's sensory capacities in relation to radiation, it also devoted itself to the elimination of its atmospheric sensitivities. Increasingly sophisticated and complex chemical technologies were made to cope with atmospheric and climatic variations so that taking photographs no longer seemed to be a matter of air as well as light. The ideal photographic emulsion would be insensitive to changes in temperature and humidity, immune to the gases and particles contaminating the air. Its potential in this direction would ideally be entirely stifled.

Toward the end of Mary Shelley's *Frankenstein*, the tragedy of the Creature's treatment is spelled out when he says: "Once I falsely hoped to meet with beings who, pardoning my outward form, would love me for the excellent qualities which I was capable of unfolding. I was nourished with high thoughts of honor and devotion."[8] Likewise, photographic materials have other "excellent qualities" that they are "capable of unfolding." The analogy suggests that they are alive, brought to life as Frankenstein's Creature is, and I am happy to think of photographic materiality as lively if not strictly alive. I am less happy with the echoes of an older photography theory in which photography as a practice is often seen as aggressive, masculine and predatory.[9] Although, in fact, it is usually the photographer who is cast as predatory—in theoretical writings the materials and technologies of photography are imagined very differently. For example, in theories of photographic indexicality, film is imagined as a passive recipient of light: it is exposed and it changes, something is done to it, but it does not act itself.

Terms like "actant" and "agent," which are used to describe the liveliness of the nonhuman, do not seem to apply so easily to this sensitive material since they invoke action and activity. They have as their shadow-image the inactive or the latent, the passive and the inert. However, if the point of recognizing the agency of nonhumans is to challenge the assumption that historical actors are always human, and that the world is divided into subjects (humans) and objects (nonhumans), then passivity, the latent, and the dormant must be considered. Historically and etymologically, passivity relates to passions, to feeling and to suffering, and to being exposed to the actions of others.[10] If we are going to grant agency to nonhumans, we need also to grant them passivity, recognizing that the capacity to be acted on is still a capacity. Photographic materiality offers a model for passive vitality, in the form of sensitivity, receptiveness, an openness to impressions, the ability to register and record.

But photography also has a kind of collective agency. It does not just represent the world back to us but makes and remakes it. The making of photography is also part of the making of new social relationships and new ways of being in the world. Photography coproduces the reality it then sets about recording. By this I mean something more than cajoling people to pose, rearranging the furniture in a room, or moving a particularly evocative object into shot. Photographic media—still and moving images—shape the meaning of ordinary activities and influence the feelings associated with them. They produce the new visual experiences involved in looking at photographs, and new senses of self and identity through acts of photographing and being photographed. But they also go deeper, prioritizing certain ways of moving and being over others, new kinds of behaviors, feelings, sensibilities, and dispositions, reinventing and rationalizing our bodies and those of other species.

Worlds are never complete, whole and finished, but always incomplete, and continually contested. In the interwar period, photographs pictured the existing world in new ways, while, as a set of practices and industries, photography and film were altering it irrevocably. The presence of cameras transformed everyday situations and contributed to changes in how people experienced time and space. Photography was both repressive and productive: it overwrote existing experiences and ways of being, such as those of the colonized who suddenly found themselves subjected to the camera, but at the same time it expanded the thresholds of the visible world through new techniques of scientific visualization and increasingly sensitive materials.

04 A DAUGHTER OF COAL

► Mid-twentieth-century social histories of photography tended to begin with a paradox: Why was it, they asked, that photography had not been invented or discovered sooner? (Indeed, the very question of whether it was an invention or a discovery was vexing.) They pointed to the early knowledge of photochemistry and photosensitivity in China and other countries, to the long familiarity with the basic chemicals of photography such as silver halides and nitric acid, and with the workings of the camera lucida and camera obscura. The answer tended to be that "the time was ripe" in nineteenth-century Europe due to a coalescing of social factors: class mobility and a new demand for portraits among the rising middle classes, increased literacy accompanied by the growth of the press and lithography, and so on.[1] This answer suits a Marxist social history, because it suggests that developments in culture and technology are dependent on the historical development of social institutions and relations between classes, explaining why photography emerges in the context of capitalism. But such explanations miss the crucial fact (which Marx himself might not have missed) that photography developed out of the Industrial Revolution and out of empire, and was dependent for its materials on the intensive exploitation of coal.

In their 2022 book *Residues: Thinking Through Chemical Environments*, Soraya Boudia and her five coauthors make the point that we tend to think about the environmental impact of industrialization in terms of the burning of fossil fuels, whereas in fact coal, oil, and gas have had just as much impact, if not more, as "the feedstocks for chemical industry, which synthesizes a massive variety of compounds used in every sector of the economy."[2] On Barak similarly argues in *Powering Empire* that by putting too much focus on the burning of coal, we miss its significance in shoring up the British Empire through the establishing of coal bunkers across its domain, not just for fueling but for territorial expansion; he notes that by the turn of the century British coal accounted for 85 percent of the world's export trade.[3] In the years between 1840 and 1914 the volume of coal mined in Great Britain grew tenfold.[4] The grounds for this huge exploitation of coal had been laid a century earlier, due not just to the islands' large and accessible coal reserves, but also to an early market economy, a weakened relationship of peasantry to land, and established overseas colonies.[5] Nevertheless, as Barak summarizes, "Britain's industrialization and imperialism were not separate processes: both—not only the former—were predicated on coal."[6]

Photography was also predicated on coal. Even the very first photographic experiments in Europe, circa 1800, were reliant on coal-fired industrial developments as well as on colonial trades in nitrates, silver, and other materials. By the late nineteenth century, photography studios and factories had heating and cooling systems fueled by coal and gas. They also made use of the by-products of the coal and gas industries, which they recirculated into the soil, the waterways, and the atmosphere in the process of producing photographs and photographic materials. Among these by-products was coal tar, which resulted from the destructive distillation of coal (i.e., heating the coal in the absence of air) to make gas.

From the 1870s, photographic material manufacturers and chemists started to make use of photosensitive dyes derived from coal tar. These coal tar dyes, which altered the light sensitivity of emulsions, had been discovered by accident in the 1850s, in the search for synthetic quinine to treat malarial outbreaks among British colonial troops and settlers. In the decades that followed, dye manufacture was financially supported by European governments. Germany, especially, was keen to reduce its dependence on supplies of madder (red) and indigo (blue) from India, which was under British rule.

Despite an early head-start, the British dye industry lost momentum and Germany became the principal supplier of photosensitive dyes before the Great War. War accelerated British production of new dyes for photography, often imitating the now unavailable German ones. As in Germany, the photographic industry in Britain was connected to the artificial dye industry, the wider chemical industries, and the military. Together they were economically and chemically dependent on what the ecologist and geographer Andreas Malm usefully calls the "fossil economy."[7]

The Industrial Revolution did not introduce coal as a source of energy—Britain had long been the world's largest consumer of coal for heat, and China's coal mining reportedly dates to the Song Dynasty (960–1279)—but it marked the transition to a full fossil economy when steam (from burning coal) was used by industrialists to drive machinery.[8] Malm argues that the fossil fuel economy is attributable to the social relations of capitalism, in which "steam arose as a form of power exercised by some people against others."[9]

As Steve Edwards argues in *The Making of English Photography*, scientists in the 1830s attributed an almost magical liveliness to steam as a way of suppressing the agency of skilled workers or artisans.[10] Nineteenth-century scientists and capitalists thought of coal as a resource, translated into power through technology. So, when they attributed autonomy to steam and coal and downplayed human agency, they were not being animists alive to the sensitivities of the nonhuman world, but reflecting a dominant Western worldview that had already driven a wedge between human beings and their lifeworlds.[11] This way of thinking was rooted not only in a capitalist attitude to nature as resource but also in an older, colonial attitude that disavowed settlers' responsibility for colonial violence by attributing it to God or nature. For example, in *The Nutmeg's Curse*, Amitav Ghosh discusses how English settlers in the Americas justified their violence as less cruel than that of the Spanish because it was meted out through "nature" and "material forces," effacing their own role in "setting environmental changes in motion; it is as if they occur independently of human intentions."[12] As Ghosh argues, this justification required that the settlers had already partitioned off the human and the natural, while in fact, in settler-colonial conflicts, human and

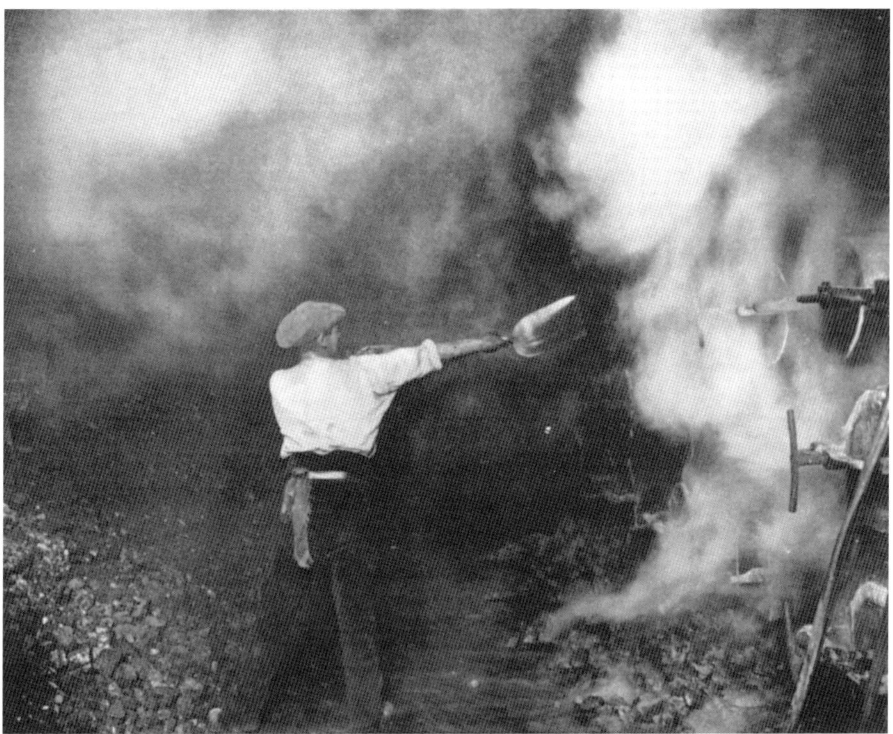

FIGURE 8. Fred Marsh, "Gasworks, Charging Retorts." From *The Photographic Journal* 22, no. 1 (September 1897), Illustrated Catalogue of the Royal Photographic Society's Exhibition, September to November 1897, plate 32. Reproduction courtesy of Michael Pritchard.

more-than-human agency were both at play. Similarly, 1830s scientists assumed a prior separation of human, "natural" and technological agents, even as industrialization bound these ever more tightly together.

In the age of steam, the interrelated world of coal, gas, electricity, and steam was of course not self-propelling. It was driven, not solely but not least, by the vast numbers of men who endlessly shoveled filthy coal into furnaces. In 1897, the Royal Photographic Society awarded a prize to Fred Marsh's photograph of a stoker at a coal-fired gasworks (fig. 8). His back is to the camera, his legs are astride, and he has a shovel in his outstretched right arm. He stands on a

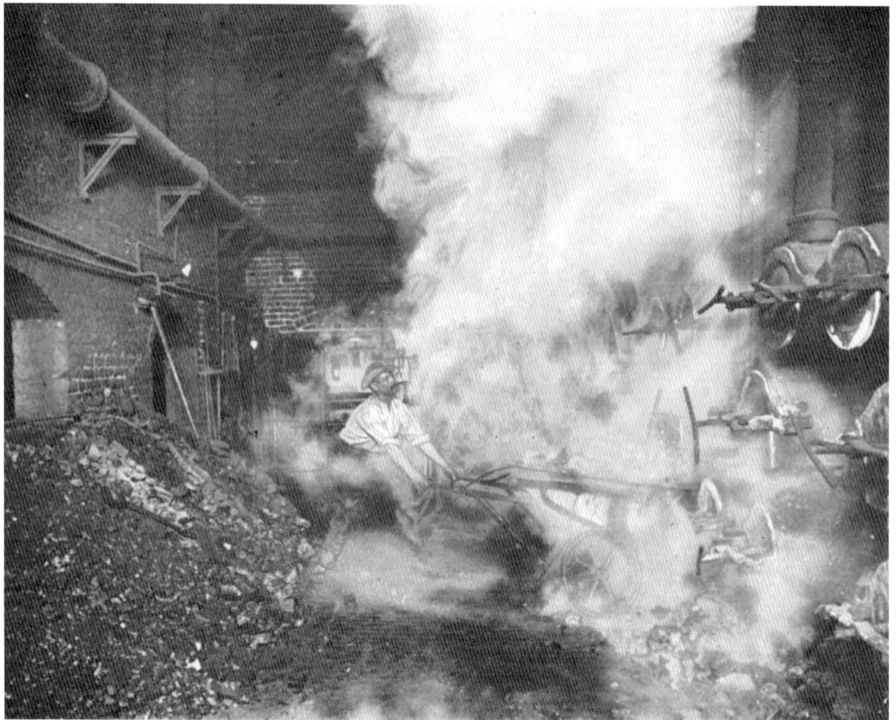

FIGURE 9. Fred Marsh, "Gas-Works: Warm Work." From *Photograms of the Year* (London: Dawbarn & Ward Ltd., 1897), 57. Reproduction courtesy of Michael Pritchard.

path cut between a heap of coal on the left and boilers to the right, surrounded by smoke. Though he appears to be in motion, the rigidity of his stance suggests he may have posed for the photograph. He wears a brimmed hat and a white shirt, and he has a rag or neckerchief hanging from the belt of his trousers. Another photograph of the same man was published in *Photograms of the Year 1897* under the title "Gas-Works: Warm Work" (fig. 9).[13] This one depicts the stoker pushing a wheelbarrow and is evidently retouched. The folds in his shirt, the outline of his arms, and the wheelbarrow have been drawn in. The retouching reveals that the man and his labor would otherwise have been obscured by the smoke.

My own great-grandfather was a stoker. John Hugh Devlin, or Old Man Devlin, as we called him, had shoveled coal in the belly of a steamship during the battle of Jutland in 1916.[14] In the age of steam, stokers were the men who fed the beast, not only in ships and gasworks but in factories and on steam trains. It was hot and dangerous work, and there was an interchange between the stokers of the gasworks and those on the ships, with Admiralty schemes employing gas stokers in the Royal Naval reserve.[15] While Marsh's gasworks stoker appears almost heroic, Royal Navy stokers have been characterized in histories and memoirs as disorderly, mutinous, uneducated drunkards, although many would have been qualified mechanics.[16] Though physically strong, they were often short (John Devlin was five feet three), an advantage in the stokehole of a ship.[17] Stokers did not fit easily with the idealized image of the efficiency of modern industry or the patriotic image of the naval seaman. I found a group photograph of the stokers from my great-grandfather's ship which plays up to the stereotype of these unruly men. They are slightly disheveled, but uniformed, and strikingly young. One is drinking from a bottle, one is playing a lute, and the central figure, clearly a prankster, has a chamber pot on his head. Far from the more typical, formal portraits of Royal Navy sailors, the atmosphere of this photograph is restless and playful, with only the three figures at the back, all wearing wide-brimmed hats like Marsh's stoker, holding themselves to attention.

In contrast, Marsh's photograph renders the stoker visible but cleans him up. It conforms to the aesthetic of the industrial sublime that was taking hold in the 1890s, and to a pictorialist obsession with atmosphere that would maintain its grip on English photographers even into the late 1930s. To give photography the expressive power and status of art, pictorialists used atmospheric elements that help to soften the image and bring it closer both to human perception and painterly impressionism. Because of this, Marsh's picture depicts something that photographs of ships stokers rarely show—the smoke and coal dust that filled the air of the boiler room. Unlike other photographic approaches, his pictorialist aesthetic rendered atmosphere thick and tangible. The smoke allows us to intuit that just as the men acted on the coal, so the coal acted on them. Stokers were at risk from scalding and explosions, carbon monoxide poisoning, heatstroke, and various lung and throat diseases.[18] Though John Devlin would survive the battle of Jutland and live to eighty-nine, he was severely scarred by burns to his face, left arm, and hand.

Today, changes in light, atmosphere, weather, and the chemical makeup of our environment are loaded with meaning in the face of global ecological crisis and the dramatic weather events resulting from anthropogenic global warming. Now we understand air pollution not only as damaging the breathability of air in a locality but also as changing the balance of gases in the atmosphere and ultimately affecting the livability of the globe for humans and other animals and plants. Yet if you have the sense that coal, like chemical photography, is on the way out, I should point out that coal use is currently increasing. According to the International Energy Agency, coal was the largest growing source of electricity globally in 2021 and was behind the overall increase in emissions that year. While oil dominated global energy sources, coal came a close second. More than half of the United States' electricity was generated from coal in 2020, as was China's in 2021. In India and China, coal use is surging.[19] As the historian Dipesh Chakrabarty points out, these countries justify the use of coal by referring to "the number of people who urgently need to be pulled out of poverty," complicating arguments about climate justice and global inequality.[20] All these statistics refer to energy generation, the largest use of coal and the greatest cause of fast-rising CO_2 emissions today.[21] But energy generation is not the only purpose for which coal is extracted, and as Boudia and her coauthors show, coal and oil are used to make a wide variety of substances that contaminate the earth and contribute to an ecological crisis that involves much more than climate change. Some of these substances were and are present in the materials of photography and in photographic processes.

As nations attempt, half-heartedly and ineffectually, to move away from a fossil economy while still clinging to capitalism as the dominant economic system, it is time to consider photography's historical dependence on coal and its related industries. Photography is a "daughter of coal" and of capitalist imperialism, dependent for its materials on the extractive practices and trades of empire, a global market in chemicals, and, most of all, on the development of coal tar by-products that followed from the coal-fueled Industrial Revolution.[22]

05 HIGH STREET CHEMIST

► Come with me back to the 1920s and '30s, when Ilford Limited was in its heyday. These are decades of economic depressions and unstable governments but also a growing consumer culture, particularly in South East England and the Midlands. On the suburban high street, dispensing machines, enamel signs, sandwich boards, and elaborate window displays announce all sorts of new products. There are newly patented technical innovations subtly transforming everyday life: women are wearing brightly colored rayon dresses, neon signs light the city streets at night, and at the cinema the talkies and Technicolor make their first appearance.

Look at what they are selling in Boots the Chemist, alongside the cough medicines and the cosmetics and the cleaning fluids: bathing caps, sun goggles, and motor goggles; the Buty Wave Cap-de-Luxe (for wearing at night and preventing "permanent waves" from being crushed); Veet hair remover and elastic hosiery. There are some surprising products, such as Regesan Cigarettes for Catarrh and Asthma and radioactive bath salts. The week of July 30, 1930, is "Foot Comfort Week." Another week is "Rat Week." The Photographic Department does developing and processing and sells books on photography, alongside a wide range of cameras including Agfa Speedex, Hawk-Eyes, Ensign Carbines, Coronet Cine Cameras, Soho Cadets, Kodak Brownies, and Pocket Kodaks. They sell chemicals and papers for the home darkroom, and something called "Mighty Atom Crystals." They stock the new "rollable" snapshot films by Kodak, Rajar, Ensign, Imperial, Illingworth, Ilford, Selo, Pathé, and Agfa (though Boots's customers may not have known, Rajar, Imperial, Illingworth, and Selo were all subsidiaries of Ilford Limited by the 1930s). Most of all, Boots's salespeople "push the Kodak film" since they have an agreement with the company.[1] There are ready-made wood photo frames, photo albums, and pocket print wallets, and in the winter a whole host of other nonphotographic items appear in the big city center branches of Boots: toasters, wirelesses, electric flash lamps, and clocks.

What you can't see on the high street: incremental changes in the specialist types of photography, including the introduction of infrared and high-speed films. All the photographic firms have research laboratories now. Ilford Limited are supplying photographic products for all kinds of purposes, from aerial surveillance to dentistry, forensics, and photomicrography. The First World War is over but the "warfare state" is here, and some of Ilford's biggest customers are the various branches of the military.[2] Sales to the military, health, and aviation industries help the photographic industries to survive economic stagnation in the 1920s and the Great Slump of 1929–33, as does the fact that "rollable" photographic film belongs to the category of small consumables that continue to sell when the market in larger items declines. As well as film, Ilford Limited is still manufacturing dry (glass) plates in great volume—sales of these start to slow around 1934 but their manufacture and use continued well into the postwar period (see fig. 52 and plate 7).[3] For the visitor to Boots's Photographic Department, perhaps the most noticeable things in these decades are the rise of the "miniature" camera, which takes 35 mm format silver halide film, and the arrival of the new panchromatic film, supersensitive and fine-grained, which must be developed in total darkness. By 1935, the most exciting arrival is photographic film in "natural colors" for ordinary cameras, as opposed to specialist color cameras. Manufactured by companies such as Dufay-Chromex (under control of Ilford Limited) and Kodak, it is too expensive for everyday use and annoyingly only produces

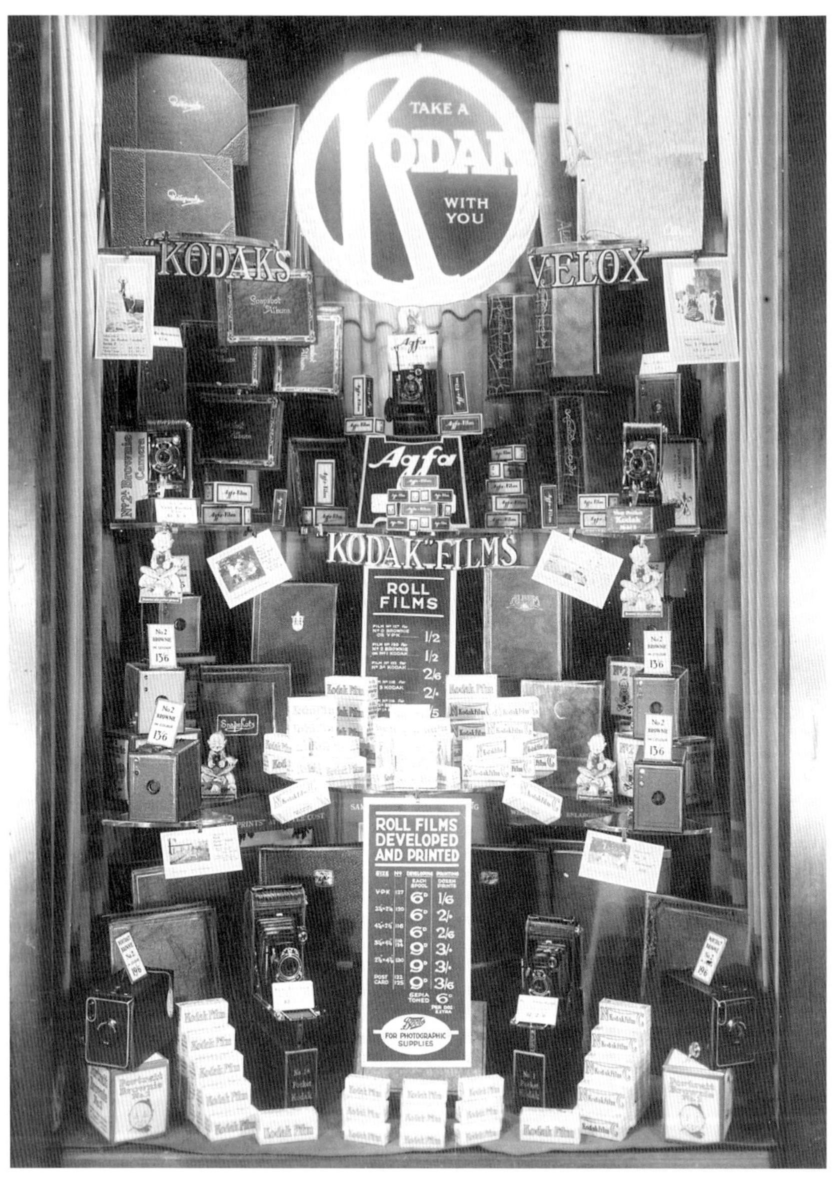

FIGURE 10. Boots display window highlighting Kodak products. 1930s. A similar window display was featured in *The Merchandise Bulletin* in 1925. The Boots Archive.

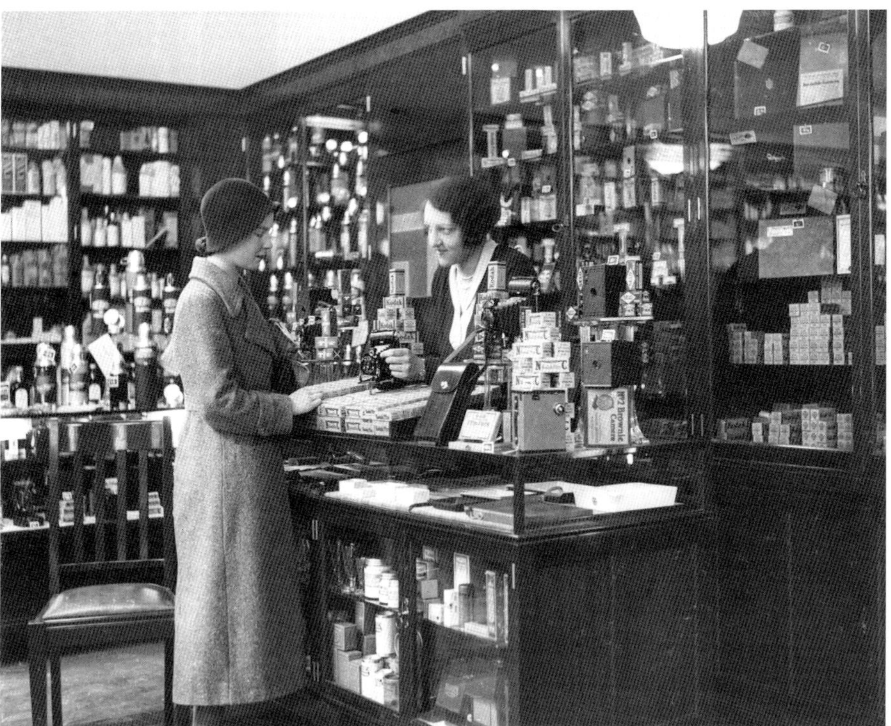

FIGURE 11. Boots Photographic Counter, Bournemouth store, 1933. The Boots Archive.

transparencies or slides, not prints. Although Boots stock Dufaycolor transparency film and Kodachrome from 1935–36, color negative film will not be available until the 1940s.[4]

Since most amateur photographers have little interest in the technical know-how needed to operate good quality cameras, by 1926 Boots is directing the majority of its customers toward cheap and simple box cameras, focusing the company's profits on developing and processing rather than camera sales. In the 1930s, most people visit the photographic department to buy black-and-white films to document their summer holidays, and then return to have them processed. But the salespeople, at the insistence of Boots and the photographic companies, are keen to get them to try new flashes and lights

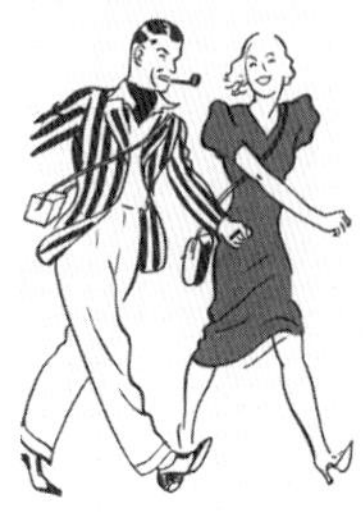

Here's sound advice, Let Boots do your DEVELOPING & PRINTING

The fun of taking photographs is made all the more delightful if you have the right equipment, if you know how to use it, and if you entrust the specialised work of developing and printing your snaps to experts who get the best results out of your efforts.

Boots are photographic specialists. They have everything you need for this fascinating hobby, Cameras, Films, etc. What's more, they have a wonderful developing and printing service which for efficiency, speed and moderate charges is unexcelled.

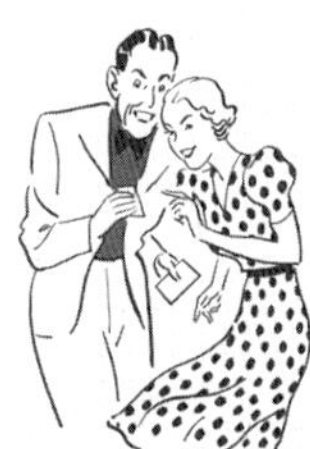

DEVELOPING CHARGES

4 exposure standard film, per spool - **4d.**
All 5, 6, 8, 12, or 16 exposure spools up to postcard size, per spool - - - - **6d.**
Fine grain developing, per spool - - **9d.**

PRINTING CHARGES

BLACK AND WHITE

All sizes up to 2½in. x 1⅝in. (vest pocket) - - - **1½d.** each
2¼in. x 2¼in., 3in. x 2in., and 3¼in. x 2¼in. - - - **2d.** each
4¼in. x 2½in., 3½in. x 3½in., and ¼ plate - - - **2½d.** each
Sepia prints 25% extra.

ENLARGING

A really happy snap deserves enlargement and Boots carry this out for you at most moderate charges as the following prices show :—

ENLARGED PRINTS approximately 4in. x 3in. according to size of negative, from sub-standard negatives, **3d.** each.

POST CARDS from the whole of the negative. From one negative, four for **1/-**.

From assorted negatives **4d.** each. **3/6** per dozen.

FILMS

Whatever the make of camera you use you will find the films for it in Boots Photographic Departments. Wherever you may find yourself on holiday in England, Wales or Scotland you will be within reach of a branch of Boots if you run out of films.

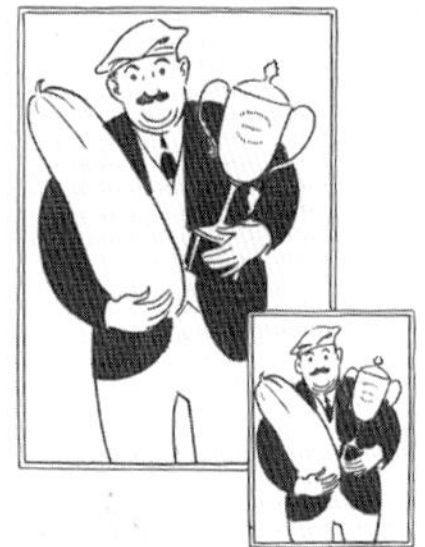

You can rely on Boots

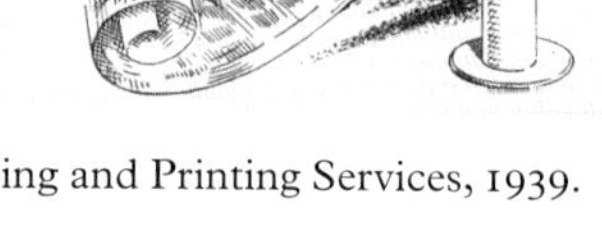

FIGURE 12. Leaflet advertising Boots Developing and Printing Services, 1939. Original in color. The Boots Archive.

for indoor, night, and winter photography—that is, to keep photographing even when the days grow shorter.[5]

Although the bloated British Empire is beginning to disintegrate, pictures of Britannia and words like "Imperial" still adorn photographic papers, films, and plates. Like other industries, photographic manufacture is becoming dominated by a few monopolies: Ilford, Kodak, Agfa (though by the mid-1930s the German Agfa film, widely considered the best, is being boycotted by Jewish photographers as well as by people who don't want to buy "foreign" goods).[6] Ilford Limited is highly dependent on its imperial markets, where it is less threatened by these larger competitors. In its products, its marketing, and its newsletters, the company takes into account the different weather and climate conditions of the empire and produces chemical products specially for tropical conditions.

In magazines like *The British Journal of Photography*, *The Amateur Photographer and Cinematographer*, or Ilford's *The Ilford-Selo Record*, discussions of "the tropics" abound. The term is used very loosely to describe all sorts of hot, humid places which are not necessarily located in the geographical tropical zones. Across the empire, articles in the photographic press about "difficult" or intemperate climates for photography echo wider discourses about difficult and unruly colonized peoples and about the place of Britain in the world.[7] Ideas about race and about climate are tightly bound. Marketing materials also reveal that concerns about culture, race, and class haunt discussions of how photographic materials and cameras respond to heat, dust, moisture, and pollution. The conversations in letters and articles repeatedly contrast the unruly tropical climate to a "normal" British one, conveniently ignoring how wet weather and poor light makes photographing in Britain, outside of the April to September "photography season," difficult for most amateurs (see chapter 15).

Even if the photographic press tries to make light of British weather, the industry does not. The variable weather of the British Isles is shaping attempts to control the sensitivities of photographic materials. Industrialization has changed the atmosphere of these islands, releasing gases and particles that subtly alter its overall balance and transform the climate but also, more immediately and locally, cause sudden, unbreathable, deadly fogs. Persistent fogs plague the capital each November, and gray smog seems to hang permanently over the northern industrial cities. In January 1931, following some of the worst fogs in England, Boots's sales staff are told that "FOG" stands for "Feature Own Goods" and that they should promote the company's own Compound Glycerin of Thymol for the treatment of catarrh and colds, and Boots's Throat Pastilles to "Impregnate and Purify the Air that you Breathe."[8]

Just as human lungs respond to these kinds of changes in the atmosphere, so do photographic materials. By the 1920s, light-sensitive materials are taking the measure of the industrialized atmosphere. Meteorological and metaphorical atmosphere meet on photographic negatives, recorded as a certain vagueness, a veil that falls over the image, softening focus and lowering contrast. The technical literature on photography treats the impact of chemicals and particles in the air as a hazard that can be overcome. Yet the factory, the darkroom, and the photography studio are beset by unwelcome visitors: by the dust that people always shed, made worse by their coal fires at home, by the greasy particles that come in the polluted air from other industries burning coal, by the general damp of the climate which seems to penetrate the thickest walls, by the filthy paraffin safety lamps (finally replaced by clean electric lights in this period). And most of all by the fog. Pollution events such as smogs directly affect the industry, whose products required clean air and bright light. In 1930 and 1931, Kodak Ltd., the British subsidiary of Eastman Kodak, even attempts to bottle the London fog in order to analyze it.[9]

In the interwar period, climate was not understood to be in "crisis" as it is now, and in the late 1930s, when industry scientists did become aware of ongoing global warming and knew that it was caused by industrial gas emissions and coal burning, they assumed temperature rises in the atmosphere would continue to remain at a relatively low level, would guard against the threat of another ice age, and would "prove beneficial to mankind."[10] But if devastating global warming was not anticipated, for more than half a century there had been concerns about the dangers of smog for human health, by reducing sunlight and forcing inhalation of toxic gases and particles. By the 1920s, it was known that industrialization changes the atmosphere, releasing gases that subtly alter its overall balance and transform the climate but also, more immediately and locally, cause sudden toxic fogs. The First World War had also changed ideas and fears about breathable air, since atmosphere had become militarized with the introduction of gas warfare in 1915. Despite the bright artificial colors and electric lighting that shone with the optimism of modernity, every generation has its fears, and pollution events threatened to revive nightmares of the Flanders battlefields.[11]

06 THE MEUSE VALLEY FOG

To understand how the meaning of fog had changed after the Great War, we need to visit a specific fog: the Meuse Valley fog in Belgium, between Huy and Liège. In 1930 this fog killed sixty people over a period of three days between December 3 and 5, and caused several thousand people to become seriously ill. Cattle also died and sickened in large numbers. Today the Meuse Valley fog is considered a landmark in the understanding of air pollution. An investigation of the causes of the disaster in the Meuse Valley proved that atmospheric pollution could cause death and severe illness, and identified the likely causes of the fog. These findings were published in a report in 1931 (translated into English in 1936).[1] It attributed the deaths to a toxic combination of the local topography, seasonal climatic conditions, and air pollution in a densely populated and heavily industrialized area where homes were heated by coal and the coal-fueled factories along the river Meuse manufactured glass, steel, zinc, fertilizer, and explosives.

The investigation was conducted by an expert committee appointed by the Royal Prosecutor of Liège on December 6. This committee interviewed family doctors and patients to establish the symptoms brought about by the fog, and studied the conditions in the area. It concluded that the fog was at first a meteorological phenomenon produced through anticyclonic conditions and high atmospheric pressures combined with a light wind. Warm air held the cold fog near the ground but above the level of the factory chimneys, which prevented the factory fumes from rising higher and forced the gases and particles they emitted to accumulate in the valley. The Liège basin was one of the most industrialized regions of mainland Europe. Over ten thousand tons of coal were burned daily in the valley, releasing tons of carbon dioxide, carbon monoxide, nitrogen, sulfur dioxide, and dust particles into the atmosphere. There were twenty-seven factories emitting about thirty different substances, including lead and cadmium and various metal oxides.[2] Most of these chemicals were excluded as causes, but the committee's report said that "fine soot particles, onto which irritant gases had been adsorbed, had a major role in the noxiousness of the fog."[3] The particles could remain suspended in the air for a long time, entering the houses and the lungs of their occupants. The report concluded that it was in fact sulfur dioxide from coal burning, both domestic and industrial, that was the most toxic.

The Meuse Valley investigation has been celebrated by some writers as the first scientific proof of deaths and illness caused by air pollution, but others have argued that the report fudged the issue by emphasizing the exceptional topography and weather, and by bundling industrial coal consumption together with domestic. Alexis Zimmer says that the Meuse Valley fog was made to seem a singular event, almost a fluke, and plans to handle future episodes merely involved asking industry to reduce atmospheric emissions in certain weather conditions (he argues that this approach remains today and "in a way, we still live in the mists of 1930").[4] Its significance for my account here is not in how it proved the deadliness of air pollution, but in what the press coverage of it reveals about perceptions of fog at the time, particularly in Britain.

The fog was described as gray in areas and yellow in others, heavy and slippery, smelling of sulfur. It scorched throats and lungs, stung eyes, tightened chests, turned saliva black, made people retch and vomit and caused the cattle to wheeze. Zimmer describes how the inhabitants of the Meuse Valley experienced the fogs as unnatural, with a distinctive smell and taste, so that they did not doubt its industrial source. Even so, before the release of the report, scientists and other commentators offered the media a range of speculative explanations.

As Zimmer summarizes, they suggested "special microbes carried the southerly winds; gases and dust from a recent volcanic eruption; a 'slower drowning'—a less academic version of hydrodiffusion—the saturation of air with water vapor causing spasms from the muscles and glottis leading to asphyxia."[5]

The report dismissed these theories along with others that were circulating, such as the notion that airplanes had sprayed poisonous gas on the valley, or that the factories had deliberately taken the opportunity of poor visibility in the fog to release unusually toxic fumes.[6] Nevertheless, these ideas had already been spread in the extensive press coverage in Belgium and abroad. The event was widely discussed in the British Isles, where the newspapers of December 6 carried sensational headlines such as "The Mist of Death," "Grim Fog Horror," "The Death Vapor," "The Trail of Death," and "Fog Horror on the Meuse." Many invoked the Great War with headlines like "Scenes Reminiscent of the War" and comparisons to wartime gas attacks, as well as repeated rumors of poisonous "War Time Dumps." Words like "poison fog" and "poisonous gas" were repeatedly used, implicitly comparing the event to the poison gas attacks at Ypres in the First World War. At the same time, journalists speculated that this was a "Form of Plague," a revisitation of "Black Death," or at least some kind of "Deadly Germ."[7]

Interest was heightened by the fact that England, and particularly London, was unusually foggy in November and December 1930. About a week after the Meuse Valley event, the *London Evening Standard* was reporting "the day that never came" due to a "black smoke fog" that lingered above the streets, cutting out the light below, forcing people to walk with flares and lights, and slowing traffic to a snail's pace.[8] Within a few days, articles began to appear that raised fears of a "peril" in Britain similar to the Meuse Valley fog.[9] The newspapers shifted their accounts as the investigation in the Meuse unfolded, one day attributing it to the freezing fog alone, another to the germ theory, wartime dumps, or the various emissions from the factories. Most of all they emphasized the mystery of the fog and the lack of a definite knowledge of the causes, and they seized on moments when the cause seemed to have been definitively identified.

The Meuse Valley fog also gave commentators a chance to pontificate on the health effects of fog in general. The fog was a familiar character in the British press in this period, particularly in local papers during the winter, when it was credited with stopping play at football matches and causing train derailments and road accidents. Newspaper coverage played on existing attitudes toward fog and emphasized the exceptional nature of this fog. Take, for example, an article entitled "Why Danger Lurks in Fog: History Repeats Itself in Belgium." Its anonymous author, described as "A Doctor," links the Meuse Valley fog to ongoing "widespread and continuous" fogs hanging over Britain. They then move on to the subject of events in Belgium, events which "in their suddenness and symptoms resembled those of poison gases as used in the war," although, unlike mustard gas, this fog hung around for a long time and covered an entire valley.[10]

Having stoked readers' fears by suggesting the parallel with gas warfare, the writer then makes a comparison which draws attention away from the idea that the fog might be anthropogenic: the Meuse Valley fog resembles "The unusually severe winter of 1916–17" in which respiratory diseases killed some of "our men in Flanders." The article muddles old and new theories of contagion, making claims about the unhealthiness of "mists and dank air" that draw on the Hippocratic theory of miasma, which attributed contagious diseases to vapors from putrefying organic matter or stagnant water. Since the sixteenth century, as plagues swept across the continent, European medics had thought of contagious diseases in terms of poison, and had revived the Hippocratic theory, understanding miasma as a kind of aerial or atmospheric poison, and using the term until the 1880s. Though outdated, in the interwar period this theory persisted in popular ideas about health, alongside newer chemical or toxicological accounts of the effects of the gases now being produced industrially.[11] The article by "A Doctor" blurs together various impacts of fog, such as depression and chills, stuffy interiors, lack of sunlight, and "soot, dust and particles of poisonous organic matter" combined with "the various waste gases of life and civilization." The overall argument is that fogs, however polluted, are not dangerous to people who are healthy, so long as they breathe through their noses when outside and ventilate their homes, but that they can be deadly to anyone suffering from respiratory diseases.

So, while "history seems to be repeating itself in the ill-fated district around Liège," it is to winter illnesses, not poison gas, that the reader is directed. The writer mentions, but plays down, pollution as a cause of death, and invokes the modern understanding of infectious diseases as caused by viruses and bacteria alongside the older miasma theory. This maneuver, in which public fears are raised and then dismissed, can be seen in other coverage of the Meuse Valley fog. In their determination to minimize the possibility of air pollution as a cause, news articles like this drew attention to the anxiety among the British that the deadly fog in Belgium might trigger.

07 THE TERROR AND THE BEAUTY

► It is curious to see how the British press covered the Meuse Valley disaster photographically in the absence of photographs of this specific fog. *The Illustrated London News* of December 13, 1930, uses a photograph of one of the villages in the valley "in which there were numerous victims" (fig. 13). It depicts a picturesque scene of a cobbled street along the edge of the river. It is a clear day and the village is quiet, populated only by a cyclist, a pedestrian, and one car in the distance. On the river there is a procession of four boats; the first appears to be a tug, steam and smoke rising from its tall chimney. Perhaps these exhaust fumes stand in metonymically for the fog itself. The brief text below the image dismisses the idea that industrial pollution might have contributed to the fog: "It was first suggested that the heavy fog which had been hanging over the district for some days contained poisonous fumes from neighboring zinc works; but it was found that the works in question had been closed down for some time. . . . It now appears that the glacial character of the fog in the river valley was alone responsible." A second photograph is captioned "The Queen of the Belgians in the Area Stricken by the 'Fog-Plague': Her Majesty Talking to a Mourner." The queen is shown in profile at the left of the image, surrounded by a crowd of men. It is unclear to whom she is directing her gaze, or who in the image is the mourner. The two photographs are late to the event; they seem to depict an

1074 THE ILLUSTRATED LONDON NEWS Dec. 13, 1930

EXPLOSION; TORNADO; DEATH-FOG: DISASTERS OF THE SEA AND OF THE LAND.

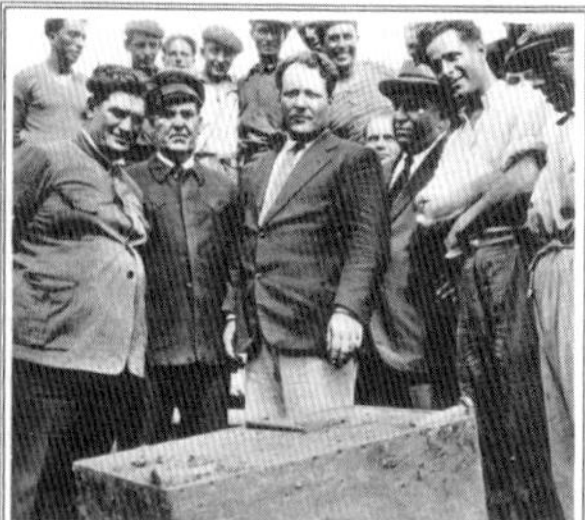

THE SALVAGE-SHIP DISASTER: THE "ARTIGLIO," WHICH WAS DESTROYED BY THE BLOWING-UP OF THE MUNITIONS-SHIP "FLORENCE," WHICH SHE WAS ATTEMPTING TO DEMOLISH.

VICTIMS OF THE "ARTIGLIO" DISASTER: CAPTAIN BERTHOLOTTO (IN CAP); AND DIVERS FRANCESCHI, GIANNI, AND BARGELLINI (L. TO R.).

The Italian salvage-ship "Artiglio," famous for her successful location of the wreck of the P. and O. liner "Egypt" this summer, and for other feats of deep-sea salvage, was destroyed, off Quiberon, by a violent explosion on December 7. She had remained in the Atlantic after suspending work on the "Egypt," and had undertaken to clear away the wreck of the steamer "Florence." The "Florence" carried a cargo of munitions. To demolish her, charges of explosives were placed round her as she lay in only 50 ft. of water, a little north of the Island of Houat. The "Artiglio" was about 100 yards off when the charges were fired, and, it is assumed, by some miscalculation the cargo of the "Florence" blew up. Eye-witnesses describe an enormous column of smoke and water 900 ft. high. When this cleared away, the "Artiglio" had gone, and only seven of her crew were picked up alive.

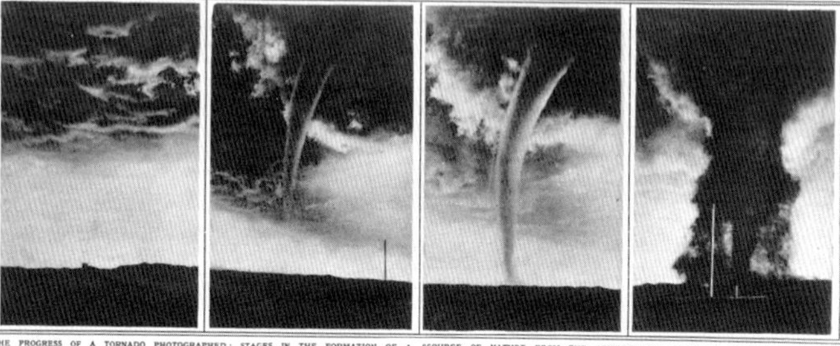

THE PROGRESS OF A TORNADO PHOTOGRAPHED: STAGES IN THE FORMATION OF A SCOURGE OF NATURE FROM THE ACCUMULATION OF BLACK CLOUDS TO THE POINT AT WHICH A FARM-HOUSE WAS STRUCK.

Readers will remember that we have from time to time made a feature of remarkable photographs from America illustrating the formation of big tornadoes there. The series reproduced above (taken by Mrs. Roy Homer, of Gothenburg, Nebraska) is of peculiar interest in that it shows the successive stages of a tornado's progress—(l. to r.) firstly, black clouds rolled together; secondly, the formation of the curious funnel pointing downwards "like a radish"; thirdly, the tornado reaching a small pond, which it sucked dry; fourthly, striking a farm-house.

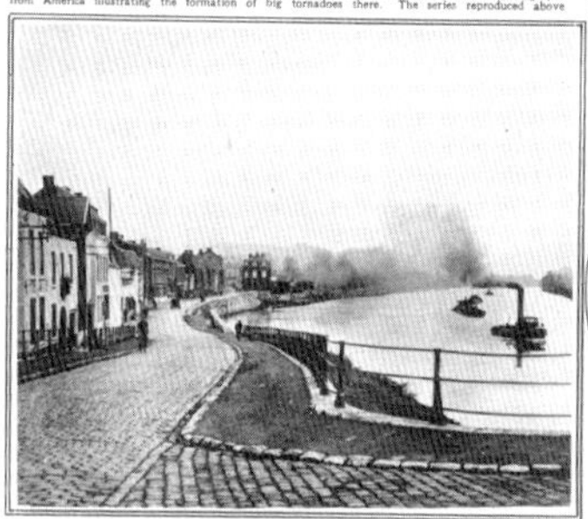

THE MYSTERIOUS, MUCH-DEBATED BELGIAN "FOG-PLAGUE": A LITTLE MEUSE VILLAGE IN WHICH THERE WERE NUMEROUS VICTIMS.

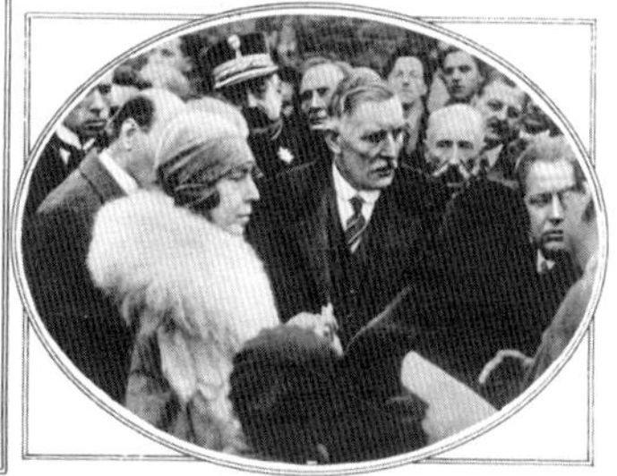

THE QUEEN OF THE BELGIANS IN THE AREA STRICKEN BY THE "FOG-PLAGUE": HER MAJESTY TALKING TO A MOURNER.

Some sixty persons died suddenly in three or four villages situated on the banks of the Meuse, in Belgium, about ten miles above Liège, on December 5, and, as a number of cows and other animals died at the same time, a rumour quickly spread that some sort of epidemic was raging. It was first suggested that the heavy fog which had been hanging over the district for some days contained poisonous fumes from neighbouring zinc works; but it was found that the works in question had been closed down for some time. Another explanation advanced was an escape of poison-gas left buried by the enemy during the war. It now appears that the glacial character of the fog in the river valley was alone responsible. The victims were all aged persons, men or women suffering from asthma or other bronchial affections. At the instance of the Queen of the Belgians, a scientific commission is to enquire into the precise cause of the deaths.

FIGURE 13. "Explosion; Tornado; Death-Fog," *The Illustrated London News*, December 13, 1930. © Illustrated London News Ltd/Mary Evans.

aftermath, a return to order—although it is possible that the first photograph was taken prior to the disaster.

The appearance of orderliness is disrupted by the overall narrative of the whole page of photographs, with the header "Explosion; Tornado; Death-Fog: Disasters of the Sea and of the Land." Two photographs at the top of the page illustrate a story about the accidental explosion of the Italian salvage ship *Artiglio*, one showing an intact *Artiglio* prior to the explosion, another a group portrait of crew members who were killed. Again, the actual incident is not depicted, and can only be imagined. The center of the page shows a sequence of four photographs of "The Progress of a Tornado." Here the story is about the "remarkable" photographs themselves, which portray four stages: the clouds gathering together, the tunnel of the tornado forming in the sky, the tornado "reaching a small pond, which it sucked dry" (although the pond is not visible in the reproduced photograph), and finally "striking a farm-house." We are not told if anyone was killed or injured.

The placing of the Meuse Valley fog story directly below these photographs invites readers to visualize fog and explosion as something like the photographed tornado: gray and black clouds moving irresistibly and inevitably toward destruction. The victims and mourners are bit players in a larger drama in which storms, explosions caused by munitions, and industrial smog are equivalent—all disasters, tragic accidents, fatal and fated and spectacular even if they cannot be photographed, only imagined. Through this arrangement of photographs and stories, *The Illustrated London News* sets out both to shock its readership with stories of terrible events and to persuade them that these disasters are of the same order. There is something disconcerting in the way that horror and attraction sit side by side, in the celebration of the extraordinary ability of photography to capture the tornado, even as this stands in for its failure to have captured the other events as they happened. The coverage reflects an ambivalence particular to the interwar period in Britain, in which an older aesthetic of pictorialism with its interest in picturesque atmosphere coexists alongside a new high-speed technical photography that can arrest the most spectacular sights of science and nature.

A similar fascination with disaster and destruction can be seen in an earlier issue of *The Illustrated London News*, also from 1930, that brings together fire, fog, and gas attacks in the United States (fig. 14). A floodlit, nighttime scene of a fire at the White House is accompanied by an image of workers clearing the detritus in President Hoover's office. In the middle of the page an image of an explosion on a research yacht captures the moment of destruction as a plume of dark smoke envelops the rear deck; next to it is placed a photograph of the tops of New York buildings emerging from a fog. Bottom left is a staged hold-up of a bank to demonstrate the use of tear gas for defense. The page ends, as the Meuse Valley one did, with an image of authorities apparently in control: a photograph of police handling a replacement supply of tear gas canisters, following a prison mutiny that had been suppressed using the gas. Real disasters entailing loss of life (the yacht's explosion) merge with staged ones and events alluded to but not seen. The page reveals an aesthetic interest in the formless qualities of clouds, smoke, and fog, and a fascination with the disaster and danger they signify. Tear gas had not yet been used in Britain or the British Empire, with popular opinion set against it following British First World War propaganda condemning the Germans' use of gas warfare. It was considered particularly exotic and American, associated with the United States thanks to its appearance in American gangster films such as *The Big House* (1930).[1]

The obsession with the visual effects of smoke and fog continued into the late 1930s. On January 21, 1939, the magazine *Picture Post* published a selection of London fog photographs under the heading "Foggy Morning" (fig. 15). The accompanying text refers to an earlier article by Edward Hulton bemoaning the destruction of "the natural beauty of Britain" through building developments, then adds: "On these pages you see another kind of natural beauty which man has so far not been able to destroy—the beauty of atmosphere." It is hard to imagine that the author did not realize that the London fog was related to "man's" destruction of the atmosphere through pollution. Ironically, concerns about urban industrialized atmosphere, along with overcrowding in inner cities, were helping to propel the building of very suburban ribbon developments that

JAN. 11, 1930 THE ILLUSTRATED LONDON NEWS 61

FIRE; FOG; AND A "SYNTHETIC" GAS-ATTACK: UNITED STATES NEWS ILLUSTRATED.

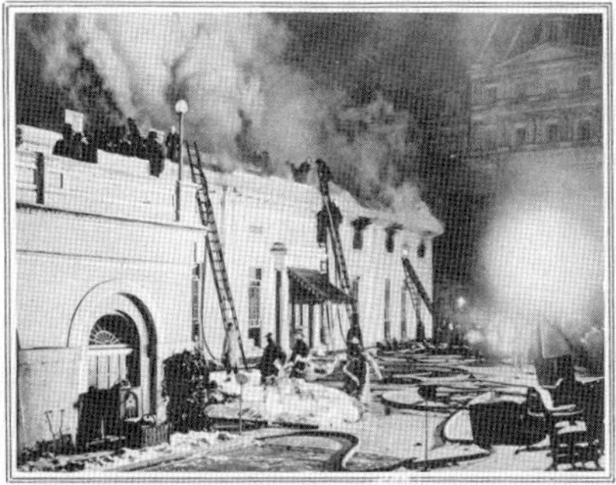

THE CHRISTMAS EVE FIRE AT THE WHITE HOUSE EXECUTIVE OFFICES: FIREMEN AT WORK DURING THE OUTBREAK, WHICH MADE IT NECESSARY FOR PRESIDENT HOOVER TO MOVE INTO THE HISTORIC STUDY OF PRESIDENT LINCOLN.

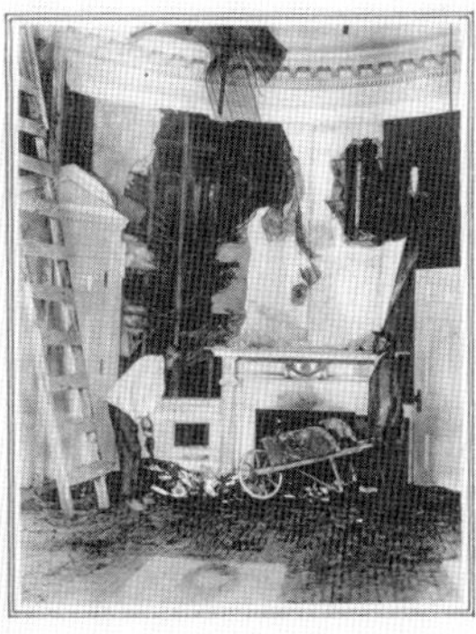

WHY MR. HOOVER HAD TO MOVE HIS WORKING QUARTERS! THE PRESIDENT'S ROOM BURNT OUT IN THE WHITE HOUSE EXECUTIVE OFFICES, WASHINGTON.

The Executive Offices of the White House, a long and low building some two hundred feet from the White House proper, were on fire last Christmas Eve, and two hours passed before the blaze was extinguished. It was possible to save the President's personal files, but his work-room, with others, was practically gutted. Many historic public documents were destroyed, though, luckily, many of their kind were removed long ago to the Congressional Library or to the State Department. As a result, Mr. Hoover had to transfer his work to the study President Lincoln used to occupy in the White House itself.

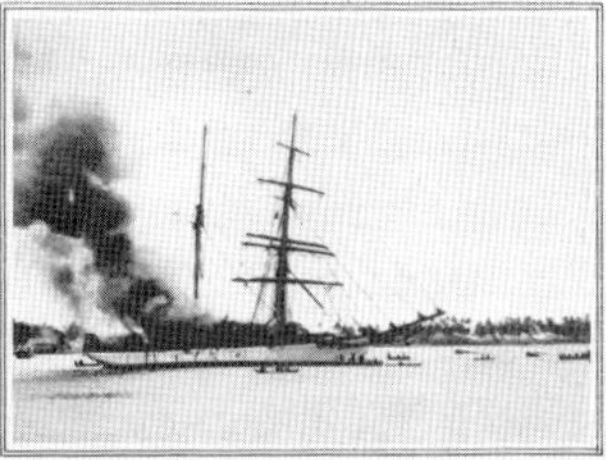

THE FIRE ON THE NON-MAGNETIC SCIENTIFIC YACHT "CARNEGIE," THE RESULT OF A PETROL-EXPLOSION WHICH HAD FATAL CONSEQUENCES AND DESTROYED THE VESSEL: THE SHIP ABLAZE IN APIA HARBOUR, SAMOA.

On December 1, a wireless message reached New York, from Apia, stating that the non-magnetic yacht, "Carnegie," of the Carnegie Institute of Washington, which was specially built some years ago for the purpose of scientific research, had been burned in the harbour there, as the result of a petrol-explosion which had occurred on the Saturday. Captain Ault, the commander, died of burns; and the cabin-boy was reported missing. Other members of the crew were injured. The ship had on board a party of seventeen scientists, who were cruising to study atmospheric electricity and magnetism.

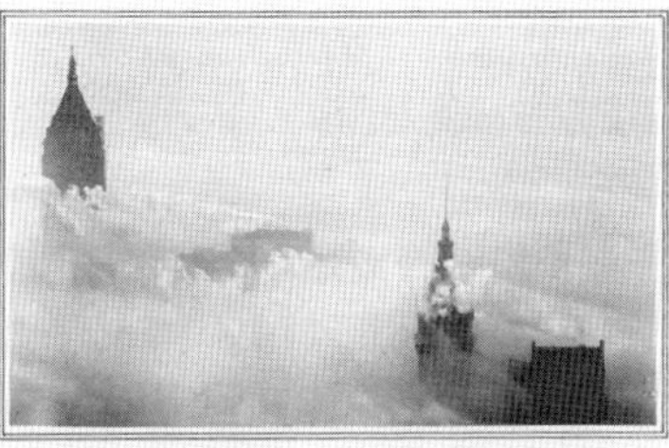

SKY-SCRAPERS IN THE CLOUDS! AN ODD PHOTOGRAPH TAKEN FROM THE TOWER OF THE WOOLWORTH BUILDING, NEW YORK; SHOWING THE TOPS OF BUILDINGS IN THE CLEAR BELT ABOVE THE BANK OF FOG ENVELOPING THE CITY.

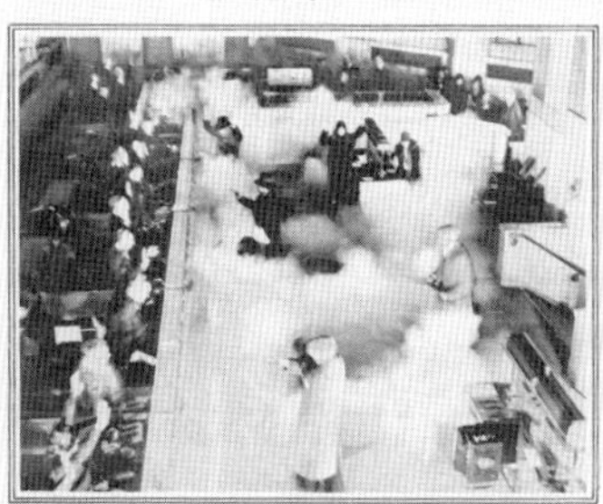

A "SYNTHETIC" GAS-ATTACK: A BANDITS' RAID STAGED IN A PITTSBURG BANK TO DEMONSTRATE A SYSTEM OF DEFENCE BY BLOWING TEAR-COMPELLING FUMES TOWARDS THOSE ATTEMPTING TO HOLD-UP THE CASHIERS AND CUSTOMERS.

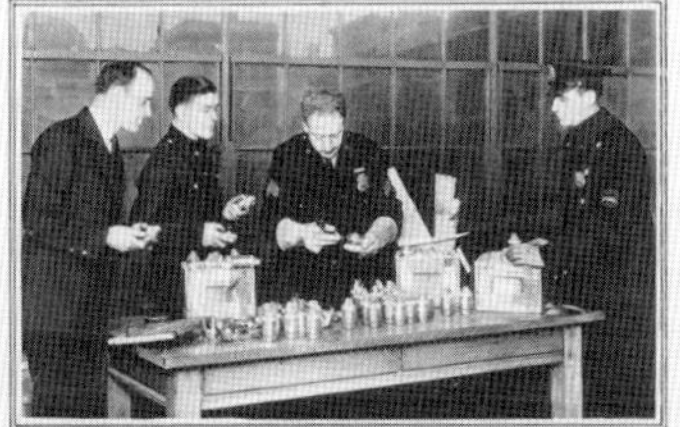

TEAR-GAS AS PART OF THE EQUIPMENT OF THE UNITED STATES POLICE FORCE: RELOADING TEAR-GAS BOMBS IN NEW YORK, TO REPLACE THOSE SENT BY AEROPLANE FOR USE DURING THE AMAZING MUTINY AT THE NEW YORK STATE PRISON AT AUBURN LAST DECEMBER.

The "gunman" is so active in the United States, and so daring, that many a bank takes very special precautions. The first of the two pictures here given shows one of those initiated at Pittsburg, Pennsylvania, illustrating what a correspondent describes as a "synthetic" raid on the Monogahela Bank: that is to say, an imitation raid to demonstrate how bandits might be defeated by means of tear-gas released at a given signal and blown towards them. Needless to say, a harmless cloud of smoke was used for the test. The second picture is of a scene that was a sequel to the prison mutiny which took place at the New York State Prison, at Auburn, on December 11 last, and was illustrated in our issue of January 4. On that occasion, tear-gas bombs were sent to the "battle-area" by means of aeroplanes, and were used by the troopers attacking the building.

FIGURE 14. "Fire; Fog; and a 'Synthetic' Gas-Attack," *The Illustrated London News*, January 11, 1930. © Illustrated London News Ltd/Mary Evans.

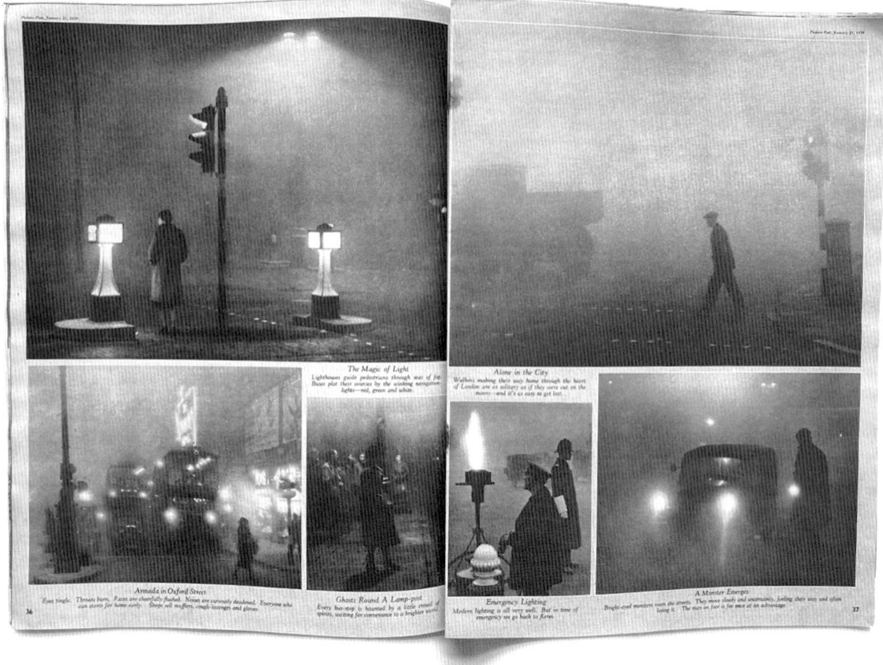

FIGURE 15. *Picture Post*, "Foggy Morning," January 31, 1939. Private collection.

Hulton had disparaged. But this writer is concerned with beauty, a beauty that "owes nothing to man" and indeed can turn humanity's "ugliest productions . . . into something lovely."[2]

In the black-and-white photographs, the amorphous fog is given form by lighting: without the presence of bus headlamps, navigation lights ("miniature lighthouses"), shop signs, and illuminated advertising, fog would be nothing other than a gray smudge, the photographs indistinguishable from failed or "fogged" photographs. Oddly, the captions do not entirely steer away from the health effects of the London fog or the difficulties of navigating the fog: "Eyes tingle. Throats burn. Faces are cheerfully flushed. . . . Shops sell mufflers, cough-lozenges and gloves." At bus stops people wait "for conveyance to a brighter world," and in the foggy city it is "easy to get lost." Yet despite the unpleasant physical effects, the overall impression is a romantic one of a city transformed into a mysterious and magical place.

At least one reader objected to the article. Ernest Restell wrote to *Picture Post*: "I am sorry your pictures were so beautiful, for fog is an ugly, harmful thing. In its commoner form of smoke (produced by the inefficient combustion of raw coal) it is a nuisance. In its concentrated form—fog—it is a menace."[3] He goes on to describe the negative effects of smoke and fog on buildings, on sunshine levels, on health, and on women (by increasing their load of housework), concluding that "'Beautiful' hardly seems the epithet." Like the literature that mythologized the London fog, photographs of fog contributed to the impression that fog was extraordinarily picturesque, even beautiful, and detracted from the actual experience of being in its choking, filthy air. Restell's complaint contrasted the brutal reality of the fog and its picturesque representation. Though we do not know Restell's views on photography in general, he echoes older concerns about how an attachment to the picturesque aesthetic of atmospheric soft focus undermined the core purpose of photography, understood as the detailed, sharp, and ideally "objective" representation of the world as it is.

Restell's letter also reflects the conflicting meanings that fog had come to hold. It had acquired new, sinister connotations that coexisted with the perception of it as picturesque. If, by the interwar period, people recognized that the London fogs could be deadly, they had also become newly anxious about toxic atmospheres thanks to the use of poison gas in the Great War (see chapter 6). Additionally, decades of clean air campaigns had worked hard to give the dirty and smoke-filled fog negative meanings. It was associated with the destruction of the city environment, especially historic buildings. Because of the additional housework that soot entailed, coal smoke was seen as a nuisance to women in particular. Smoke abatement campaigns drew on emerging discourses about women's liberation from household drudgery. Funded by the gas industry, clean air campaigners pushed for gas heating to replace coal from the late nineteenth century onwards. In the 1930s, Gas Association advertising promoted the National Smoke Abatement Society, emphasizing the filthiness of coal and its effects on health.

By aligning coal smoke with unnecessary housework, the Gas Association ads seemed to place themselves on the side of women's liberation and introduced a form of gender politics into the debates about the dangers of fog. For example, an advertisement that ran in *Picture Post* on November 19, 1938, (fig. 16) emphasizes that coal smoke is a cause of women's overwork because of the enormous amount of soot produced by coal fires: "All over the food. All over the furniture. All over the walls and covers and curtains." Women are described as "soot slaves." The advertisement suggests replacing coal fires not only with gas but also with coke—a coal-derived fuel that was less smoky and conveniently produced together with coal gas. Other ads in the same campaign suggest that the dirt produced by the coal fire may be imperceptible: "Thousands of tiny invisible particles floating in the air."

The ad parodies two common features of women's magazines, photographically illustrated recipe pages and photo-story narratives, the latter constructed through sequences of captioned photographs that draw on the lighting and framing techniques of cinema. In this extremely limited vision of women's liberation, housework remains unquestionably a woman's task and an untroubled marriage her principal goal, but the thing that causes disharmony is the coal fire. The inset photo-story-style pictures depict a before and after—in the first, the man is a threatening figure looming over his wife, who looks as if she is preparing to defend herself with the object she holds in her right hand. The photograph is lit from below and to one side, in the manner of a scene from a thriller. The second photo, lit and framed as a close-up in the style of a romantic film, presents a scene of happiness as, the woman's work done, the couple head to the cinema. The implication is that women have only themselves to blame for overwork, and even for domestic abuse, if they have a coal fire.

The emotional repertoire of this advertising campaign is striking. Images of soot in custard and text about "tiny invisible particles" are designed to evoke disgust, and the image of the man looming over the woman implies the threat of violence. An ideal life is suggested,

Picture Post, November 19, 1938

Week-end recipe

CUSTARD WITH SOOT SAUCE

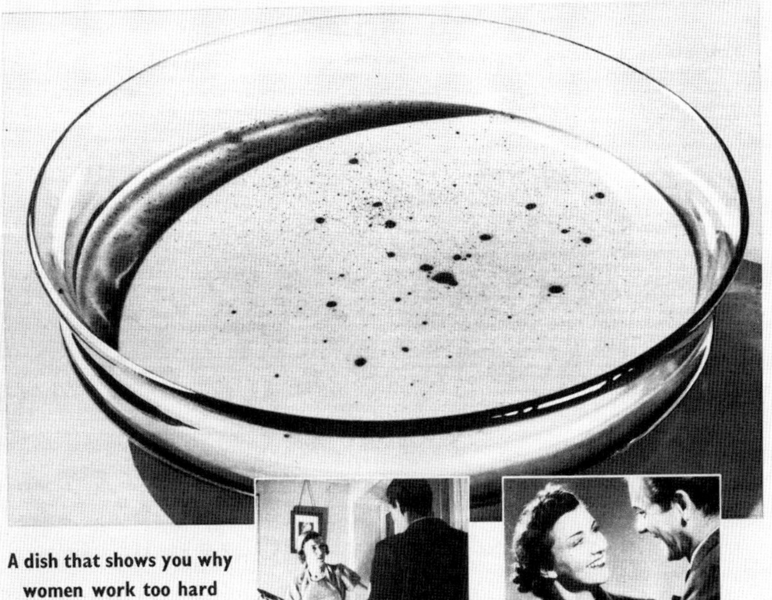

A dish that shows you why women work too hard

When you leave a custard uncovered, and find it nicely decorated with smuts like this, you know at once what the trouble is. The fire's smoking. But did you know that ordinary, old-fashioned fires are *always* smoking? They're *always* sending tiny particles of soot and dust and smuts into the room. All over the food. All over the furniture. All over the walls and covers and curtains. That's why so many women have to slave much harder in the living-rooms, washing paint and fabrics, than in the bedrooms. They have a clean and efficient gas or coke fire in the bedrooms, and yet stick to their old-fashioned smoky fire downstairs.

Heartbreak House: 'Are you home already for your supper? I haven't finished cleaning the place yet . . .' Old-fashioned sooty, smoky fires are starting trouble in this home.

Happiness House: 'Darling, work is so EASY nowadays! The place almost keeps itself clean since we've put in the gas fire and the coke fire.' And so . . . off to the pictures.

GAS AND COKE free soot slaves!

Stop eating soot sauce—and stop working as hard as two washerwomen! Why don't you join the sensible modern housewives who heat all their rooms by gas and coke—fires without smoke? There's nothing as cheery as a glowing coke fire for the living-room. There's nothing as handy as a handsome gas fire for the dining-room. Work—and washing bills—are cut down weekly. And you'll be prouder than ever of your warmer and *cleaner* home. Pop into the local Gas Showrooms when you're out to-day.

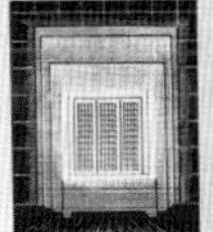

For rooms where you want warmth quickly, if only for short periods—a gas fire. Choice of many colours; quick, clean heat. It ventilates a room automatically

NO SMUTS
NO SMOKE
with
GAS
and
COKE

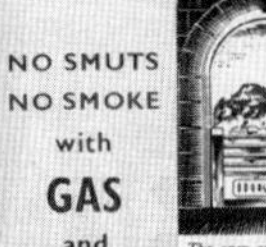

The open coke fire burns without dirt or smoke. Saves 2d. in the 1/- on your fuel bills. Gas-ignited on tap. Very convenient, very modern.

Advertisement issued by the British Commercial Gas Association, Gas Industry House, 1 Grosvenor Place, London, S.W.1.

BRITAIN'S BURNING SHAME
GAS and COKE can clear the Air!

70 per cent. of all city fog and grime comes from smoke in home chimneys—'Aerial Sewage' that blots out health-giving sunshine. Why would the same family spend 1s. less on washing bills in Harrogate than in Manchester? What is the Black Smoke Tax? How can you fight smoke? Get the illustrated magazine-book, 'BRITAIN'S BURNING SHAME,' issued by an independent body—the National Smoke Abatement Society. This is of such importance that supplies have been made available at your Gas Showrooms. Ask for a free copy. It will make you angry. It will make you think!

60 PICTURE POST

FIGURE 16. Gas Association Advertisement in *Picture Post*, November 19, 1938. Private collection.

too—cleanliness, romance, leisure time, and respite from work—but feelings of fear, disgust and anxiety are prevalent. The campaign effectively reinforces the negative emotions attached to the smoke-filled fog, with its threat of suffocation, contamination, and bodily harm.

08 TROUBLE WITH FOG

Fog—i.e., a general deposit of silver all over the plate, including those parts that should not have been acted on by light. This may arise from accidental exposure to light before development; from exposure to active light during development . . . ; from the use of a too energetic developer; from too prolonged development; from over-exposure in the camera, in which case the edges of the plate that are protected by the rebate of the dark slide will remain clear, except near to dense parts of the picture; from imperfectly blackened camera bellows, from which oblique light, which should have been absorbed is reflected on to the plate; from dust on the surface of the lens, causing the light in its passage to be partially diffused; possibly, though rarely with modern plates, from defective emulsion.

C. H. BOTHAMLEY, *The Ilford Manual of Photography*, 1921

► Photographic practice in London is from the beginning inhibited by the city's persistent fogs, which become more frequent and more dense just as the "mania" for photography starts, in the 1840s. This is the decade in which London, always fog-prone, gains its long-standing image as the city of fog, the "big smoke." If the throats and lungs of city dwellers register it, so do photographic materials. On June 15, 1841, following experiments with his newly invented calotype process, William Henry Fox Talbot writes to his wife, Constance, from London: "My windows in Cecil St command a good view of the river but unfortunately I find that the London atmosphere prevents a good result, even when the fog is hardly visible to the eye."[1] Photographers would tend to assume that it was the lack of light caused by fog that made it difficult to photograph in a period when long exposures were required, even in bright sunlight. But Talbot's letter seems to hint at something else—that it is not the absence of light or the low visibility but something in the atmosphere of the city that affects his process. From its inception, photography has been affected by atmospheric conditions. Photographic materials behave like barometers or weathervanes—they take the measure of the surrounding air.

Some eighty years later, on January 19, 1923, at the British company Thomas Illingworth, a staff chemist writes a memo: "We have definitely decided that in future coating of any kind of emulsion must not be commenced or proceeded with during a fog as the risk is not worth the small gain we aim at in coating during a fog."[2] There had been fog in the London area on January 17, 1923, and probably the "risk" refers to contamination from atmospheric pollution, while the "gain" might be the added moisture in the atmosphere that helped prevent film from curling during the coating process.[3] In the archive, the memo is glued to a page inside one of the Illingworth experiment books, under the heading "Trouble with fog." The experiment books are handwritten notebooks that include formulas dating back to the early 1900s, but which continued to be added to as late as the 1930s. The purpose of the experiments described in these books was to identify and resolve common technical issues, one of which was fogging.

"Fog" in photography does not usually refer to the weather. It describes a distinctive appearance of a print or a negative, an overall and uneven, muddy, or lackluster appearance, in which the range of shades between black and white is reduced so that everything in the image seems flatter and lacking in contrast. It is as if something—a veil or a mist—has got in the way between the viewer and the depicted scene. In 1928, Olaf Bloch of the Ilford Research Laboratories published an article called "The Chemist in the Photographic Industry," in which he described how, in cases of fogging, the difference between the unexposed and light-exposed halides (silver salts) in an emulsion is reduced, giving less gradation in the image.[4] The causes vary: contaminants in water, in the photographic paper, or in the atmosphere; light leaks in the camera or during storage of light-sensitive materials. Bloch listed the London fog among the potential contaminants that posed a hazard to the sensitive photographic emulsions.[5] The 1923 note warned against "coating in a fog" because the London fog, laden with contaminants including smoke particles and sulfurous compounds such as hydrogen sulfide and sulfur dioxide, caused photographic fogging.

Before reading this memo in the archive, I had not made any connection between actual meteorological fog and the photographic term "fogging." As someone who had grown up with photography, and who has developed and printed my own photographs, I am familiar with the photographic sense. I had just not connected it with the weather. Or at least, I had assumed that "fog" applied to photography is a metaphor because a fogged image looks *as if* a fog or a mist has been cast over the image. After all, we also use the term "fog" figuratively to describe a confused or unsure mental state, as in "brain fog" or "fogged memories," and variations of this have apparently been part of the English language since around 1600, only half a century after the first recorded literal use.[6] But in photography, fogging is a technical term without an alternative, which means that images are literally rather than figuratively fogged. In the context of chemical photographic practice, terms like "fog" and "fogging" are examples of what literary theorists and linguists call catachresis, or false metaphor, in which a word attached to one concept, such as a person's arm, is attached to another similar concept, such as an arm of a chair.[7] The term "fog" in photography derives from the experience and appearance of fog, just as the term "arm" in relation to furniture derives from a resemblance to human arms and the experience of resting human arms on chair arms. However, this little note pasted into the Illingworth experiment book acts as a reminder that actual fogs plagued the production of photographic materials.

It may seem, from Talbot's experience with the London atmosphere—or from the numerous damaged, faded, or decayed examples of old processes such as daguerreotypes, salt prints, and calotypes—that fogging might always have been an issue in photography. Yet as far as I can tell, the term was not used to describe photographs until the advent of the collodion process introduced by

Frederick Scott Archer in 1851.[8] Perhaps this is because that process was more complex and involved more steps than most earlier processes. In any case, from the 1850s onward the technical literature of photography is full of descriptions and distinctions between different causes and types of fog. In his 1854 book *The Collodion Process on Glass*, Archer attributes fogging to contaminated water, incorrect pH of the "exciting bath" (the silver nitrate bath used in the collodion process), or light leaks during development. Archer's discussion makes it clear that by 1854 fogging was already a familiar term: "If the surface of the whole picture assumes a dull, brownish gray color, (fogging as it is so commonly called), it arises from the precipitation of silver itself, which may be entirely in the body of the collodion, or partly as a deposit on the surface, veiling the developed picture."[9] This explanation chimes with later descriptions of fog on gelatin-based plates. *The Ilford Manual of Photography*, in a section called "Negatives and Their Defects," defines fog as "a general deposit of silver all over the plate, including those parts that should not have been acted on by light" and also describes partial fog "from the action of an impure atmosphere" or contaminated paper (containing metals or other substances).[10]

In 1928, Olaf Bloch offered an explanation of fogging in gelatin emulsions. To make emulsion, the silver solvent (silver nitrate) is added to a mixture of gelatin, water, a soluble bromide, iodide, and ammonia, and this mixture is warmed. Microscopic grains of halide begin to grow into crystals. Bloch acknowledged that how this happened, or what caused it to happen, was not fully known. After this controlled crystal-growing process, termed "ripening," the emulsion is set (to a jelly) then broken up or shredded and washed and then "further digested" in another warming process to develop the capacities of the emulsion as required. "If, at any stage, the processes are pushed too far, the emulsion tends to give fog during development, i.e., the difference between the unexposed and light-exposed halides grows less."[11] Here, fogging is not a result of contamination but a matter of timing. There is more risk with fast (more sensitive) emulsions, as small variations in production and development time have greater effect: "in the case of a fast emulsion, we are working upon the border-line of fog; and it is not a little remarkable that a product of such character should retain its properties unimpaired over a long period of years."[12] Fog always lurks as a possibility, and as faster emulsions are produced, the risk of fogging increases.

In the case of wet plate collodion, Archer had gone so far as to distinguish between the appearance of different fogs. Fogging that results from the precipitation in the nitrate bath has a specific color cast—"a dull, brownish grey color"—and silver "veils" the image just as a fog veils a scene. Fogging produced by light leaks has "a certain indistinctness" or "haze." He was one of several writers who began to set out a taxonomy of photographic fog.[13] They practiced methods of reading photographs symptomatically, which is to say, looking for causes. Because photography was already conceived of as a means of making light-pictures, its other sensitivities were understood as fallibilities, listed and studied in great depth in order to eliminate them.

Fog was something that could happen, not just in the darkroom but inside the camera itself, or during the process of putting the film into the camera. Additionally, cameras and films, generally designed to approximate the vision of the human eye, share our eyes' inability to see in the fog. The worst London "pea-souper" or "London particular" (as it was often called) could penetrate both cameras and rooms. A man named Louis Katin, in a letter to the *Palestine Post* published on February 28, 1934, described how he yearned for the blue skies of Palestine but found himself in the "fog-bound" colonial capital: "You shut the windows up tight and the fogsmoke comes in through the keyhole. It makes your eyes water, and your nose smart. It makes your throat and mouth feel like the morning after the night before. You ram the keyhole with paper, but still it doesn't take the hint. It comes in through the chimney instead." This is the world turned upside down. Surely smoke is supposed to exit via the chimney, not enter the house by this means? A similarly topsy-turvy thing happens on the street when near becomes far and the blind become the only ones able to "see": "With a great fear, you realise the way is blank and lost in the fog. Thirty yards distant is home, a cheerful fig [*sic*], and a loving family. It might just as well be thirty miles, or thirty thousand. You are exiled in a London fog. For hours you grope around strange streets, knocking at strange doors, accosting strange people. If you are lucky you may meet with a blind man and he will lead you safely home."[14]

Outside on the street, fog-blindness afflicted eyes and cameras, making the world a strange and unfamiliar place. Since cameras are not necessarily airtight, the film or plate in the camera could receive the fog, not only as a visual thing mediated by the lens, but as something which touched the photographic surface itself, just as it touched Katin's eyes, nose, and throat. If photographic emulsions respond to contaminants in fog when wet in the darkroom, might they also respond when dry, inside the camera? This raises the tantalizing possibility of understanding the fogged image not just as a failed image but as a self-portrait on the part of the fog. Alongside a symbolic "reading for atmosphere," in which the flaws, marks, and inconsistencies of an image contribute to the affective or aesthetic atmosphere it conjures, I am suggesting the possibility of a literal one—a diagnostic reading in which the marks and blotches of an image speak of the very air that surrounded it.

09 HOW TO PHOTOGRAPH FOG

At present, people see fogs not because there are fogs, but because poets and painters have told them the mysterious loveliness of such effects. There may have been fogs for centuries in London. I dare say that were. But no one saw them, and so we don't know anything about them. They did not exist till art had invented them.

VIVIAN in OSCAR WILDE, "The Decay of Lying" (1891)

► Drawing on long artistic and poetic traditions giving symbolic meaning to fogs, mists, and smoke, in the first half of the twentieth century photographers and writers continued to represent the city of London as a city of fog. London was defined by its atmosphere: in his 1909 book *London: A Book of Aspects*, Arthur Symons wrote that "English air, working upon London smoke, creates the real London. . . . The English mist is always at work like a subtle painter, and London is a vast canvas prepared for the mist to work on."[1] Fog was useful to pictorialists because it made atmosphere visible and photographable while rendering objects and scenes indistinct and mysterious. London's fog seemed to have a "peculiar picturesqueness": in 1903, the pictorialist Walter Benington was praised for the way that his photograph of St. Paul's Cathedral captured the "beauty and romance" created by the "murky air of the metropolis" (fig. 17).[2] Press photographers were also drawn to the fog and similarly emphasized

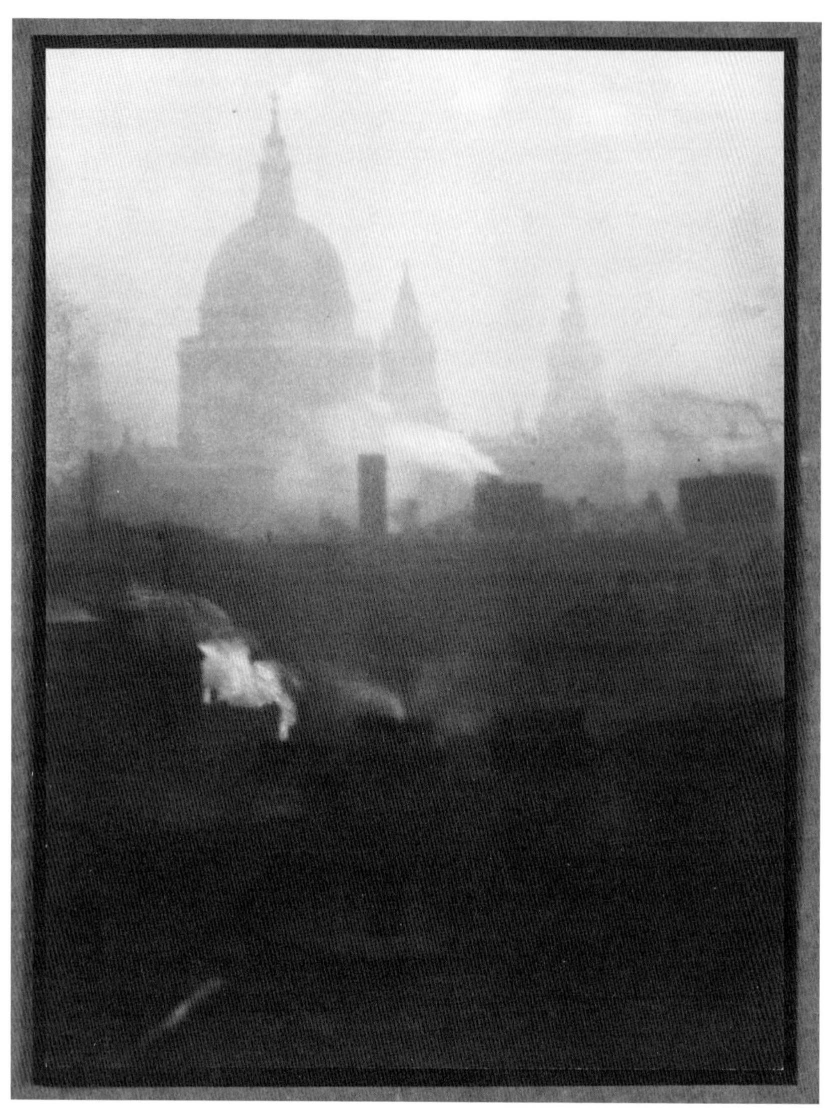

FIGURE 17. Walter Benington, *The Church of England*, platinum print, 1903. © Victoria and Albert Museum, London.

its picturesque side, despite the suffering and endurance that living with recurrent fogs required.

The London fog shaped perceptions of the city as the center of imperial power in the interwar period, even as the overstretched British Empire was straining under the pressure of independence movements and insurgencies from Ireland to Egypt and Punjab to Palestine. In the face of such challenges to the legitimacy of colonial rule, imperial propaganda intensified, and the fog, rather than undermining the image of British superiority, paradoxically supported it by signifying Britain's industriousness, heightening the mystery of the imperial center, and emphasizing its distinctive silhouettes.[3]

In press photographs, fog gave a new twist to an iconography of the city that was already becoming hackneyed—double-decker buses, traffic policemen, horse guards and beefeaters, and the neon advertising at Piccadilly Circus. It also introduced new subjects—the linkmen and their flares used to light the streets, bus conductors leading their buses on foot through the gloom, men and women walking in front of cars, using handkerchiefs to show the way. The darkness brought by fog evoked criminality, secrecy, and eroticism. Some London fog images resemble nocturnal photographs of the city, especially the books of night photography published in the 1930s. These books celebrated what the French-Hungarian photographer Brassaï described as the "magic of the lower depths."[4] If Brassaï's night photographs unveil the "eros of the city," as one writer puts it, 1930s press photographs of fog at least hint at it.[5]

The fog might leave its own photographic traces, regardless of the intentions of the photographer. Just as the sun drew its own picture, according to the early Victorians, so too did the fog, though its marks are less easy to identify. In a reproduction of a photograph by Monty Fresco, taken during the Great Smog of December 1952 for the Topical Press Agency (fig. 18), I thought I had found such a picture. It depicts a lamppost on the bank of the Thames at Blackfriars, the branches of surrounding trees, and a couple in hats and warm coats, leaning on the wall and looking out over the river, although the river itself is not visible. In the digitized copy, the fog is so thick that only objects in the mid-ground and foreground are identifiable, beyond a second lamppost little is visible. The image is completely covered

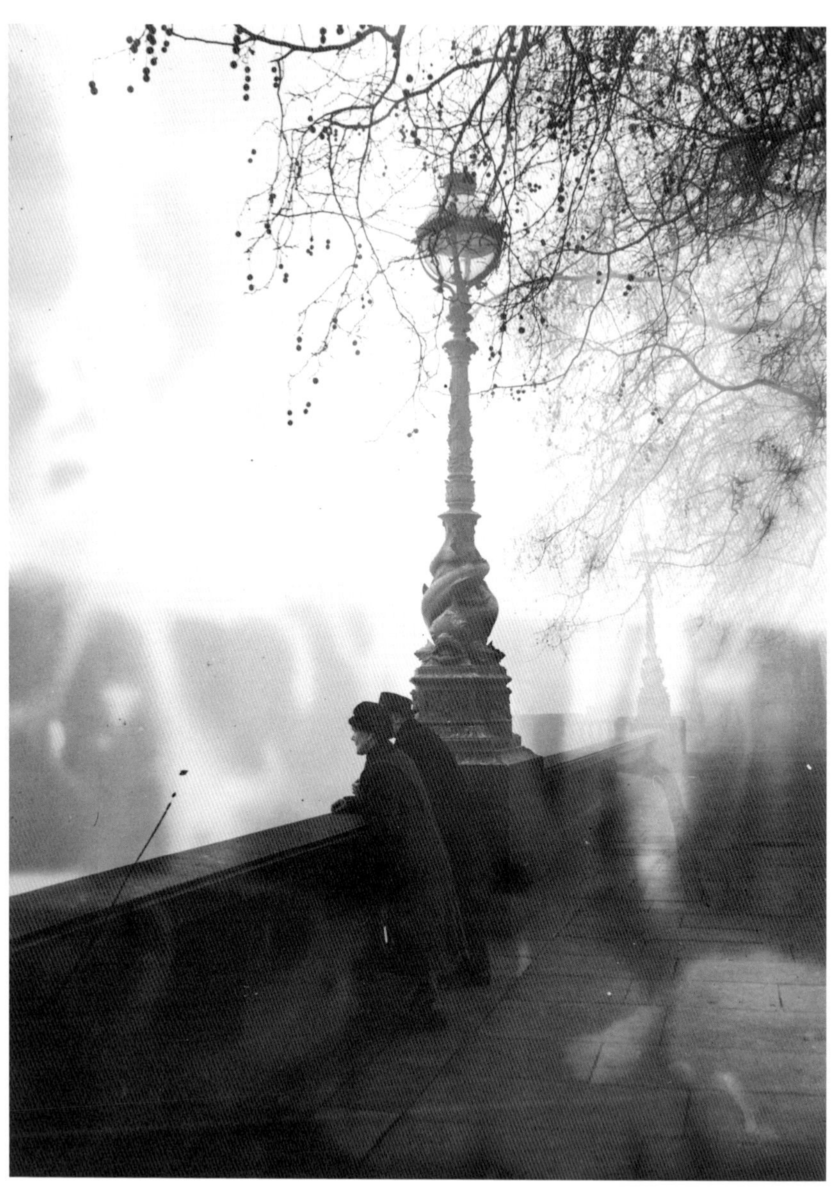

FIGURE 18. Monty Fresco, *Embankment at Blackfriars*, London, December 5, 1952. © Getty Images, Hulton Archive.

in blotches, white amorphous shapes which seem to explode in the sky above the couple as well as bubbles, smears, and dribbles marring the paving stones in the foreground. These uneven patches do not belong in the scene but are clearly chemical effects that result, for example, from a poorly fixed, poorly developed, or damaged image.

Why was this photograph considered good enough to keep rather than destroy? Was it good enough to publish? Perhaps these marks would not be seen in halftone reproduction or could be eliminated through retouching. Or perhaps they were the result of the image being recently scanned from a damaged print or negative? Or perhaps—indulge me here—it was the fog itself that touched the image. Did the fog get into the camera or the darkroom? Is this the result of atmospheric contamination, the fog portraying itself?

Consider another photograph, taken a day later, by a different photographer, Edward Miller (fig. 19). It depicts a streetlight and a number 9 double-decker bus, but little else is visible except great washes of gray, almost as if the surface of the print had been painted with large brushstrokes. There are blotches, blobs, splashes, and crazing on different areas of the image, plus a visible fingerprint. Were some of these marks deliberately added in the darkroom to give a pictorialist, painterly look? Or again, was this the fog, painting its own picture?

At Getty Images' Hulton Archive, I found several different prints made from the same two negatives, most (if not all) produced by the picture library decades after the photograph was taken. The prints from Fresco's negative show little sign of the streaks and blobs so visible in the online version, and less fog: the paving stones receding into the distance are clearly visible. But the negative, a five-by-four-inch glass plate, does have the marks that are visible on the digital scan, appearing blue-gray and sepia in patches and dribbles (see plate 9). None of the other glass negatives in the same box, taken in the same week, are affected.

If time has just been cruel to the negative, why was this effect only on this photograph, the sole one in that set that depicted fog? Alternatively, if the marks had been there since the negative was developed, why do the analog prints not reproduce them? Would the marks show when digitally scanned but not when printed traditionally with

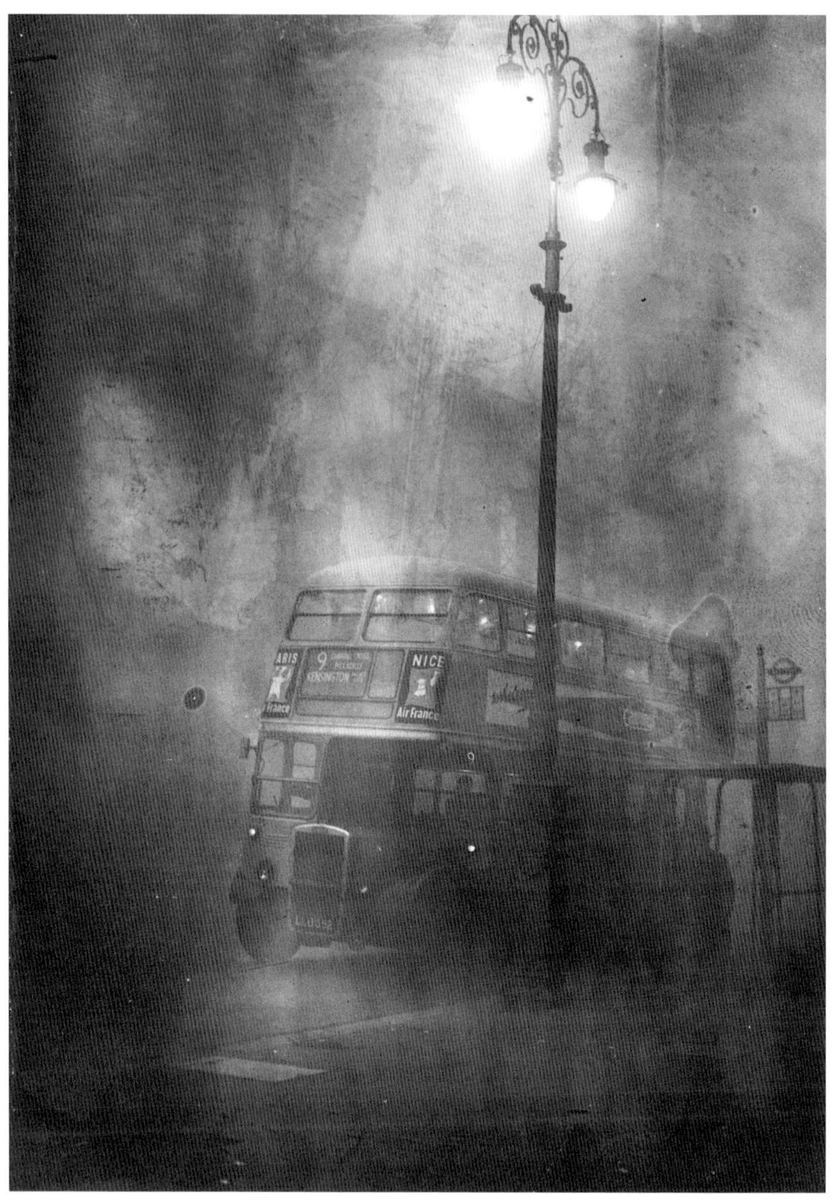

FIGURE 19. Edward Miller, *Fleet Street*, December 6, 1952. © Getty Images, Hulton Archive.

an enlarger? Experts at Getty Images explained that there could be many causes of the marks: contamination during development might reveal itself only after years or decades; there could have been a flaw in the manufacture of the plate; or contact with a destructive substance. While atmospheric sulfur contamination might produce these marks, it is impossible to know for sure.

In the case of the Miller photograph, all the painterly marks and patches are on the original glass plate negative, including the fingerprint at the base of the picture. It did not appear to have been manipulated, and the prints correspond to the current state of the negative, although processed to increase the contrast and tonal range. This negative is poorly developed, flat, and thin; I was told that this meant that all the marks of sloppy processing would be more evident. Fog made photographic practice difficult in many ways: difficult to get a good exposure because of low light, difficult to get a good range of tones, and inevitably harder to protect the sensitized plate, exposed negative, or print from the particles in the air. I wanted to read these marks symptomatically, to diagnose a definite cause, yet I could neither prove nor disprove the notion that the fog had made its mark on these pictures.

In one sense, paying attention to atmosphere entails considering the range of ways that photographs might speak of the air and the weather.[6] Literature scholars have argued that novels and poems can act as sensitive registers of weather and climate. Jesse Oak Taylor says that a novel can register the atmosphere "in a manner akin to a meteorological instrument."[7] I am making a similar claim about photography, but unlike novels, a photograph might register atmosphere chemically, without the mediation of its author. Tobias Menely compares poems to tree rings and ice core samples, as "particularly sensitive records for tracking the history of climate."[8] Photographs track climate in ways even closer to trees and ice since photographic materials are literally sensitive to radiation, chemicals, minerals, and moisture.[9]

Although it is tempting to treat a photograph as a barometer or air-quality sensor, the question of what becomes understood as "atmospheric" is interesting. I have yet to find Fresco and Miller's photos in reproduction, but it seems they were intended for circulation

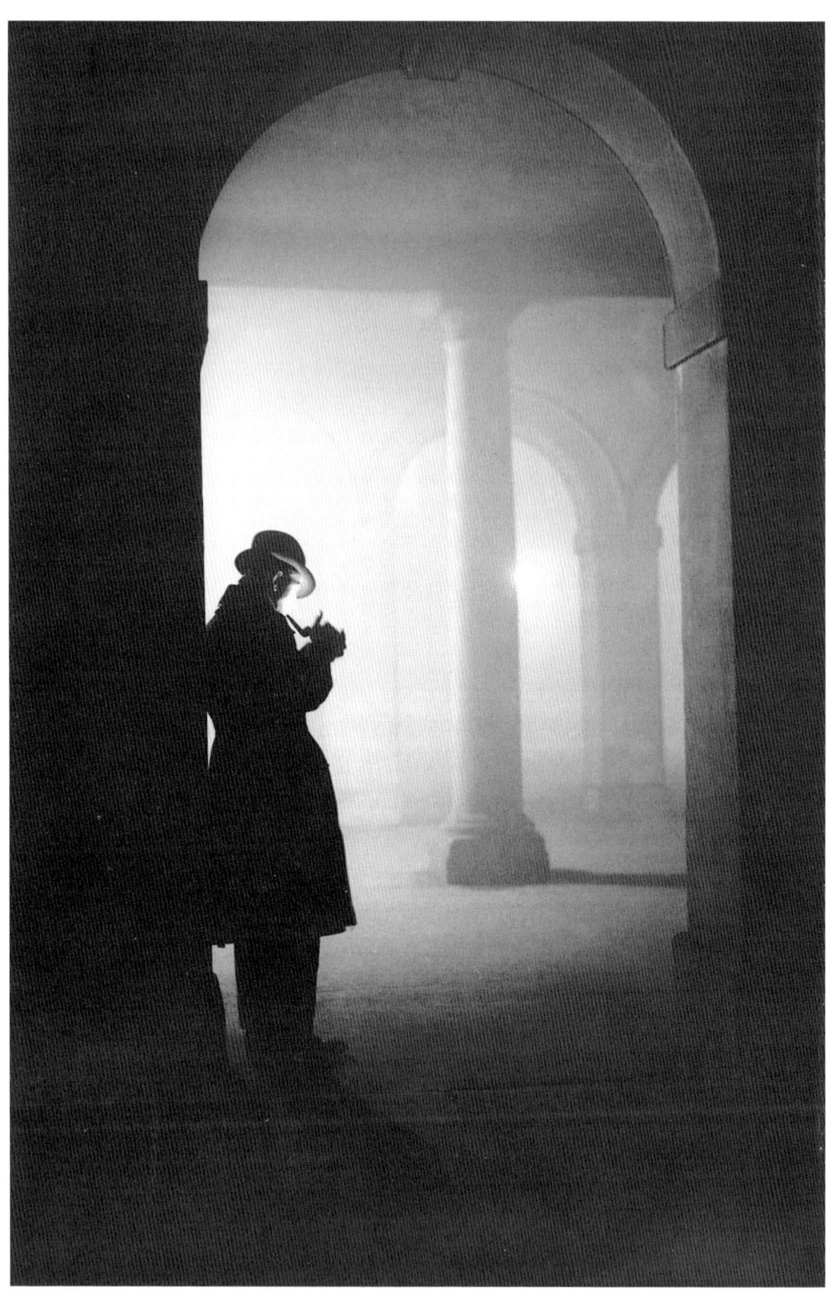

FIGURE 21. Arthur Tanner, *A Man Lighting His Pipe in Thick Fog Under the Arches at the Temple*, London, December 23, 1935. © Getty Images, Hulton Archive.

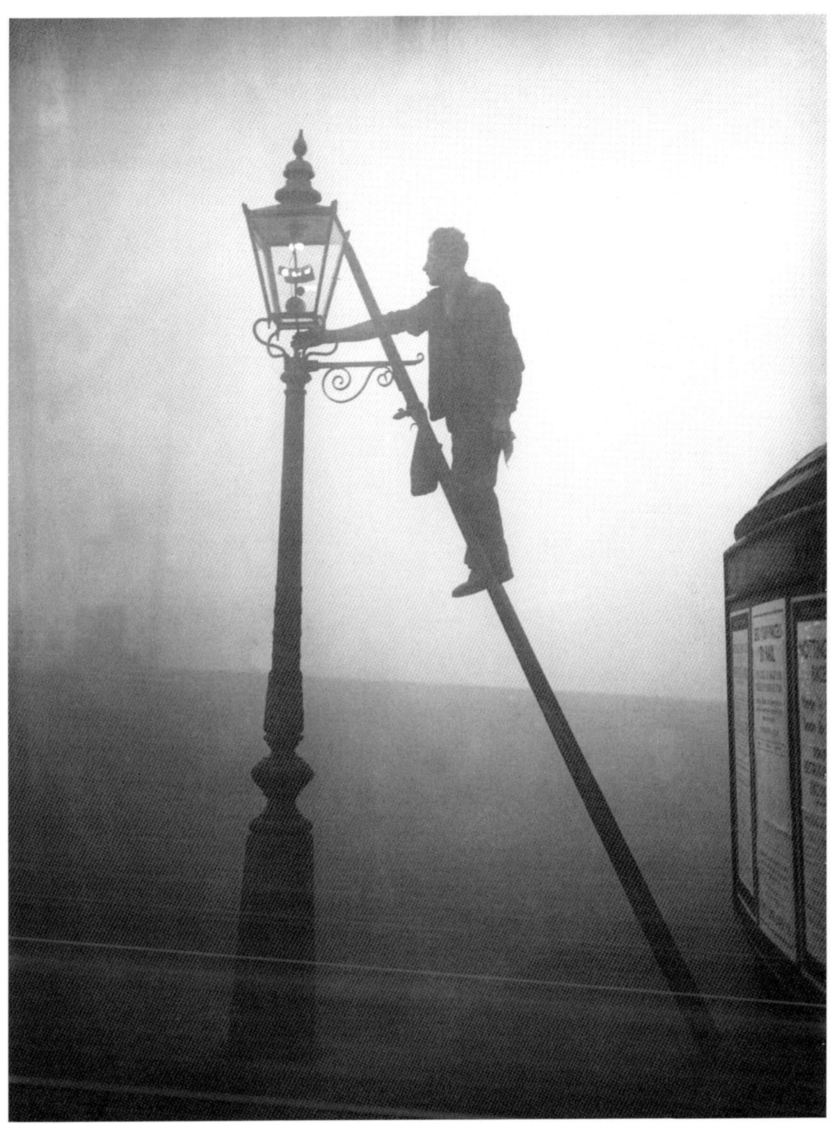

FIGURE 22. E. Dean, *Lamplighter*, Finsbury Park, London, October 17, 1935.
© Getty Images, Hulton Archive.

read it that way, receiving German movements like New Objectivity and New Vision as misguidedly favoring facts over feeling.[13] While British amateurs were encouraged to seek out atmosphere, New Objectivity was characterized by an "aversion" to atmosphere.[14] Katerina Korola has written about how the champion of New Objectivity, Albert Renger-Patzsch, carefully removed from his backgrounds any elements that might suggest less than pure air, even in his pictures of the mining plants and coal industry of the heavily polluted Ruhr valley.[15] He expunged both the flaws that intervened in printing and developing (such as streaks and bubbles from development, fogging, or dust) and the extraneous details that might intrude into a pure sky (such as cloud and smoke, wires, or birds).[16]

Pictorialists could also be virtuosos at controlling the appearance of their prints, but the darkroom technicians of the press, working with images destined for halftone newspaper reproduction and needing rapid turnover, took much less care, and the fog's continuous gray tones made vivid any dust, scratches, or signs of poor processing. Yet in some of these photographs, the net effect is to create a kind of flatness and a graphic, stylized impression different from both the subtle tonal range sought by the pictorialists and the crisp clarity of Renger-Patzsch's images. In a 1935 scene of a man lighting his pipe, the background appears flat like a stage or film set (fig. 21). It recalls film noir, partly because of the staged appearance, though it predates that genre. A silhouetted figure framed by an archway was a common trope in interwar photographs of fog, creating a contrast between the black foreground and the shallow gray scene beyond. Recognizable silhouettes were preferred: lamplighters, the bobbies (policemen) with their distinctive helmets, the porters at Billingsgate market, men with top hats. Fog appears as a dense, almost opaque backdrop, like the flat areas of gray paint used on prints by newspaper retouchers to isolate objects prior to reproduction. If Renger-Patzsch's prints cleaned up the filthy Ruhr valley air and rid the atmosphere of any materiality, the press photographers disarmed the murky London fog by turning it into an opaque surround against which the city's familiar "characters" become legible.

10 THE PATHETIC FALLACY

► In photographs, as I have suggested, effects of light and air are mobilized for meaning, to construct and convey moods and to communicate specific feelings. The means by which a photographer and a photograph convey mood, the material used to do so, include the objects in a scene—the expression on a face, the slump of a shoulder, the dirt on a street—but also the light and air, conveyed for example through shadows, reflections, vapor, and smoke.

A highly controlled deployment of the expressive elements of photography can be seen in Masahisa Fukase's photobook *Ravens* (*Karasu*, sometimes translated as *The Solitude of Ravens*), published in 1986 and renowned as a masterpiece of Japanese expressionism. Fukase uses techniques that another photographer might consider mistakes: underexposure, blur, very visible grain, multiple exposure (using flash and long exposure together), blotches, and distortions that come from looking through glass windows or from moisture on the lens of the camera. These effects lend a deep sadness to the images, as if they are regarded through eyes filled with tears. They combine with effects of light, weather, and air: darkness, fog, looming clouds, snow lying on the ground and swirling in blurry flakes, mist on the surface of the sea, steam and smoke from industrial chimneys and small fires, bright streetlights, moonlight, the breeze lifting a woman's hair, sunlight reflecting on the metal body of an airplane, an explosion of dust and detritus in a waste disposal machine.

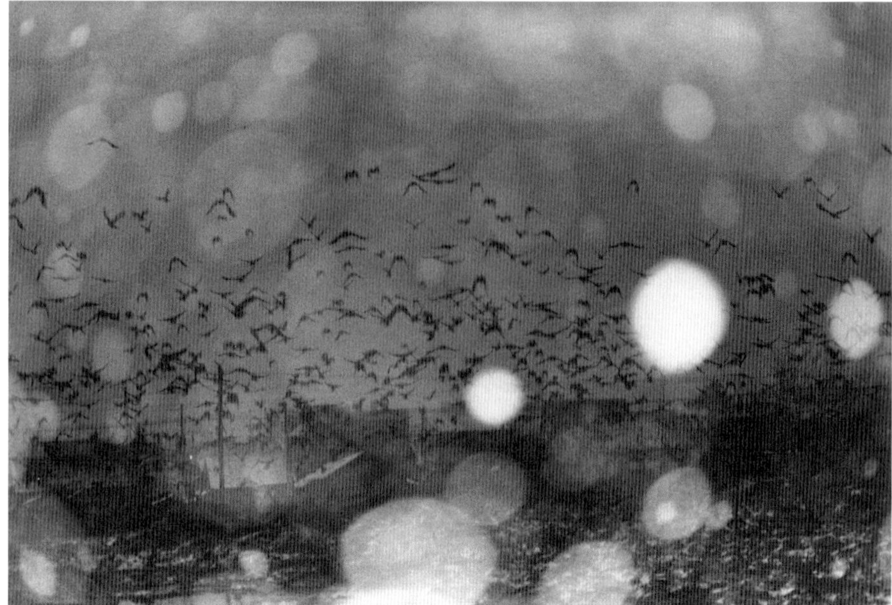

FIGURE 23. Masahisa Fukase, *Nayoro* from the series *Ravens*, 1977, © Masahisa Fukase Archives.

While the central characters of Fukase's series might be considered to be the ravens that he follows, which appear in many of the photographs and have their own explicit symbolism, the birds are invariably set against and viewed through this atmosphere that emerges from both meteorological conditions and photographic technique. Atmosphere mediates between feeling and meaning; we interpret the photographs not by decoding but as if by breathing them in. In a photograph, atmosphere can be the singular and ungeneralizable feeling of a unique moment, and at the same time suggestive of a wider state of affairs beyond the frame. As journalist Sean O'Hagan wrote in a review of Fukase's book, it "merges the deeply personal—his forlorn and obsessive state of mind—and the allegorical—the collective trauma of postwar Japan."[1]

Fukase's dark and personal expressionism, produced via a handheld camera, is very different from the more stately and static pictures of late nineteenth- and early twentieth-century pictorialism,

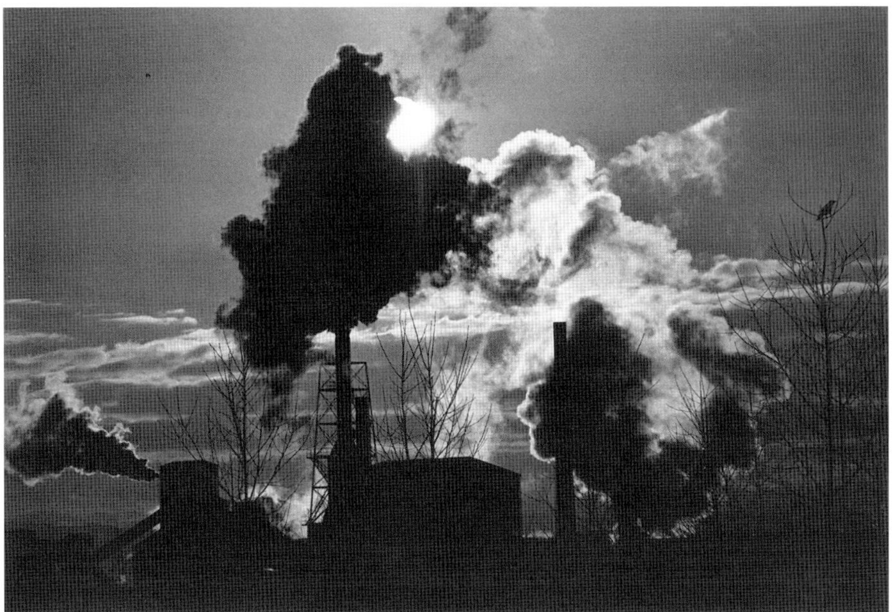

FIGURE 24. Masahisa Fukase, *Nayoro* from the series *Ravens*, 1978, © Masahisa Fukase Archives.

in which techniques of blur and grain generally serve the ambition of giving the photograph an aesthetic quality thought to raise it to the status of art in line with the painterly conventions of the time. For some pictorialists, fog, mist, and smoke could have symbolic meaning. Robin Kelsey, writing about the pictorialism of P. H. Emerson, says that mist and fog were used metaphorically by the Victorians to stand for "the filtering effects of emotion, communication, or memory."[2] Smoke from factories and houses could mean something more down-to-earth: for Walter Benington, set against the "spiritual" dome of St. Paul's cathedral, the surrounding London buildings "with their sordid smoky chimneys" stood for "bodily matter" (see fig. 17).[3]

But pictorialists also used atmospheric effects such as rain, fog, mist, and smoke to give emotional atmosphere to their pictures and to add a desirable vagueness and mystique that aligned their images with human vision and painterly impressionism. During the interwar period, promoted in the pages of *The Amateur Photographer and Cinematographer*, pictorialist aesthetics continued to be popular among amateurs who found that weather conditions helped to bind the details of the photograph into an "expressive unity."[4] One of the appeals of fog for them was the way in which it gave an immediate softness to the image and created effects akin to shallow depth of field. Critics of pictorialism, who often saw the job of photography as being to record the world objectively and realistically, had long derided the amateurs who went out looking for "atmosphere" on foggy days, calling them "mud-and-slush" photographers.[5] They attacked them not only for rapidly reducing the idea of photographic atmosphere to a set of clichés but also on moral grounds as threatening the truthfulness and realism of photography.

Shallow depth of field separates foreground, middle ground, and background and it ameliorated the much-criticized tendency of photography to "flatten" reality by rendering all things equal (and equally sharp).[6] Similarly, fog limits our vision to what is closest to us, obscuring objects in the distance and isolating objects in the foreground. Fog photographed is atmospheric in the aesthetic sense because instead of everything depicted being plainly visible, there is a sense of obscurity, presences are suggested without being fully delineated. The viewer must fill in the gaps, imagine what lies beyond the visible. Vagueness, softness, the impression of a veil over parts of the image: all gave fog a meaning in photography that it did not necessarily have in daily experience.

Writing about *flou*, a French term that encompasses blur and visual softness, the French photography theorist Pauline Martin talks about the long tradition of blurring and softening techniques in painting, claiming that "painters had already mastered blur for centuries, well before photographers entered the scene."[7] The two kinds of photographic blur—motion blur and the blur resulting from shallow depth of field—resulted from the specific technical design of cameras, lenses, and emulsions, and they had entirely different connotations. Motion blur seems more obviously a technical product of photography, because it does not correspond to the ways that humans see. Shallow depth of field lends a painterly quality that also seems to correspond to human optical experience, resembling perceptual blur (the defocusing we experience as part of human vision). Even so, as Martin points out, there is no direct relationship between painterly softening, photographic blur, and blurred vision, although people tend to conflate them. Nevertheless, she argues, in the nineteenth century "a game of back-and-forth" existed between the different kinds of blur.[8] I would add that this is complicated by an additional kind of *flou* in photography, the softening and obscuring effects of atmospheric conditions such as mist, rain, sea spray, and fog.

The process of projecting human states of mind onto the external world is sometimes called "the pathetic fallacy," a term coined by the British critic John Ruskin in his 1856 book *Modern Painters*. Ruskin thought that poets who project human qualities onto nonhuman things are falling for a fallacy. It was a sign of moral weakness, in his view, and of a "morbid state of mind."[9] He argued that reality needs to be seen for itself, not as something reflecting our own fantasies and affectations. I wonder what Ruskin would have made of Fukase's photographs, which depict a world saturated with the photographer's own grief. In Ruskin's view, the acceptability of projecting emotion or pathos onto nature hinges on whether a truth is being told about a state of mind such as grief, or if anthropocentrism or solipsism stands in the way of direct perception of the world.

Ruskin was famously observant of the weather, clouds, and fog. For example, in his 1884 lecture "The Storm-Cloud of the Nineteenth Century," he read extracts from his own diaries to show how the weather was changing, describing his first experiences of what he called a "plague-cloud" and "plague-wind," a "dry black veil" and a "deep, high, *filthiness* of lurid, yet not sublimely lurid, smoke-cloud; dense manufacturing mist."[10] This commitment to precise observation underpinned his earlier criticisms of artists and writers: they misrepresented nature by humanizing it, while Ruskin's observations allow him to see the human or anthropogenic element already present in "nature."

However, other writers have since argued that the pathetic fallacy is not a fallacy at all, but rather a use of metaphor to connect with the world. The environmentalist Neil Evernden writes that metaphor is driven by the desire and ability of the human mind to identify with that which is not human, saying "the Pathetic Fallacy is a fallacy only to the ego clencher."[11] Another writer, Bernard Dick, argues that the pathetic fallacy can be traced back to ancient texts like *The Epic of Gilgamesh* and understood as part of an older tradition of correspondences or correlations between humanity and nature.[12] When weather conditions become metaphors for emotional states, the effect is not necessarily to turn us away from our surroundings but to establish a relationship between interior and exterior, inner life and environment. In this way, a metaphor might bring the viewer or the reader into that "dense manufacturing mist," to help them get to know it and breathe its thick and choking air.

In my view, one of the most weather-sensitive of photographers, who also made pictures of the London fog in the 1930s, was Wolfgang Suschitzky (1912–2016), a Viennese Jewish émigré who had a successful career in photography in the UK. Technically trained at the Bauhaus, though not sharing a Bauhaus aesthetic, Suschitzky did his own developing and printing and made skillful use of atmospheric effects in his work. His photographs from this period are full of long shadows, reflections on wet pavements, and the glare of electric lights as well as fog and steam. He was not opposed to soft focus or motion blur if it proved atmospheric. In another, more realist photographic practice, Suschitzky's atmospheric effects might be regarded as "mistakes," especially the use of various kinds of blur, but also the conscious choice to shoot toward the sun, carefully controlling its effect through framing, or the filtering through smoke and steam.

There is a great deal of precedence for this, especially in Britain, in the tradition of pictorialism, but Suschitzky put these devices to work for a form of socialist realism. In doing so he went against the sharp clarity of New Objectivity, the style of photography dominating German-speaking countries in the 1930s. In its place he introduced a kind of romanticism (some would say sentimentality). Examples include a photograph of roadworkers spreading steaming tar (1935), a man pushing a milk cart early on a rainy morning on the Charing Cross Road (1936), a woman jumping a puddle in the same road (1937), and a steam train on the Oban line in Scotland in 1944 (shot directly into the sun). As Amanda Hopkinson wrote

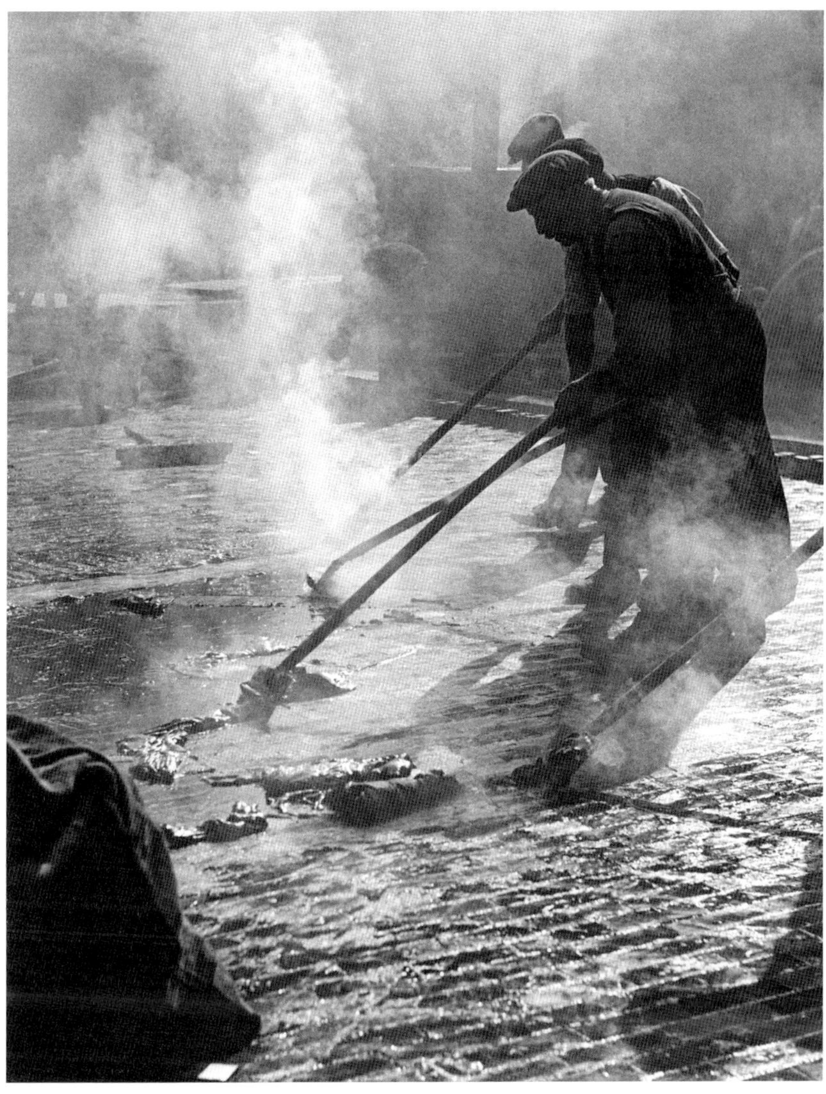

FIGURE 25. Wolfgang Suschitzky, *Woodblocks being tarred by roadworkers on Charing Cross Road*, ca. 1935. © The Estate of Wolf Suschitzky, courtesy of The Museum of London.

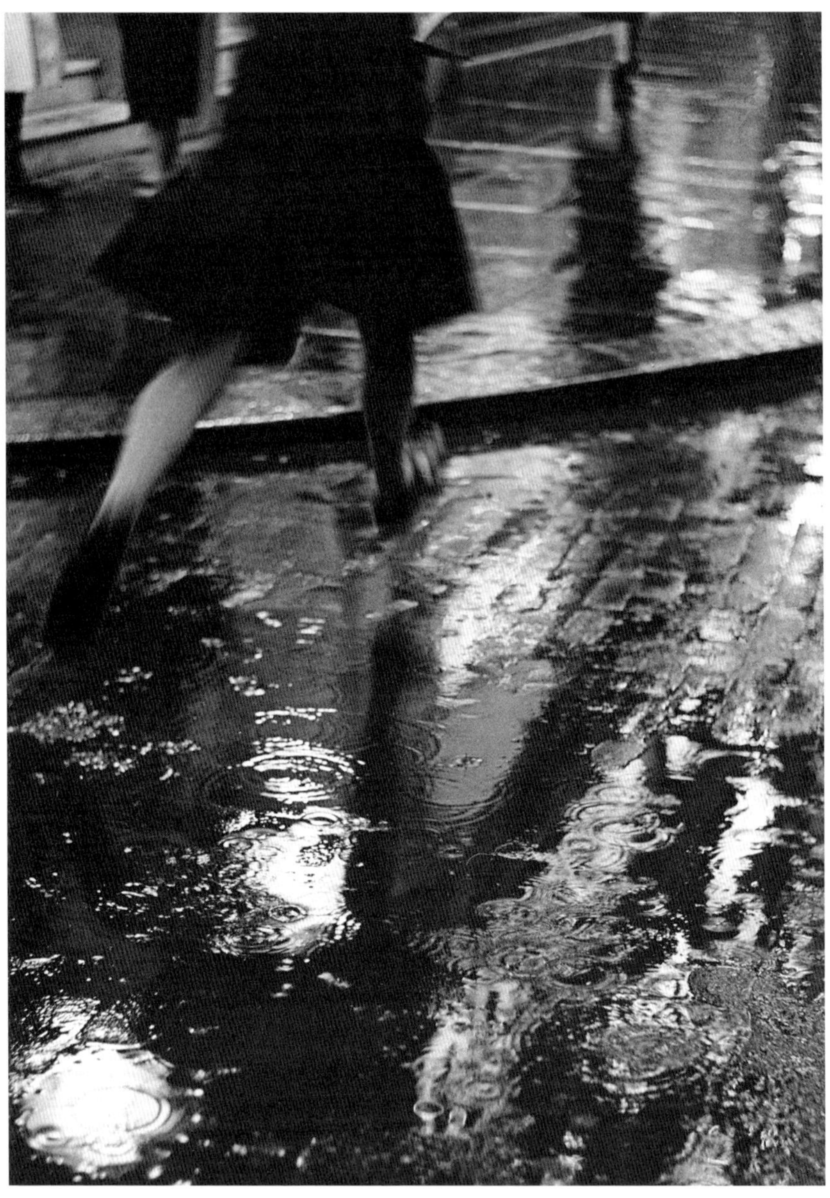

FIGURE 26. Wolfgang Suschitzky, *Puddlejumper, Charing Cross Road*, 1937. © The Estate of Wolf Suschitzky, courtesy of Fotohof.

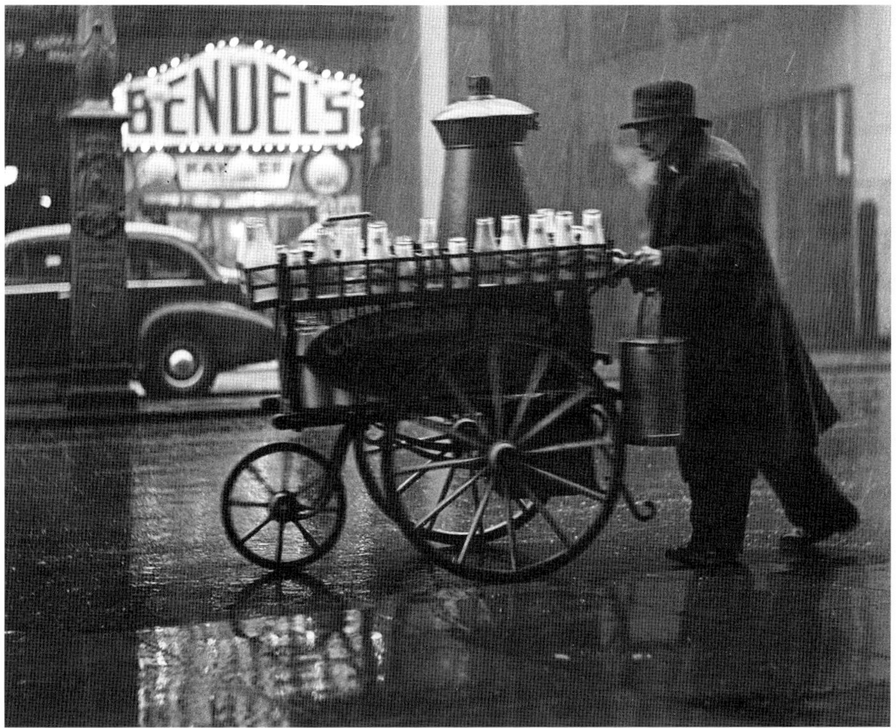

FIGURE 27. Wolfgang Suschitzky, *Milkman (Coliseum Dairy)*, Charing Cross Road, 1936. © The Estate of Wolf Suschitzky, courtesy of Fotohof.

in Suschitzky's obituary for *The Guardian*, the Charing Cross Road photographs in particular are "eerie period essays in black and white, a paean to the dignity of labour."[13]

If the pathetic fallacy is essentially anthropocentric, making use of the weather to reflect a mood or inner state of a character, Suschitzky's weather-infused photographs are anthropocentric in a different way, showing meteorological conditions as both the setting for, and consequence of, human action; a thick environment (of rain and dark, fog and steam) through which humans must struggle. The emphasis (in these Charing Cross Road photographs at least) is on human labor—workers tarring the road, the milkman pushing his cart. Writing not of Suschitzky but of Alfred Stieglitz, Kelsey makes the point that, in Stieglitz's pictorialist photographs of workers on the streets of New York, "the ephemeral vapors enfolding these workers spoke both to the fleeting quality of urban experience and to the precarious circumstances of their labor."[14] In Suschitzky's case, steam, darkness, fog, and rain make labor visible as such, but rather than conveying an impression of ephemerality and precarity, the atmosphere adds a kind of density. It is something these workers toil *against*, something tangible rather than just visible, which allows the viewer to imagine being in the dark, the wet, and the cold.

11 BREATHING THE LIGHT

► This photograph *Sienne* (fig. 28) is by Hervé Guibert, a French photographer and writer. It was made in 1979. Taken from a bed, or one side of a bed, it shows a room with an open window and closed shutters on the outside. Two beams of light enter the room sideways from between the shutters, their edges sharp and diagonal, made more visible as they slice through white swirls of smoke. The smoke seems to emanate from a valley between the shoulder blades of a naked man who is slumped on a desk below the window. We can see his back, the fluid curve of his stomach, his left knee, and his left arm to the elbow, but the rest of his body is concealed. The light illuminates the surface of the desk and traces the top of his arm, his shoulder, his bony shoulder blades. It makes a bright, hard rectangle of white on the top of his back, trickles down toward his waist, and outlines the intricate wood of the chair behind him, then lays itself out in a thin strip across a blanket neatly folded along the end of a bed in the foreground. It also slips down between the desk and the wall, counts out three bars of the radiator below, and rebounds to form a patch on the wall itself.

On the left another light seems to come straight toward the viewer, soft-edged and overexposed. It forces its way between the slats of the outside shutter, below a horizontal bar or transom, then between the metal grid on the open window, and illuminates a further, interior wooden shutter. The room is like a camera—a *camera obscura*—filtering the light through layers of

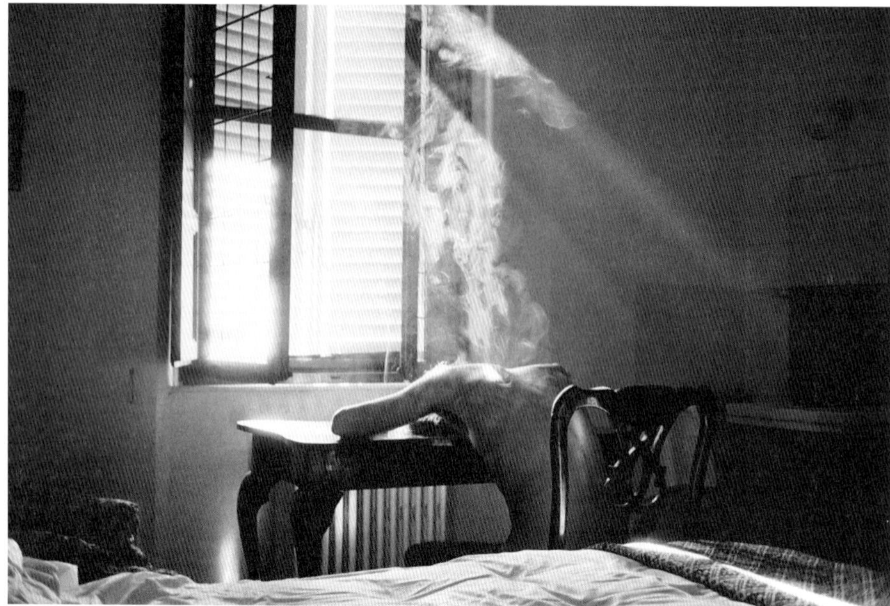

FIGURE 28. Hervé Guibert, *Sienne*, 1979 © The Estate of Hervé Guibert.

mechanisms, lenses, shutters. Cameras began as rooms, and since it doesn't take much to turn a room into a photographic apparatus, it is apparent that this photograph is showing us a room behaving as a camera.

Although this is a photograph of an interior, it tells us that somewhere, outside, the sun is shining, and this exterior enters the room, while air and smoke are breathed, in and out, by this man in this room, and by a second man, the photographer. The photographer and the camera are inside, sharing with this man the space, the light, the smoke, and the various pieces of furniture, yet (one might argue) they do what all photographers and cameras do: they posit an outside for themselves at the point of making the image. This outside or exterior posited by the camera and the photographer in the act of taking the photograph is the position we occupy as observers of the image, which allows us to regard photographs in the same way as we might regard a scene through a window. But if the room the photograph shows is (analogically) also the interior of a camera, we are also placed inside the apparatus.

I am reading this room as a camera and reading this photograph as an image of photography itself—hardly a stretch when Guibert himself was both a photographer and a thoughtful writer on the nature of photography. In his book *Ghost Image*, published in 1981, he talks about what photography is and can be from the point of view of a photographer. The book is partly a response to Roland Barthes's *Camera Lucida*, published the previous year, which looks at photographs from the perspective of a viewer or a photographic subject but not from the point of view of the person taking the photograph. Guibert was very taken by *Camera Lucida* and corresponded with Barthes about it.

This photograph is a photograph of photography, but it is also representing a moment: the moment when Guibert saw the smoke dance, his lover slumped at the desk, light like wings on his back. And it represents things he may not have seen or paid attention to, such as the washbasin in the corner of the room, the indentations on the sheet from where it had once been folded or imprinted by the springs in the mattress, or the pattern on the blanket at the end of the bed. Since the photographer is another man within the room, also possibly naked, on a bed, breathing the same air and basking in the same light, this moment was not only seen but felt and inhabited by him.

We don't see the air in photographs very often, but smoke and fog are useful to the photographer because they make air visible. The breath exhaled from the body mingles with the light entering the room, solidifying that light. In this photograph, light and air become visible as objects in themselves, rather than just as the media through which we see. The smoke comes from a cigarette, we can assume, but one might imagine it as the spirit of a man leaving his body. This is an especially tempting interpretation in the light of Guibert's own preoccupation with death, and with the hindsight of his sickness and eventual death in 1991 as a result of AIDS. In 1979, at the age of twenty-three or twenty-four, and before his HIV diagnosis, Guibert was already thinking about death; in his first book, *Propa-*

ganda Death, he describes his own corpse on a laboratory table. But in *Ghost Image*, he describes taking and viewing photographs as acts of love and desire.[1] If death is present, so is intimacy, an intimacy made stronger by bringing the observer inside the camera itself and making the trajectory of light through air visible and trackable, allowing the light to explicitly delineate the space containing both the photographing and the photographed, subject and object.

Those of my students who are not familiar with chemical photography are often astounded to find there is nothing inside a camera before you load it with film. They exclaim, it is empty! But it is not: like our own lungs, it is filled with air. Light passes through air both outside and inside the camera. To think about photography atmospherically might be, in part, an exercise in imaginatively slowing down this process, thickening and solidifying the air, making visible what is invisible.

The photographic effect of halation, visible in this photograph as the light from the window permeating the edges of the window frame, offers one way of thinking about the interrelation of light and air. Halation describes light behaving like a breath on glass or a mist across a landscape. Though the word sounds like it might relate to breathing—inhalation and exhalation—it is actually derived from the word "halo" and was reportedly named by astronomers who saw its effect as a circular bleed or bloom in photographs of celestial bodies. *The Ilford Manual of Photography* describes it as "A spreading of the high lights beyond their proper boundaries, with consequent blotting out of the details of the surrounding parts."[2] Early versions of the manual subsumed it within the category of "fog" and under "Defects in Negatives." Ilford company historian and sales manager A. J. Catford describes it as a "violent spreading of the highlights."[3]

Halation is caused by pointing a camera directly toward a light source (such as a bright window), especially where, as the manual says, "the presence of deep shadows in the same subject as brilliant high lights necessitates a somewhat long exposure."[4] Though it persisted into the 35 mm film photography of Guibert's time, it was a particular issue with glass plate photography (wet collodion or gelatin dry plates), where the light, instead of being absorbed by the emulsion, would pass through the plate and rebound inside the camera, back onto the underside of the emulsion coating. For several decades photographers painted backings onto the plates to prevent light bouncing back through the glass. These could be made of Indian ink, burnt sienna powder or soot with gum and glycerin, even caramel or jam.[5] These backings then had to be sponged or scrubbed off before development—a process that could produce marks and splashes on the exposed plate.[6] From 1897, plates were sold with dissolvable opaque backings. Later, on roll and flat films, halation was reduced using synthetic dyes suspended in a gelatin layer on the reverse of the film. Backed films and plates were sometimes marketed as "anti-halo" (see plate 3).

The French word for halo, *aureole*, reveals the proximity of "halo" and "halation" to "aura." Some films (especially infrared films) without antihalation layers were marketed as "aura films," making a virtue of highlights that appeared to glow. Aura self-toning paper was manufactured by Houghton and Son circa 1900. Walter Benjamin noted there were attempts to reinstate aura in photography artificially, for example in the pictorialist techniques of soft focus, shallow depth of field, and the use of mist, vapor, and fog, or in the techniques used by Hollywood photographers to make film stars seem to radiate light.[7] But there were also accidental effects of aura: the flaws caused by contamination, the light leaks from a badly made camera, or halation caused by inadequate backing. Though regarded in the technical literature as a defect and an accident, even in the early twentieth century halation was beginning to be recognized as an aesthetic effect that could be capitalized upon.

These connections of halation with aura seem to do with the way it affects the clarity and sharpness of the image, obscuring parts, increasing ambiguity, undermining a kind of visual perfection that had become associated with mechanization, and creating the impression of other presences haunting the image. But in *The Ilford Manual of Photography*, the description of the causes of halation also recalls that strange quality of aura identified by Walter Benjamin, which is its simultaneous distance and proximity.[8] For the light that forms the "secondary image" of halation has traveled, passing through the emulsion and the sensitized surface, and is reflected back again at an angle. As with the scattered light caused by various kinds of atmospheric haze, this light has "been turned aside and then turned back again, finally arriving from a direction very different from that in which the distant object really lies."[9]

Halation, understood as a flaw or a mistake but also as an aesthetic effect, makes light in a photograph not just a means to see objects, or a property of objects, but something that simultaneously constitutes and disrupts the photograph.[10] In Guibert's photograph, it collaborates with the smoke to solidify the air through the window. Halation and exhalation work together to produce something not altogether removed from what the English Romantic William Wordsworth called a "glory." Some writers see Wordsworth's use of glory as synonymous with Benjamin's aura, but a glory also meant a polychromatic effect of light diffracted in fogs and cloud that suggested angelic or spectral presence.[11] Glories like the Brocken specter, an optical illusion seen from mountains in certain atmospheric conditions, were both scientifically explicable and loaded with mystical significance. Guibert's photograph, though monochrome, seems to depict a glory, even if at the same time it reveals the mechanism by which the illusion is constructed.

In 2023, I visited the exhibition *Exposé.es* at the Palais de Tokyo in Paris. The exhibition included a work called *Visitor* by the Canadian artist Moyra Davey. *Visitor* references Guibert's work at the same time as it meditates on the American healthcare system through text and images (Davey cocurated the first exhibition of Guibert's photography in the United States). It is composed of eighteen prints made of photographs and handwritten text, some of which reflects on Davey's anxieties about her son being treated with the opioid fentanyl. The fourteenth print is a photograph made by her son while still at school, a recreation or transcription of Guibert's *Sienne* (plate 8). At the Palais de Tokyo, Guibert's own photographs were displayed opposite. Looking at *Sienne* together with *Visitor*, I realize for the first time that the smoke and the slumped body imply sedation.

Davey's son's photograph has been folded and posted from New York to Paris, the image now disrupted by tape, postage stamps, handwritten and stamped text, and the marks of having traveled. On the left frame of the picture is written the word "air," referring to airmail, and there are tears where it is folded. Mailing photographs is a technique Davey uses repeatedly, folding and sending pictures to the shows in which they are to be exhibited so that they arrive in a condition in which the artist has not seen it, referencing both the artist (the sender) and the gallery where it is exhibited (the recipient). In turning her photographs into pieces of correspondence, Davey layers another level of intimacy onto work that is already intimate and concerned with interiors and interiority. At the same time, the technique slows down the process of transmitting images and reasserts their thingliness (one could send a digital file from New York to Paris in an instant, after all). Her and her son's reworking of *Sienne* loses the sense of being inside a camera (gone are the shutters and the halating light) but gains folds and tape and marks. It reminds us that photographs are made of light and air, but also, like human bodies, sustain damage precisely because they are substantial and material.

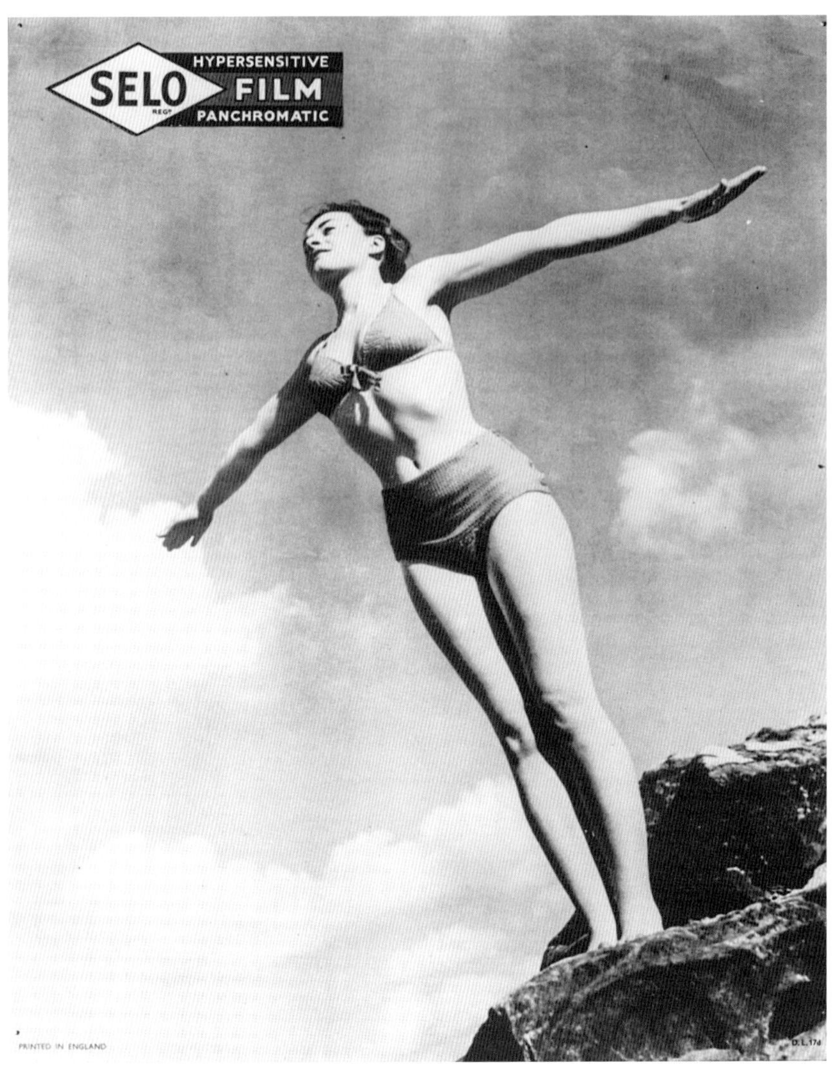

FIGURE 29. Selo Showcard, 1939. Ilford Limited collections, Redbridge Museum and Heritage Centre. Courtesy of Redbridge Museum and Heritage Centre 2025.

12 BODIES IN FLIGHT

► Look at the photograph of the diver (fig. 29). It is an Ilford show card for Selo panchromatic film, used in promotional displays. The woman stands rigid, on tiptoes, apparently tilting, about to fall. Ilford Limited's fastest sensitized materials were perfectly capable of recording divers in midair. But this photograph is intended for a particular consumer, perhaps someone with a simple box camera working in the soft light of a British summer. The woman represents a friend or a relative who is not an expert diver. She is making a good show of it—perhaps she will execute a perfect dive—but this is fundamentally about play, not sport. At the same time, she really is diving; she is already committed, there is no backing down. Diving like this is something you do when being watched: it is a moment of showing off, something done for other people's eyes and cameras as much as for your own pleasure. Diving platforms and boards, even high rocks, are stages for performance.

Note her outspread arms. This takeoff position is not very conventional among competitive or leisure divers today, but it was common in the period. As a springboard diver in the 1980s, I rarely began a dive like this, but my grandfather, who learned to dive on a hardboard in the late 1920s or early 1930s, did. He was a master of the swallow (or swan) dive, admired in the lidos (outdoor swimming pools) of south London. When I learned to dive, the swallow dive was not taught by my instructor. Neither I nor my grandad were athletes, but through this experience and from watching competitive diving today, I can see how dives are historically specific and go in and out of fashion.

The swallow dive, as it is known in the UK, was introduced in the early 1900s. It begins with outstretched arms and is arguably the dive that most feels and looks like flying. It is named for that upward, bird-like swoop as you leave the board and arch your back. The effect (for both viewer and diver) is that you seem to hover for a moment, midair, before bending your torso, straightening your whole body, and plummeting downward. Interwar photographs and films of people diving emphasize this sense of flight repeatedly. In the 1920s and '30s, sometimes described as "the golden age of swimming," movie stars such as Esther Williams and Johnny Weissmuller showcased their aquatic skills, with the swallow or swan dive, in particular, epitomizing "the graceful elegance and freedom that these swimmers embodied."[1]

This dive suggests human flight at its most pure, unencumbered by technology, akin to the flight of a bird. It is also the most obviously photographable dive. It lends itself to still photography because of the moment in which the diver brakes, and their body seems to pause momentarily before dropping. This is, I would argue, a built-in photographic moment, amenable to a medium-fast film and shutter speeds. It seems appropriate that, in the interwar period, this particular dive was repeatedly performed and photographed.

In the 1920s and '30s, holidaying was a new experience for a sizable portion of the British population. At Ilford Limited there were no holidays for workers before 1913.[2] In the interwar period, photographs showed the kinds of behavior that might be expected or enjoyed while at the seaside. Gone were the bathing machines and the sex segregation of the Edwardian period, replaced by crowded beaches and skimpy swimsuits. The summer editions of photographic periodicals and illustrated magazines are full of photographs of people exercising—dancing, playing ball, jumping, and diving—on beaches and at lidos. Leisure was being invented and reinvented, photographically.

Photography and film had long been participating in the reinvention of human mobility, famously used by Frank and Lillian Gilbreth to rationalize factory labor. By the 1930s, sports physiology was using photographic techniques to enhance athletic performance, too.[3] This rationalization of bodily motion was done in the name of efficiency but also for the purpose of spectacle. Like the synchronized parades of the period, sport and dance were increasingly designed for mediated vision—that is, designed to be filmed and photographed. Likewise, photography did not just record summer holidays and beach bodies in this period, it helped to shape them.

The photographs used in the Selo advertising campaigns and promotional material of the late 1930s repeatedly feature children and adults at the seaside, dancing and cartwheeling in summer clothes and swimwear, throwing balls and leaping off dunes. They are presented as ordinary people on vacation, often with towels and shirts held up like sails. Even though these are not individuals who have been subjected to an explicit, technologically enhanced body training, and while people did certainly jump and dive before photography, these photographs do not simply represent something already existing: they are eliciting these activities, bringing them into being, and circulating them.

A central purpose of these promotional pictures is to emphasize the speed of the new "hypersensitive panchromatic" (HP) film and how it could successfully record movement by facilitating short exposure times. In fact, Selo film was not fast by later standards: it was probably equivalent to 100 ISO by today's measure.[4] It could only freeze motion in bright sunlight when used in a camera with a fast shutter. Thus, these are not full action shots. Even so, the gymnastic and aeronautic bodies suggest that photography is part of a transformation of everyday life, part of a new, modern, light-footed, and spontaneous world. They imply an active use of leisure time and collaborative play between the photographer and the subjects who are cartwheeling and leaping and diving on request. They epitomize a desire for a kind of weightlessness and suggest an aerial potentiality, the possibility that human bodies might fly, and the capacity of the film to produce flight and dynamism. If the most immediate and straightforward message is "this film is fast and sensitive and can capture movement," it is overlaid with something else, something like "we are free and modern people, and we express our modernity

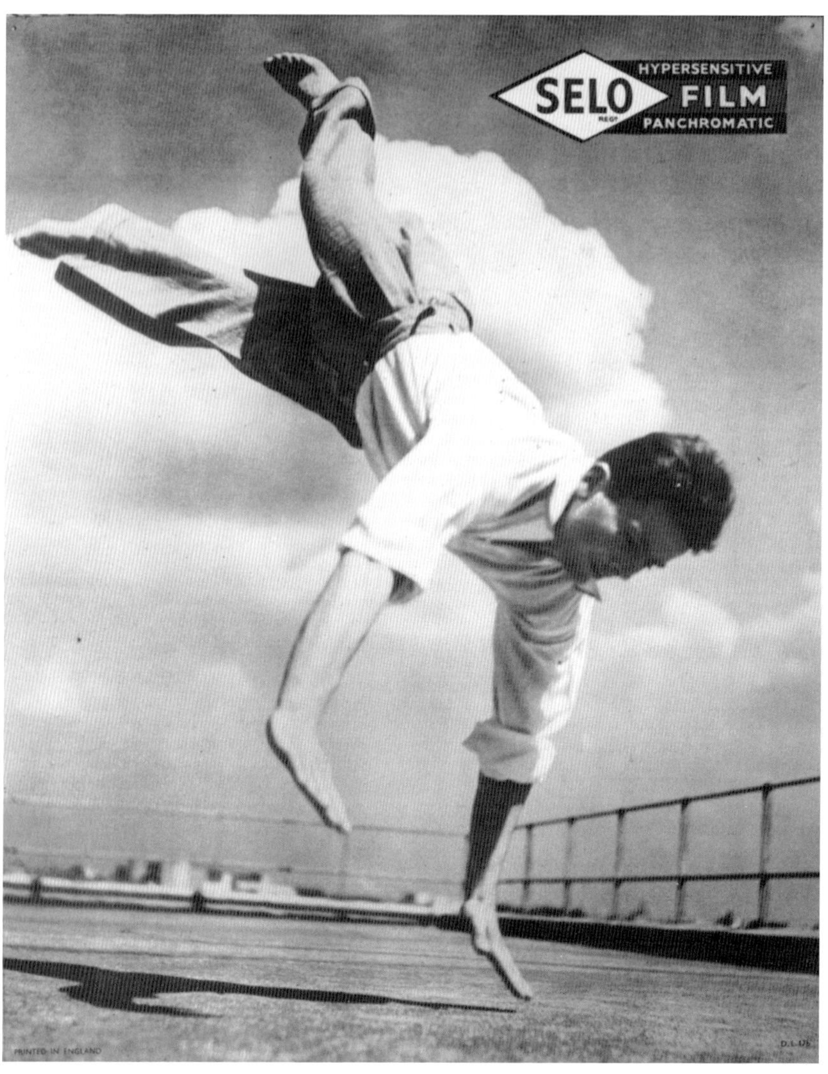

FIGURE 30. Selo Showcard, 1939. Ilford Limited collections, Redbridge Museum and Heritage Centre. Courtesy of Redbridge Museum and Heritage Centre 2025.

through dynamic, gymnastic movement." (see plate 2) In this way, Ilford implicitly presents itself as a driving force in technological progress, dynamic and forward looking.

The Selo advertising campaigns belong to a broader 1920s and '30s trend of picturing jumping and diving figures. In the 1920s, new lightweight and fast handheld cameras such as the Leica had enabled the body of the photographer to move in concert with the subject, opening up a new kind of play (although the attraction to leaping bodies in flight begins more than a decade earlier, famously in Jacques Henri Lartigue's childhood photographs of his flying, leaping relatives). Images of diving and jumping became part of the "New Vision" in photography, included, for example, in Franz Roh and Jan Tschichold's *Photo-Eye* and László Moholy-Nagy's *Painting, Photography, Film*.[5] They were popularized across the visual culture of the time, ubiquitous, for example, in dance magazines, where, as cinema scholar Michael Cowan notes, "the reader could observe photographs of dancers caught hovering in midair for page after page."[6] As the Selo promotion suggests, this mobility was facilitated by new emulsions as well as cameras, by the wet technology of photographic chemistry as well as the dry hardware of lenses and shutters.

These campaign images also owe a debt to the wider body cult that was circulating in the 1920s and '30s. In this period, participation in physical culture and sport became associated with women's liberation.[7] Just as women's clothing in the period suggested a new freedom of movement, so images of women exercising seemed to express a new mobility and lightness associated with modernity, sloughing off the weightiness of nineteenth-century culture and patriarchal tradition. Freedom here is imagined as freedom of movement, as a joyful leap into space.

The interwar body culture also carried connotations of national and racial identity, though the precise form these took varied between countries. Physical fitness and especially outdoor sports were aligned with ideas of progress, modernity, and freedom, understood as reflective of the health and vitality of the nation. Toward the end of the 1930s, the British government recognized bodily fitness as a national issue, setting up a National Fitness Council and a National Fitness Campaign from 1937–39. In Britain, the "body beautiful" offered an image of national regeneration in the face of a declining empire and the impact of industrialization on health. It could also conceal something more sinister, as in the Soviet Union, where famine hid behind the socialist realist iconography of ideal bodies, or in Nazi Germany, where the Aryan body ideal was celebrated while other kinds of bodies were being annihilated.[8]

New health practices such as heliotherapy (sun cures) drew on a growing understanding of the processes by which human and animal bodies could obtain vitamin D from ultraviolet radiation. Sunlamps and sanatoriums, lidos, fitness classes, and physical education literature linked the health benefits of sun and exercise to the body of the nation.[9] In Britain, organizations such as the Women's League of Health and Beauty and the Sunlight League considered sunbathing and taking the sea air as antidotes to the city fog.[10] Sunlight campaigns were associated with smoke abatement and slum clearance. Campaigners saw sunshine and pure air as having health benefits and moral value, since they viewed urban pollution as causing stunted growth and ill health among the working classes, and contributing to moral degeneration. In the 1920s and '30s, suntanned skin, previously associated with manual labor, became increasingly valued among the white middle classes. Sunbathing was even linked to ideas of imperial regeneration, in arguments that contrasted the poor physiques of British city dwellers with the healthier bodies of other "races."[11] These racial comparisons were intended to stimulate a revival of British masculinity, which was equated with a revival of imperial power.[12] Sunlight, swimming, and beach life became tied to the eugenic ambition to shore up "the race" (often meaning, rather parochially, the English) against degeneration.

The Selo campaigns are not the only examples of Ilford promotional materials that aligned the company with the moral qualities of sunlight, beach, and fitness culture. Advertisements for Dufaycolor cine film (also an Ilford product) promised to "bring the pink and rose flush of perfect health faithfully to your screen," while its roll film would record "faces tanned by the open air."[13] The ads implicitly set the bodies they depict against two other kinds of body: the naturally brown-skinned and the sickly, rickety, vitamin D–starved urban working class. Similarly, images of ideal aerial bodies gained desirability through their opposition to the earthbound, as the cultural theorist Mary Russo has shown in her writing on American images of aeronautic, streamlined female bodies (wing-walkers, divers, dancers), which project an ideal of the liberated woman: "The very idea of freedom as a flight from the bodily takes wing, so to speak, from the very shapes and movements it would leave behind."[14]

Yet those whose bodies are most stigmatized as heavy, earthly, or grotesque are even more likely to imagine emancipation in terms of flight, weightlessness, or disembodiment. As Dipesh Chakrabarty writes, this causes a difficulty for an environmental politics meant to emphasize our mutual dependence on other earthly beings.[15] The power of images of bodies in flight does not reflect the degree to which viewers identify or recognize themselves in such pictures but rather the extreme stakes involved in the desire to transcend the bodily and earthbound. These advertisements are not necessarily addressing the kinds of people they depict; after all, advertising speaks to desires and aspirations, regardless of whether these can ever be fulfilled. The image of freedom and success expressed in images of leaping and flying bodies depends on the exclusion of the more uncomfortable and negative aspects of modernity—including, in this case, the dangers of urban pollution and fog for health, and the working-class bodies that betray such facts.

13 NIGHT PHOTOGRAPHY

► Although snapshot photography was strongly associated with bright, sunny weather, from the late 1920s, low-light photography was made possible by the development of faster films for amateurs, new kinds of flashlights, and indoor lighting equipment. These allowed photographic companies in Britain to promote the use of their materials indoors, at night, and in winter. The firms set out to increase consumption of photographic materials and cameras by suggesting amateur photographers go out with their cameras in the evenings after work or take family photographs indoors, in winter as well as on summer holidays. Photographic subjects could now include private moments like the baby's bathtime—a theme frequently promoted in the various photographic booklets published by companies like Kodak and Ilford—or subjects previously restricted to more expert and well-equipped photographers, such as the lamp-lit architecture of nighttime London. As these examples suggest, both amateurs who wanted to document family life and those who had more artistic and pictorial ambitions were being encouraged to take photographs in low light.

The new films and lighting sets promised to release amateur photography from being a seasonal practice. In the British Isles, which were (and are) so often rainy and overcast, the photography season was short, lasting from April to September for all but the most enthusiastic and well-equipped. In the interwar period, photographic manufacturers and retailers tried all sorts of measures to persuade amateur photographers to maintain their hobby outside the season. In 1927, for example, the *Merchandise Bulletin* of Boots the Chemist bemoaned the "rather long period of stagnation in purely photographic materials" that continued through winter.[1] A decade later, it was encouraging its salespeople to "keep the photo business going all through the winter."[2] In the intervening years much had changed: color film (Kodachrome and Dufaycolor) was now available in its stores along with photoflood lamps for indoor photography and new sensitive, panchromatic black-and-white films. These films included Zeiss Ikon Panchrom ("the ideal film for night photography"), Exakta, Kodak's SS Pan film (for "every kind of after-dark photography") and Ilford Limited's Selo Hypersensitive Panchromatic Roll Film. They were suited to low-light photography because they were "faster" than other films and because they were panchromatic—sensitive to a wide spectrum of light. As Ilford's promotional material announced, in combination with newly available fast lenses, these films made it possible to photograph in "the yellow light of winter and the artificial light used for taking night subjects."[3]

At Boots, the ambition to increase consumption was transparent. A price reduction on roll films in spring 1930 saw the company exhorting its salespeople to quickly sell the existing stock, hoping that the price reduction would "result in a considerable increase in sales during the season" while providing "a special incentive to push the Photographic business from the branch point of view, and encourage the customer to use more film."[4] The photographic companies used a quasi-moral discourse of thrift and productivity. For instance, Ilford Limited's free *Night Photography* booklet asked, "Why waste half your life (so far as photography is concerned) by shelving your camera from October to April?"[5] A camera not in use is a waste, we are told, as is a life unrecorded. Even during the photographic season, the firms worried that photographers ought to take more pictures—an Ilford Advertising Department Report for the year ending October 1933 attributed a drop in roll film sales to (among other things) "too hot weather—a great many people bathed and lay around and were too slack to take photographs."[6] As the word "slack" implies, the company wanted people to work at their hobby.

The photographic companies and retailers promoted low-light photography through advertisements, articles in the photographic press, shop displays, and free booklets. From about 1936, Ilford Limited's booklet *Night Photography: Picture Making at Night—Indoors and Out*, and the near-identical *Winter Photography and Night Photography*, advertised Selo Hypersensitive Panchromatic Roll Film, a Selo lighting set for indoor photography, and Ilford Hypersensitive Panchromatic Plates. Kodak had a similar booklet to promote their SS Pan film, Photoflood lamps, and Sashalite flash bulbs, which were also advertised via the monthly *Kodak Magazine*.

Encouraging photographers to continue to take photographs at night and in the winter meant trying to change their habits and their perception of what might be appropriate and attractive subjects for photographs. The new products, materials, and technologies for low-light photography were not enough to make this change—it also required new aesthetic standards and criteria. The photographic companies and the retailers clearly had to work hard to expand the photographic season. Not only was it much easier to take photographs in the longer days, brighter light, and fairer weather of the summer months, but also years of publicity had cemented the association of photography with summer holidays. Although the Ilford *Night Photography* booklet optimistically asserts that "the hibernation of the camera in Winter is a thing of the past," changing habits was not easy. The *Ilford Selo Amateur Photographic Handbook* (undated but probably from 1940) also pushes the idea that people ought to be taking photographs throughout the year: "Many people with cameras do not get as much enjoyment from their hobby as they should. They confine photography to the holiday periods, whereas it is a hobby which can be enjoyed all the year round."[7]

In trying to undo the strong seasonal associations of photography, the industry was also transforming the meanings and contents of amateur photography. Expanding the practice meant extending the vast photographic archive of everyday life that had grown as snapshot photography was popularized with the widespread use of roll film and Box Brownie–style cameras following the Great War. Increasingly it seemed that there were few unsuitable times for photography (even while there remained many unsuitable or unmentionable

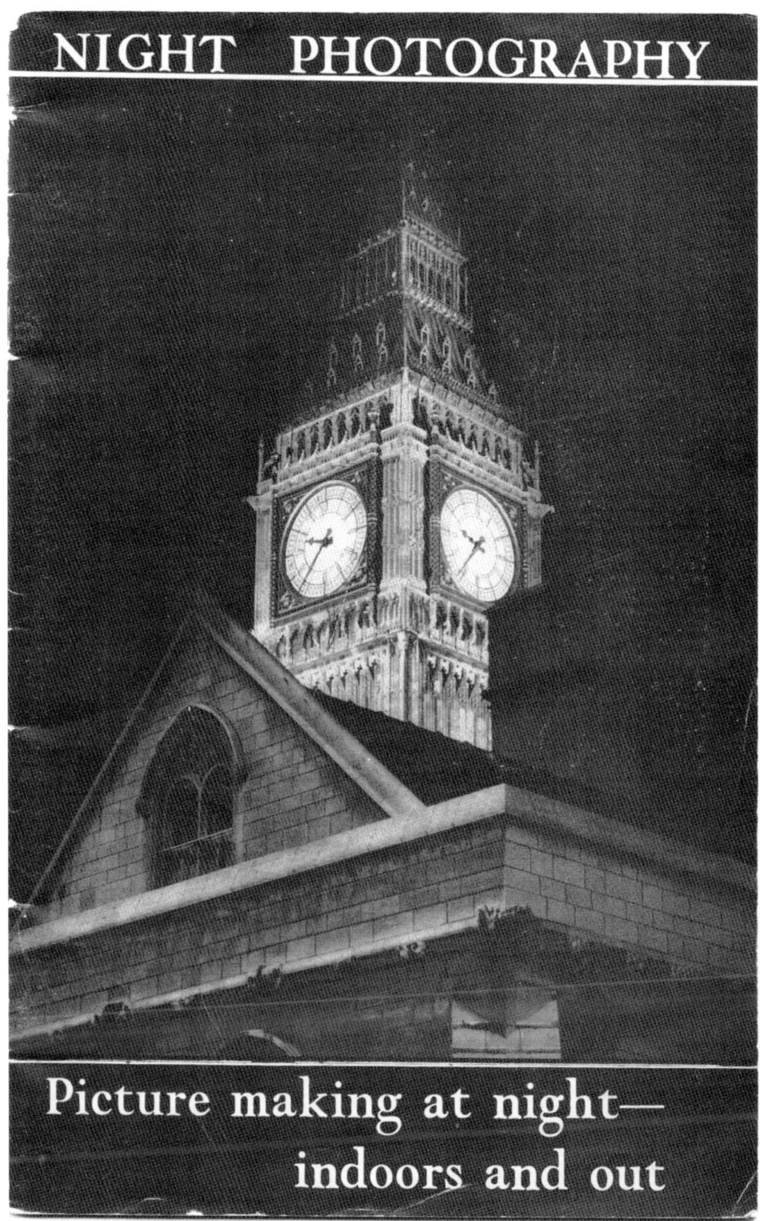

FIGURE 31. Ilford Limited "Night Photography" booklet (undated but probably ca. 1936), private collection.

subjects).[8] More and more aspects of public and domestic life became photographic opportunities.

Both the text of these booklets and the example photographs that illustrated them suggested a repertoire of appropriate subjects while also implying that the new photographic materials could appeal to a wide range of photographers and a diversity of tastes. Kodak's advertisement in the January 6, 1937, issue of *The Amateur Photographer and Cinematographer* reads: "Fireworks or firesides, floodlit buildings or baby in the bath—it's all one to Kodak S.S. Pan. Every kind of after-dark photography, outdoors and in, is made easier by the lightning speed, extremely high sensitivity to artificial light, and really effective anti-halation qualities of this Super film."[9] In Ilford's *Night Photography* booklet, the writer asserts that the new developments in photography, which make night and indoor photography possible, introduce "new subjects"—"the subjects you always longed to take." These include winter scenes previously "missing" from the photographer's album. The images illustrating the booklet are offered as a sample of an "immense variety" of subject choices. Additionally, *Night Photography* suggests one might photograph "ice hockey matches, swimming galas, cabarets, train and bus snapshots" (presumably for the headlamps and lit interiors) and "stage and circus subjects." At home, in the evenings, photography is a fun activity in which "the whole family can take part."

Strategically and deliberately, the photographic industry in interwar Britain extended photographic labor and the activities of amateur photographers into the nighttime, into the off-season, and into spaces that had previously been unvisited or at least unphotographed at such times. This is a story of capitalist growth, of companies expanding their market by introducing new product lines and attempting to increase the pace and regularity of consumption. The impact is far-reaching. The market annexes times and places where consumption had not happened before.[10] Not only do sales of materials increase, but the photographic universe expands. The growth of the market and the proliferation of images are in turn part of wider ongoing changes in social experience during the interwar period. As capitalist expansion speeds up the rhythms of production and consumption, it colonizes time and space. Immutable and supposedly natural phenomena such as the seasons and the divisions between night and day are up for grabs.[11]

Yet the interwar push toward night photography cannot be understood purely in terms of changes in photographic technology. When Ilford's *Night Photography* booklet asserted that "sunset no longer marks the end of the photographic day," it was also reflecting larger changes, especially in urban centers, that were making nights out in the city more available to a wider population, and creating new forms of nighttime labor.[12] Urban nightlife was flourishing due to changes in transport, the relaxing of wartime licensing restrictions for restaurants and pubs, and the arrival of the picture palaces and dance halls.[13] By the late 1920s, city centers were brightly illuminated by neon signs and floodlit architecture as well as street lighting. The lights of London's West End were an attraction in themselves, represented in advertisements and posters as one of the reasons to visit.[14]

A 1937 article in *The Amateur Photographer and Cinematographer* noted that night photography benefited from recent changes in public lighting: "Lighting schemes of buildings, parks, promenades, and so on are more brilliant and elaborate than ever; and what is known as flood-lighting has introduced some very striking and impressive effects."[15] Not only the floodlighting, but also the uneven illumination from streetlighting allowed for an aesthetic of dramatic contrasts of light and shadow. And since, despite the faster films, rapid "snapshots" at night were rarely possible, amateurs were encouraged to use longer "time exposures" which softened and blurred moving figures, created light trails from moving vehicles, and added mysterious atmospheric effects.

Yet there is quite a difference between the version of London depicted in Ilford and Kodak's promotional materials and that depicted in written texts of the 1930s. While the photographic companies promoted night photography with images of imposing architecture and statuary (floodlit Westminster or the atmospherically gaslit Royal Courts of Justice), popular written accounts of London at night were dominated by the genre of "slum travel writing," which described the nocturnal city in colonial terms, as if setting out into the less affluent and more ethnically diverse parts of London at night were an expedition into the wilds of Africa.

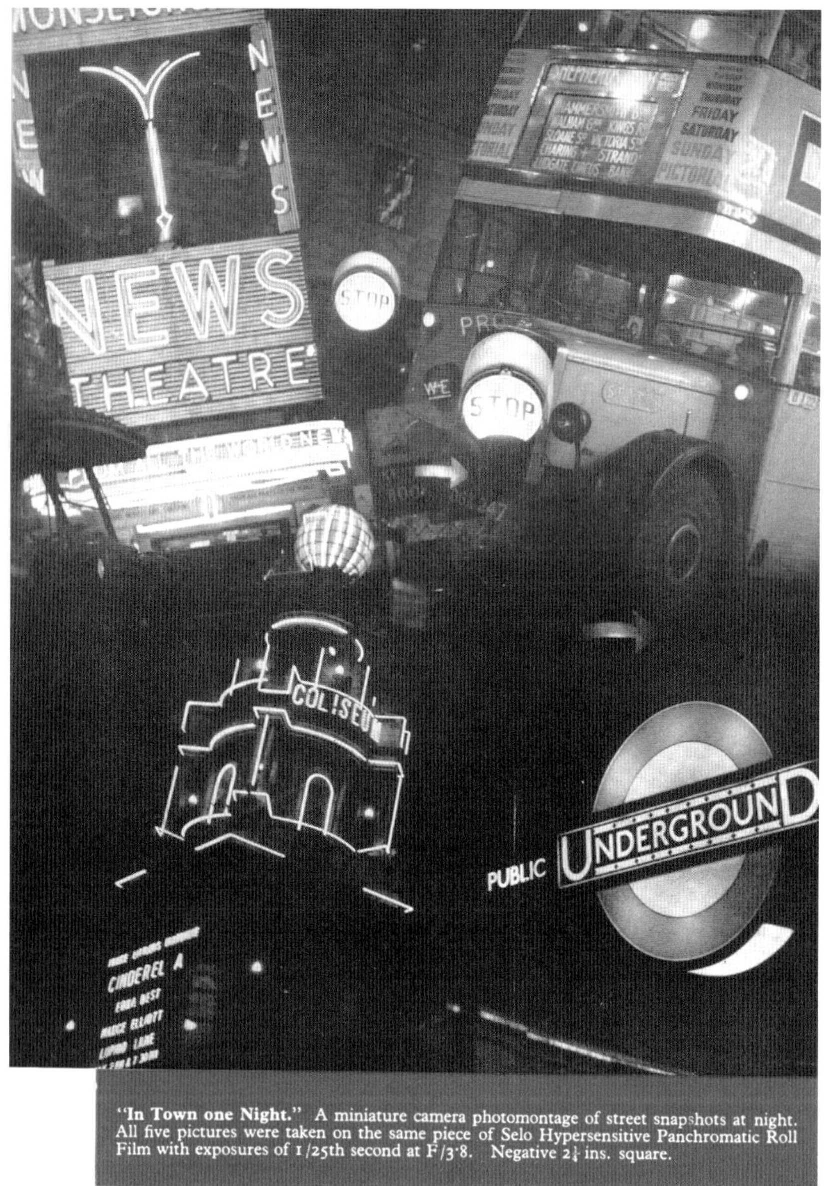

FIGURE 32. "In Town One Night," Photomontage from Ilford Winter and Night Photography Booklet, (undated), private collection.

The right-wing travel writer H. V. Morton's popular *The Nights of London* (first published in 1926 and in its fifth edition by 1932) compared the streets at night to a jungle, suggesting that the nightclubs of Soho or the gambling dens of the East End, with their migrant communities (which he described in hideously racist terms), offered an exhilarating break from "civilization." The night is something into which one "plunges."[16] Sociologist Mariana Valverde describes a "traffic in metaphors" between this kind of "slum travel writing" about the city and imperial travel writing about Africa and India.[17] Morton and other authors referenced Joseph Conrad's 1899 novel *Heart of Darkness*, the title of which had become shorthand for a racist notion of a dangerous, exotic Africa and its "primitive" peoples.[18] Like colonial narratives of far-off places, these writings about nocturnal London may have given their (presumed white) readers a vicarious thrill in imagining places they would never actually visit.

By contrast, the photography companies addressed an intended readership that was respectable and middlebrow, and who they hoped would actually go out into the city after dark. They promoted the lights of the West End, and relatively safe, staid districts and sights such as Westminster and the Houses of Parliament. Even so, this more genteel vision of the metropole is not as far removed from the literary "slum travel writing" as it appears. It too is shaped by an imperial vision, in which the city appears not as jungle, but as the locus of imperial power.[19]

When photographers headed out into London at night, they participated in the transformation of the "photographic day." But they also reaffirmed the claims of empire. This reassertion of the imperial imaginary was set against the backdrop of imperial expansion after the Great War but also in the context of anticolonial resistance and growing independence and nationalist movements like that led by Mahatma Gandhi in India. Interwar literary and photographic representations of London at night reprised in miniature the thrills of nineteenth-century imperial adventure and reaffirmed the might of the old imperial order, even while radical technical changes in lighting, transport, and leisure heralded a new and exhilarating modernity.

14 WORLDINGS

► Photography does not simply represent reality; it coproduces it. Photographic practices and images facilitate among people certain kinds of orientations and dispositions, particular sensibilities and feelings, different rhythms of life. They introduce new material artifacts and material transformations; they provide contact with other people and places. Photography is a process of gathering stuff together, classifying, using, and making the raw material of the world meaningful. Except that material was never raw, never just given to us as a resource but always already a world—perhaps someone else's world—being trampled on by this new one.

In an essay called "The Rani of Sirmur," Gayatri Chakravorty Spivak writes about colonial or imperial worlding as a violent process which gathers together and makes sense of tangible and perceptible things without regard for prior meanings and practices or for pre-existing worlds.[1] The term "worlding" comes from Martin Heidegger's essay "The Origin of the Work of Art," which contrasts the meaningful "world" with the unprocessed material "earth," but Spivak is more interested in how the process of worlding might treat an already existing world as if it were empty and unmapped.[2] Where Heidegger imagines bare earth, Spivak sees another world, impossible to reconstitute. Worlding, for her, names the process by which colonizers overwrite the world of the colonized, treating colonized lands as "uninscribed earth," on which they can inscribe their own meanings, and to which they give their own names.[3] As well as superimposing meanings and suppressing or destroying others, worlding is materially transformative—the colonizer rewrites the colonized world with maps and documents, roads and buildings, institutions, laws and regulations that discipline and control the bodies of the colonized.

In the case of photography, it's not armies of photographers nor the ever-growing mass of photographs that do the worlding but the "photography complex."[4] This is a concept from historian James Hevia, who deliberately evokes the term "military-industrial complex" (coined in US president Dwight D. Eisenhower's "Farewell Address to the Nation" in 1961), and indeed the photography complex both intersected with the military-industrial complex and rivaled it in size. It included photographers, journal editors, shop workers, industrial workers, chemists, financiers, and sales agents. It encompassed organizations—the large photographic manufacturing firms, lens and camera companies, publishers, the press, picture agencies, photographic societies, journals, galleries, museums, and stores. It involved a trade in materials and artifacts such as cameras, lenses, papers, glass plates, and chemicals, and global communication and distribution networks including the railways, canals, and steamer routes.

Spivak's concept of worlding helps focus attention on the ways in which the photography complex operated across empire. The photography complex's role in worlding is evident first in the circulation of photographs that helped to shape perceptions of the places and peoples of the colonies. As historian Paul Landau has suggested, photographs were in this sense "indispensable colonizing tools," reducing Indigenous peoples to "archetypes" and obscuring the infrastructure of empire.[5] But also, by circulating raw materials and photographic products, companies like Ilford Limited participated in the managing and processing of knowledge about empire. Their products were shipped from Britain to the territories of the British Empire and the more independent states of the Commonwealth. These materials facilitated representation of far-flung colonies and were marketed to photographers in colonial administrative centers and shipping ports. With them came the expansion of the photography complex: importers, photographic dealerships, camera repair shops, studios.

The British Empire provided crucial export markets for the photography industries. Eastman Kodak's British subsidiary, Kodak Ltd., owed its existence partly to the fact that foreign companies could only avoid the tariffs and taxes of trade with the empire if they had a manufacturing base in Britain. Protectionist legislation was intended to bolster British industries against foreign incursion, using a system of "imperial preference" for tariff-free export to the colonies. This had the unintended side effect of encouraging US companies to set up subsidiaries within Britain.[6] From the 1890s Ilford Limited also had a large imperial export market, especially in India and the Far East. As a British company, it had privileged access to these regions, and it emphasized its Britishness in its marketing. The main Ilford factory was named the Britannia Works, and one of its competitors—which Ilford absorbed circa 1918 though it kept the brand name for a long time afterward—was the Imperial Dry Plate Company, "Imperial" for short (see plate 3).

The export records of Imperial reveal that imperial preference was no guarantee of sales. They show that British Empire and Commonwealth countries did not necessarily represent the largest markets for all British companies. Prior to the First World War, Australia, at that time only nominally independent from the British Empire, accounted for most of Imperial's international sales, but second and third were Austria and Germany, perhaps surprising given the strength of the German photographic industry. Sales in the US outstripped those in the Cape Colony and the Transvaal, while dealers in India reported no demand for Imperial plates at all.

As competition grew from Kodak and Agfa, Ilford depended increasingly on its exports: Ilford's biographers Hercock and Jones write that otherwise "the company might not have survived the difficult period between 1899 and 1914."[7] This dependence on empire was part of a wider British tendency characterized by Marxist historian Eric Hobsbawm as the country's response to its "loss of dynamism" and industrial competitiveness.[8] Ilford Limited gained a reputation for the superior suitability of its plates for tropical climates. Hercock and Jones claim that "Ilford materials behaved better than most of the competition in the tropics," though it is unclear how deserved this reputation was.[9] Certainly, some photographers swore by the Ilford materials. For example, Dr. A. T. Schofield, a missionary in Uganda, in a 1934 lecture to the Royal Photographic Society in London, asserted that he had difficulty with all plates except Ilford's: "their materials, both plates and papers, somehow stood up to tropical conditions remarkably well."[10]

Imperial also exported to West Africa, and though their trade there was small, sales records from 1911 reveal that it was dealing exclusively with African photographers.[11] The census for the Gold Coast that year showed thirty-one working photographers.[12] Daguerreotypists had probably been operating along the West African coast since the 1840s, though few if any of these early images survive.[13] By the early twentieth century there were well-established photography studios across the cities of the region and a network of African professionals with both local and European clientele. In the interwar period, the photographic companies were advertising to photographers in West Africa through publications like *Nigeria* (fig. 33).

One of the earliest and most prominent West African photographers was Neils Walwin Holm, reportedly the first photographer to introduce dry plates to Lagos.[14] He used Ilford's plates in the 1890s, though in February 1893 he wrote to the London-based journal *The Practical Photographer* complaining about a drop in quality of the plates available in West Africa.[15] According to Charles Gore, who has researched the work and life of Holm, Holm kept track of photographers using dry plates (with gelatin-based emulsion on glass), counting "some eighty professional and amateur photographers in West Africa" using the technology.[16] Many of these were Africans with experience of European and North American culture: some previously enslaved people who had migrated back to the continent, and some, like Holm, the mixed-heritage descendants of African women and European colonists.[17] Historian of African photography Jürg Schneider comments that "The cultural proximity of early African photographers to Europe is evident not least in their names—Grant, Decker, Holm, Lutterodt, Sawyer and . . . Joaque."[18] Some of them participated in the British photographic community, writing letters to the photographic journals and being elected as members of the

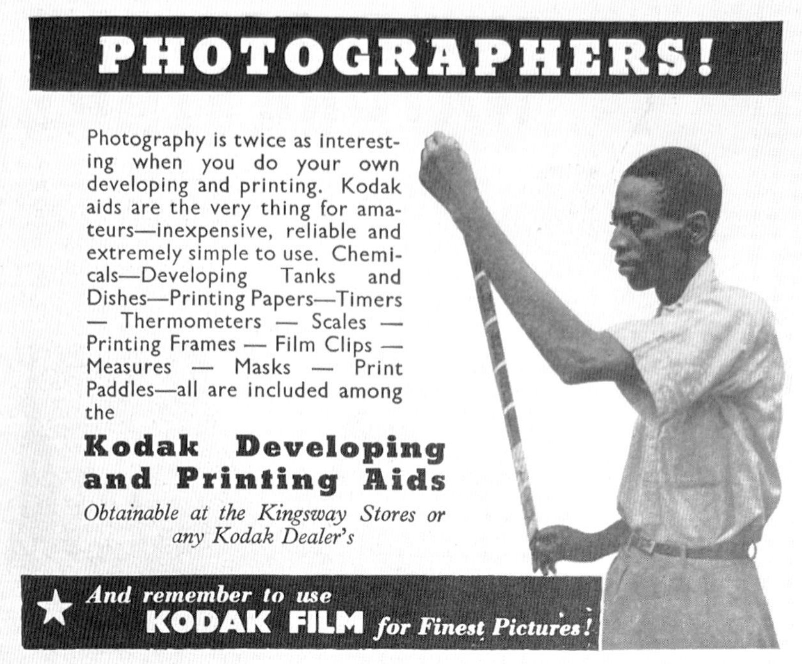

FIGURE 33. Kodak advertisement in *Nigeria*, 1939, Q2, viii. British Library.

Royal Photographic Society (RPS). Holm imported *The Practical Photographer* to West Africa and advertised his own business in it. Holm was elected a member of the RPS in 1895 and inducted as a fellow in 1896, and Francis W. Joaque was awarded a bronze medal at the 1889 Exposition Universelle in Paris.[19] But there were many less prestigious photographers: the art historian Olubukola Gbadegesin writes that "in the 1880s in Lagos, the most ubiquitous photographers were itinerant, small-scale, lone operatives in makeshift studios—often located outdoors or in busy marketplaces—which attracted working-class indigenous clients who could scarcely afford the luxury."[20]

From the late nineteenth century, photography thrived along the West African coast. Local papers advertised photographic studios for a growing and increasingly diverse clientele, and some photographers offered apprenticeships and training.[21] Scholars of West African photography emphasize the mobility of the photographic trade in this region, which was linked to urbanization, the three-way contact between European, African, and American ports, and inward migration as the transatlantic slave trade was ending. African photographers were often itinerant, moving between the shipping ports and, as Schneider remarks, "It was common for photographers to advertise in local newspapers that they would be available for work for a given period, generally for several weeks, a certain place."[22] As photographer and historian Vera Viditz-Ward notes, these mobile photographers were likewise involved in the circulation of photographic materials, distributing them to local photographers in the different cities, and the imported dry plates "made the itinerant photographer's life considerably easier."[23] However, the photographic technology could not compensate for other difficulties, such as discrimination and antagonism from white colonials.

The clientele of these photographers were often recent migrants, including people of African descent, among them previously enslaved African Americans and captives who had been "disembarked at Freetown by the British naval blockades following the British abolition of slavery in 1807."[24] These people adopted British and American cultural practices and used British and American citizenship to protect themselves against the risk of re-enslavement and increase their security. Like Europeans and Americans, they deployed carte de visite and cabinet card portrait photographs to establish and enhance their social status. Other customers were photographers themselves; among Holm's clientele, for instance, were the colonial administrators of Lagos Colony and elite African and European amateur photographers who wished to use his darkroom. From at least 1897 Holm also sold photographic materials to amateurs and offered developing, printing, and retouching services.[25] His son, Justus A. C. Holm, joined him as a partner in the early 1900s, took over the business when his father became a barrister-at-law, and began his own photographic business in Accra in 1919.[26]

African and European professional photographers shared ideas and technical tips and competed for the same clients and markets. From its base in London, the RPS spread awareness of photographic aesthetics and new technologies. With its membership including photographers from West Africa, South Africa, and India, it reflected the geographic reach of British colonial and mercantile interests.[27] West African photographers became, in Gore's words, "agents and conduits for the dissemination of new technological innovations and visual practices, including art movements such as pictorialism."[28] The photographs they produced contributed to emerging visions of African modernity, although they operated in a market that prioritized European exoticizing views of Africa. Within the RPS, in the pages of *The Practical Photographer*, and through their work, West African photographers were involved in the photography complex, producing and circulating photographs, contributing to the discourse around practice, buying photographic materials, providing intermediary services (such as darkrooms), and creating and shaping a wider photographic market.

The materials supplied by companies like Ilford Limited were used to construct versions of West Africa and of African peoples that suited the purposes of the colonial administration, but they were not, as Landau seems to claim, solely tools of colonization. On the surface of the gelatin-coated dry plate, and later on the roll film, complex negotiations could take place, identities could be asserted, and status could be established. If the photography complex was an agent of worlding, African photographers, dealers, and studio and shop owners were nevertheless part of the complex, inscribing meaning through their own photographs. Their position could be ambivalent: Photographers like Holm and John Parkes Decker received the patronage of the British colonial classes, and were called upon to document colonial military and governmental activities.[29] Yet at the same time, Holm appears to have supported the Pan-African political movement.[30] Within cultural parameters set by European institutions such as the RPS, and using materials imported by companies like Ilford and Kodak, these photographers also contributed to, and helped to shape, the technical and aesthetic discourses of photography.

15 PHOTOGRAPHY IN THE TROPICS

Although photographic businesses, studios, and importers were well established across many parts of the British Empire from the nineteenth century, supplying both local photographers and a growing tourist trade, reading the photographic press from the first half of the twentieth century, one could be forgiven for thinking that the colonial photographers who wrote articles and corresponded with these magazines were brave pioneers, the first to attempt any kind of photography in these places. They wrote repeatedly about the difficulty of photographing in regions deemed to be "tropical." Their firsthand accounts bundle together the frustrations of the climate for photography with other difficulties such as the photographer's poor acclimatization to the heat and humidity and the problems encountered in photographing native peoples. They reveal that photographic practice and technique were dependent on notions of climate and race.

Tropical climates were repeatedly classified as "trying," and difficulties of climate conflated with difficulties associated with infrastructure, culture, and the rich abundance of insect and plant life. *The British Journal of Photography* published R. Dykes's account of photographing in West Africa in 1922, complete with entertaining descriptions of preparing or developing plates in a "native hut" with no table, so that he had to squat on the floor, plagued by scorpions, tarantulas, giant flying beetles "over half a pound in weight," and lizards that dropped from the ceiling. Even the tropical night could not be relied on, as "vivid blue flashes of tropical lightning" meant the darkroom must always be totally lightproof. He advised readers to bring their own darkroom to West Africa and to send films back to England for development "packed up tightly in specially fitted tins with calcium chloride."[1] Yet research shows that cities on the West African coast had good facilities for photographers, with developing and printing services offered from the 1890s (see chapter 14).

Accounts in the photographic magazines draw on an older discourse about the tropics. Although the tropics are technically regions limited by the latitudinal circles of Cancer and Capricorn north and south of the equator, the term was loosely used to describe all hot, wet places with luxuriant plant life and exotic, often dangerous animals. According to the historian David Arnold, writing about tropical medicine, the definition of tropical was more dependent on climate than geography, inseparable from moral and cultural assumptions and claims about the "backwardness" of the tropics, which were defined as "culturally alien to, as well as environmentally distinct from, Europe and other parts of the temperate world."[2] British anxieties about the tropics grew with changes in the geographical makeup of the empire in the nineteenth century, as it shrank in North America and expanded in the hot, wet regions of India, East Asia, the Middle East, Africa, and the Caribbean.

Arnold illustrates the spread and imprecision of the category of "the tropics" using the example of India, arguing that the British increasingly perceived it as tropical, emphasizing its lush vegetation and dangerous diseases, though much of the subcontinent is not geographically located in the tropics. According to medical historian Warwick Anderson, tropical places were "both attractive and repellent," places of "luxuriance, excess, and danger."[3] Writers contrasted tropical regions and peoples to those in "temperate" regions, viewing the tropics as culturally "intemperate"—that is, lacking in moderation, control, or regulation.[4] This image of the tropics mapped abundant nature and wild climate onto native peoples, whom the colonizers viewed as indolent, wasteful, and unruly, supporting the colonizer's own self-image as "civilizing, containing, controlling."[5]

Medics continued to see these places as having “bad air,” a view based in the old miasma theory of disease.

Even as a new, statistically informed medicine emerged, based on data generated by British Empire bureaucracy, it was combined with old theories of biological racial difference to justify the exploitation of African labor and to explain the different vulnerability to sickness among populations in the colonies.[6] In the interwar period, this racialized conception of disease supported a view of the tropical climate as alien to Europeans and fueled the racist fears about white colonial degeneration that had first emerged in the 1840s. The discourse on the tropics was haunted by unease about ungovernable colonized peoples and the eventual backfiring of the imperial project. Cultural historian Rod Edmond argues that “it was becoming axiomatic that Europeans could not settle and reproduce in tropical countries, while an expanding tropical population was threatening to burst out of its zone and into the temperate civilized world.” Even the discovery of the mosquito vector of yellow fever and malaria and the advances of germ theory did not necessarily dispel such atmospheric anxieties; instead, Edmond says, the presence of invisible microorganisms “intensified the view of both tropical places and peoples as toxic.”[7]

This image of the excessive and perilous tropics is echoed in the photographic discourse. Photographers across the British Empire contrasted these hot and humid places with the England they had left, whose climate was treated as a norm and where photography seemed by contrast “a simple matter.”[8] Heat sped up development, making it difficult to control the process as the picture appeared too quickly, while fixers and salts “burst their containers.”[9] High temperatures caused the gelatin-based emulsion to swell and soften, even sliding off the plate while in the fixing bath.[10] Gelatin also proved attractive to the local fauna: photographer Frank Hurley, on a 1922 trip to Papua (on the island of New Guinea) complained of cockroaches eating the gelatin from his photographic plates, a problem also described by Dykes the same year.[11]

Photos came out “foggy, black stained or splotched” or tinted with yellow haze, and exposed negatives were covered in mold or mottled by storage in the heat.[12] A 1922 article in *The British Journal of Photography* (*BJP*) asserted that while prints and films would "deteriorate rapidly" in the tropics, those developed and printed in situ would be more permanent than those made "in the temperate zone and subsequently taken to the tropics."[13] A 1908 *BJP* article titled "Photography in the Tropics" warned of the effects of "moisture and destructive insects" on equipment, especially on the leather bellows, glue, and wood of cameras of the period.[14] Cameras went moldy, their metal parts rusted, lenses filled with condensation, shutters stuck.[15] Imperial's 1911 sales report discusses its poor trade in India, anticipating that the arrival of a new dealership will improve things since "the plates will have time to acquire a reputation before the hot weather arrives; people have got into a habit of saying Imperial plates are too soft for India."[16]

Dufaycolor, introduced in the 1930s, created difficulties since it was a reversal process requiring two stages of development. The gelatin would soften and separate from the film base so that one photographer working in "semi-tropical" conditions in Cape Town, South Africa, wrote to the *BJP* in 1937 that he "frequently had the mortification of seeing the sensitive film disappear down the sink after second development."[17] The Lumière Autochrome did not fare much better: at the Royal Photographic Society in 1929, A. Coleman described his attempts to use Autochromes in India, commenting that this work "was likely to lead in a short time to a piece of clear glass." He was more successful with Paget plates, although he "never obtained a color plate during the monsoon."[18]

Concerns about the impact of weather conditions in remote outposts were not restricted to the tropics. Writing about William Henry Jackson's 1870s survey photographs taken in Wyoming, art historian Elizabeth Hutchinson emphasizes the material obstacles, "the uncooperative nature of weather or terrain, and the inconsistent reliability of chemicals and solutions."[19] Heat and dryness caused the albumen on his papers to crack, while cold weather made the collodion solution thick and unworkable. In the photographic press, comparisons were often made between the difficulty of photographing in the Arctic and in the tropics. But the tendency to blur the distinctions between problems with materials and problems with people was characteristic of colonial photographers and associated with tropical climates. In *Photographing Papua*, Max Quanchi details the complaints of explorers, geographers, and other travelers about taking and developing photographs "in the field" in Papua. Such difficulties were conflated with a whole host of other complaints: about photographing brown skin and dark interiors, and about resistant subjects or subjects whose quickness to please and pose ruined the supposed authenticity of photographs.[20]

British photographers often found that the light and climate abroad made it difficult to achieve what they considered to be a picturesque image, their observations inadvertently revealing how much their concepts of the picturesque and the atmospheric were drawn from their experience of a British climate. For example, in 1937, one writer in *The Amateur Photographer and Cinematographer* claimed that the presence of water vapor in the English atmosphere meant "a comparatively large amount of light is reflected into the shadows," but in "tropical and sub-tropical latitudes" what he considered "good photography" was more difficult. For this writer, dryness and heat were the issue, particularly on the North African coast, where "the brilliant lighting and absence of "atmosphere" familiar to photographers at home appears to be intensified by the glaring whiteness of the buildings" as well as the deep shadows in the narrow streets.[21]

During the 1930s, Ilford Limited promoted its products to markets in the Empire and Commonwealth via a short-lived magazine called *The Ilford-Selo Record*, produced in numerous editions for different countries, from Burma to New Zealand, with only minor variations. The magazine presents Ilford products as climate resistant. For instance, when W. W. Crealock of Kidderpore, Calcutta, wrote to *The Ilford-Selo Record* describing how well his Selochrome films fared despite being left in his cameras for over four months in the heat, the editors described this as proof of Selochrome's reliability "even under unfavorable atmospheric conditions."[22] *The Record* also promoted specialist products for hot and moist climates, including slower darkroom chemicals to counter the effects of heat, and "Tropical Hardener" to enable gelatin to set. This special preparation was patented in 1918 and marketed via Johnson & Sons of Finsbury, a long-established chemical supplier to the photographic industry. It was the result of concerted effort by the firm, which had created a special tropical darkroom within the factory in early 1918, to address the complaints of photographers in the colonies. As company chemist A. P. Agnew reported to the Royal Photographic Society a couple of years later, the room was climate controlled to mimic tropical conditions.[23] At Kodak Ltd., the research scientist Walter Clark would later be skeptical of such attempts at simulation, arguing that the tropical atmosphere, its organisms, and its air could not be reproduced in a lab.[24]

In the 1930s, formulas to resolve issues with using Dufaycolor in the tropics were publicized by the Dufay-Chromex Processing House in India.[25] Other companies also emphasized the suitability of their products for tropical climates or introduced specialist products targeted at those markets. The manufacturers of Rolleiflex cameras reassured their customers that the camera was perfectly good for the tropics as it had stainless steel parts, but a leather case should always be used, the lens wiped periodically, and cautious photographers could order a "special tropical lacquer" to protect the camera.[26] Both Kodak and Ilford supplied films "with special tropical packing."[27] In the 1920s, camera makers Butcher & Sons produced a series of "Tropical Models" of their Carbine folding bellows camera, featuring insect-proof Russia-leather bellows and an unusual brass body. They were following a nineteenth-century precedent: Captain Fowke's teak camera for tropical climates, also with brass fittings, was marketed as early as 1858.[28] Such cameras were designed to resist the local fauna as well as heat and moisture. A 1925 review of the Butcher & Sons camera noted that "those who take it to the Tropics will have the satisfaction of knowing that white ants and kindred pests cannot make any impression on it."[29]

In the interwar period, the flourishing market of "tropical" photographic materials and equipment reaffirmed existing ideas about the exceptional nature of so-called tropical places, while bringing them more firmly under the grip of the photography complex. "Difficult" climates and hostile nature were now viewed as problems to be resolved technically.[30] Problems were identified, expertise developed and institutionalized, new practices and new technologies introduced. The photographic magazines were filled with tips and advice.

From Brazil, W. F. A. Ermen wrote to the *BJP* in 1934 suggesting readers avoid alkaline developers, which aggravated softened gelatin.[31] Opinions varied on the use of ice. In 1920, A. Charcois's letter to the *BJP* discussed how to develop photographs in the "tropics" of North Queensland, Australia, without ice and recommended Ilford plates that withstood higher temperatures.[32] In 1932, J. Jackson asserted in *The Times* that in the tropics "films rapidly deteriorate and development without ice is a risky process," but two years later, in an article in the *BJP*, D. M. Cuthbertson advised against using ice since "variations in temperature are much more pernicious than a steady high one."[33]

Places understood as tropical became the testing grounds for the research and development of photographic products, allowing for the diversification of the market. New equipment and chemical processes promised to address problems of "Temperature, humidity, barometric pressure, altitude, and the presence of gaseous and particulate pollutants," from instruments such as "thermometers, barometers, and hygrometers" to new chemical products, plates, and films.[34] By the 1920s, the photography industry was participating in the discourse of technical mastery of tropical conditions described by Anderson as a kind of "modern colonialism" that aimed to subdue and tame the tropics (both human and more-than-human), or at least to make them amenable to the new phenomenon of "touristic tropical chic."[35] These technical developments facilitated an iconography of the tropical, a set of visual tropes used to market a new kind of "tropical dream" to Europeans.[36]

As Krista Thompson has shown, in Jamaica and the Bahamas (British colonies) "campaigns to refashion the islands as picturesque 'tropical' paradises" began as early as the 1880s, deliberately countering British and North American fears of the islands as sites of disease and their inhabitants as threatening. The Caribbean became available for tourist consumption via photographs very early in the 1900s, part of a process Thompson calls "tropicalization," which produced a narrow set of expectations of how the islands should look, emphasizing "tamed nature and disciplined natives." Tropicalization, while appearing to attend to cultural difference, actually flattened the differences between places deemed tropical. These colonial representations, Thompson argues, had real effects, transforming the environments and societies they purported to depict.[37]

By the early 1950s, Kodak was advertising its Kodachrome color film, first introduced in 1935, in *National Geographic*, with an image of white American tourists photographing apparently compliant Indigenous people in Guatemala against the picturesque backdrop of Lake Atitlán and the San Pedro volcano (plate 5). As Thompson's work shows, inhabitants of so-called tropical paradises challenged being objectified in this way and were often "violently inscribed against their will" even as some eventually depended economically on replaying touristic clichés.[38] The advertisement dramatizes this in two ways—prioritizing the will and acquisitiveness of the tourist ("you see it . . . you can get it . . ."), and repeating the word "color" nine times, driving home a spurious relationship between people of color, "colorful" traditional culture, and the growing popularity of color film.

16 INTO THE AIR

Lady Houston: But isn't it most terribly dangerous?
Lord Clydesdale: No more than walking across Hampstead Heath on a foggy night . . .

Wings Over Everest, 1934

At the front of the 1935 *Ilford Manual of Photography* is a sepia-toned photograph of a craggy, snow-covered mountain range emerging from clouds, set against a dark sky, and with plains visible in the foreground, dappled with the shadows of the clouds. The caption reads "EVEREST RANGE (A distance of 100 miles) Negative on Ilford Infra-red Plate by courtesy of *The Times*" (fig. 34).

The same image appears in another photograph, this time printed on a large display panel (fig. 35). A man stands next to it; having just autographed the panel, he is holding the pen against his signature as he turns toward the camera. Epitomizing swagger from his smug expression and monocle down to his spats, he is the aviator Lieutenant-Colonel L. V. Stewart Blacker, who has just returned from flying over the summit of the world's tallest mountain as part of the Houston-Mount Everest survey expedition. The occasion pictured is the opening of the Ilford Exhibition of Everest Photographs at Ilford's gallery in High Holborn, London, on May 17, 1933. Next to Blacker stands a taller, older man—Sir Ivor Philipps, chairman

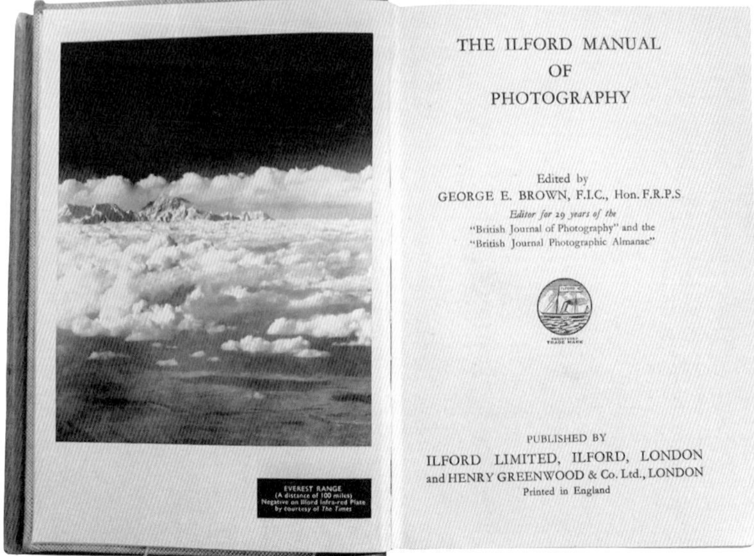

FIGURE 34. *The Ilford Manual of Photography*, ed. George E. Brown, open to show the frontispiece which is an infrared photo of Makalu titled *Everest Range*. The photograph was taken by the Houston-Mount Everest Survey Expedition. Private collection.

of the board of Ilford Limited—and beyond him Colonel P. T. Etherton, another member of the expedition.

Blacker was the initiator of the air survey which involved, as historian Patrick Zander has explained, "prominent members of the high-imperialist, often pro-fascist, far right," including its funder, the Mussolini-supporting Lady Lucy Houston.[1] The other well-known figure involved was Lord Clydesdale (Douglas Douglas-Hamilton), a celebrity and member of Parliament. Clydesdale piloted the first of the two planes, which set off to fly over Mount Everest from Purnea in India, with Blacker taking the role of "chief observer."

Mount Everest, named by the British in honor of a nineteenth-century surveyor of India, was not part of the British Empire, but it was central to the British imperial imagination. *Wings over Everest*, the film of the expedition, begins with Blacker gazing at the mountain and saying, "The pinnacle of the world, and we know no more

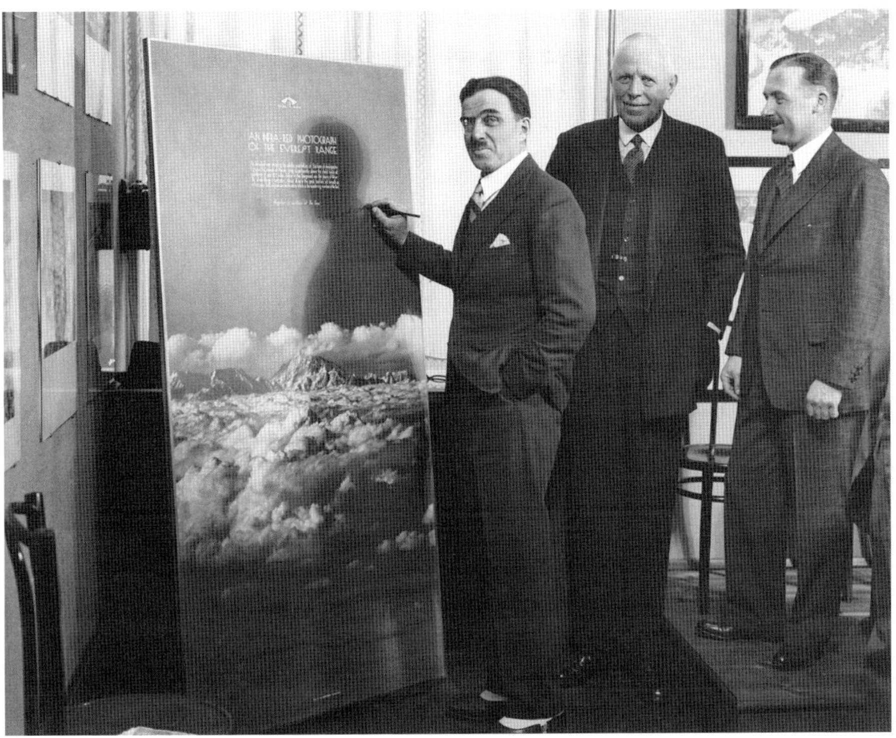

FIGURE 35. Lt. Col. L. V. Stewart Blacker with Sir Ivor Philipps, and Col. P. T. Etherton at the opening of the Ilford Exhibition of Everest Photographs at Ilford's gallery in High Holborn, London, on May 17, 1933. © Getty Images, Hulton Archive, Fox Photos.

about it than we do the moon," to which Etherton responds, "The last mystery, Blacker."[2] Though attempts to reach the summit on foot had failed, Everest's proximity to India and its height made it perfect for a British imperial spectacle.[3]

Though ostensibly a scientific survey, the Houston-Mount Everest expedition was a triumphalist propaganda exercise. This was a time of economic slump and resurgent nationalism; anticolonial movements were gaining steam in India and across the empire, and some political figures within Britain were calling for withdrawal of the British Raj.[4] Lady Houston wanted to impress "a native population in India with the courage, endurance and vigor of the new generation of Britons."[5] In a speech to his parliamentary constituents, Clydesdale proclaimed that the expedition would "have a great psychological effect in India," dispelling any impression of British "degeneration" and demonstrating that "we are still a virile and active race and can overcome difficulties with energy and vigor, both for ourselves and for India."[6] The British government was not so enthusiastic. Keen not to jeopardize plans for a land expedition to Everest, they gave assurances that there would be only limited flying, necessary for the survey, over Nepal, and none at all over Tibet.[7] Deploying cinema film and still photography, including both glass plates and roll film, as well as the latest in aviation technology, the whole expedition was, as historian Joppan George says, an "object lesson of imperial science and technology."[8] From the outset, Blacker had planned a media event. *The Times* had press and photographic rights, hence the picture credit in the *Ilford Manual*.

The pinnacle of photographic advancement was represented by the use of Ilford Limited's new infrared plates. In *First over Everest*, a book published shortly afterward, members of the expedition emphasized the novelty of using infrared photography in the air, crediting Olaf Bloch, the chief chemist at Ilford (himself a keen mountaineer), for making it possible.[9] Until 1932, long exposure times were needed for infrared daylight photography, to allow sufficient light through a deep red filter on the plate.[10] Now, "the world owes it to Mr. Bloch and his research that these exposures have been reduced to reasonable length, and even down to such lengths as made them feasible for air photography."[11]

Even so, it took great effort to use the Ilford plates. A huge camera (three feet by one foot) was suspended in the undercarriage of the plane, below the pilot's seat, where a torpedo would otherwise be. Precisely adjusted to point at the mountaintops when the plane was level, it was surrounded by horsehair-filled cushions and cradled in rubber shock absorbers to protect against vibrations. The camera operator, or "observer," would insert the dark slides (light-protected glass plates) into the camera from the back, use a mechanism to lift the foot-wide lens cap, and release the shutter. But it was the pilot who had to align the camera by maneuvering the plane to face the required view. A string attached to his wrist allowed him to signal to the observer that the plane was ready.[12] In this way, the whole plane became a camera, necessitating additional flights solely for photography.[13] These could not pass directly over Everest as there was a risk of damaging the camera above 25,000 feet, but fortunately the infrared plates were capable of cutting through the mist and haze of the freezing, thin atmosphere above the Himalayas, allowing for very long-distance images—as the caption in the *Ilford Manual* says, taken from as far as a hundred miles away.[14]

Although the expedition was promoted as an aerial survey, the way it was conducted suggests a systematic survey was not the main reason for the daring and risky flights. Rather than use the experienced British photographic surveyors already in India, Clydesdale and Blacker took a brief class at the Royal Air Force School of Photography at Farnborough.[15] The planes were well-equipped, with large electrical Williamson Eagle survey cameras for taking vertical photographs on roll film, handheld cameras for oblique pictures on glass plates, and several movie cameras loaded with 35 mm film. With no precedent for photography at such altitudes, plenty of special measures were taken. The survey cameras, for instance, were insulated against the cold and warmed with electric heating elements, which was risky since the film was nitrate and highly flammable.[16] Even so, the first flight did not yield many effective survey photographs. Blacker admitted, "The survey strips were a disappointment, and a sad one," since they had not anticipated "an amazingly high dust haze."[17] In 1935–36, employed to work with the Houston-Mount Everest photographs at the Royal Geographical Society, aerial surveyor D. R. Crone found there was no accompanying data or details of the focal lengths of the cameras. He claimed it took years for a usable map to be produced.[18]

Some decisions were made expressly for cinematic effect. For the film *Wings over Everest*, Clydesdale, Blacker, Lady Houston, Etherton, and others reenacted conversations leading up to the expedition, with other sections being filmed during the expedition itself. The filmmakers asked the aviators to drop smoke bombs on the return from Everest to heighten the drama of the cinefilm. According to George, the smoke trails were visible on the ground, leading to a "mass exodus of villagers from Lalbalu to Purnea" as they heard of "an unannounced British 'bombing practice.'"[19] Predictably, Clydesdale blamed local ignorance, but George suggests these fears were well-founded, given the RAF's aerial bombing raids carried out against of the Pashtun peoples on the Northwest Frontier (now the western border of Pakistan) in 1925, and British insistence at the World Disarmament Conference (1932–34) on their right to control British India's borders through bombing.[20] Far from impressing the Indians with British "courage, endurance and vigour," the triumphal antics intended for the film reinforced an impression of Britain's willingness to deploy technological violence.

The expedition party even prioritized the shooting of photographs and cinema film over the risk of damaging diplomatic relations with both Tibet and Nepal. Concern about the airmen's safety and about relations with Nepal and Tibet led to the aviators being recalled on April 19. Air Commodore P. F. M. Fellowes, in charge of the party, issued a statement saying, "No further main flights will be made. A local test flight to-day was used for the purpose of attempting infra-red photography of the Himalayas at long range."[21] *The Times*' editorial the next day presented the expedition as a triumph of "conquest," claiming that the "daring" and "scientific" project had undoubtedly contributed to "the progress of knowledge."[22]

Despite Fellowes's assurance, the party in Purnea decided to make another flight. This brought back some better survey photographs, though even these would prove difficult to plot properly because the pilots had flown in different directions and started and stopped the photography at random points.[23] On April 21, *The Times* admiringly reported on this second "unauthorised" flight over Everest, describing it as a "remarkable and insubordinate enterprise."[24] In an odd climbdown, a fortnight later *The Times* claimed that two flights over Everest had always been planned and that "immediately after the success of the first flight . . . the necessary sanction for the second venture . . . had been obtained from Nepal."[25] Publishing some of the pictures on May 8, the newspaper took pains to explain that the oblique photographs were not the main reason for this flight, since plenty of "splendid shots" of the Himalayas had already been taken on the first.

The infrared photographs, though purportedly "scientific" images,

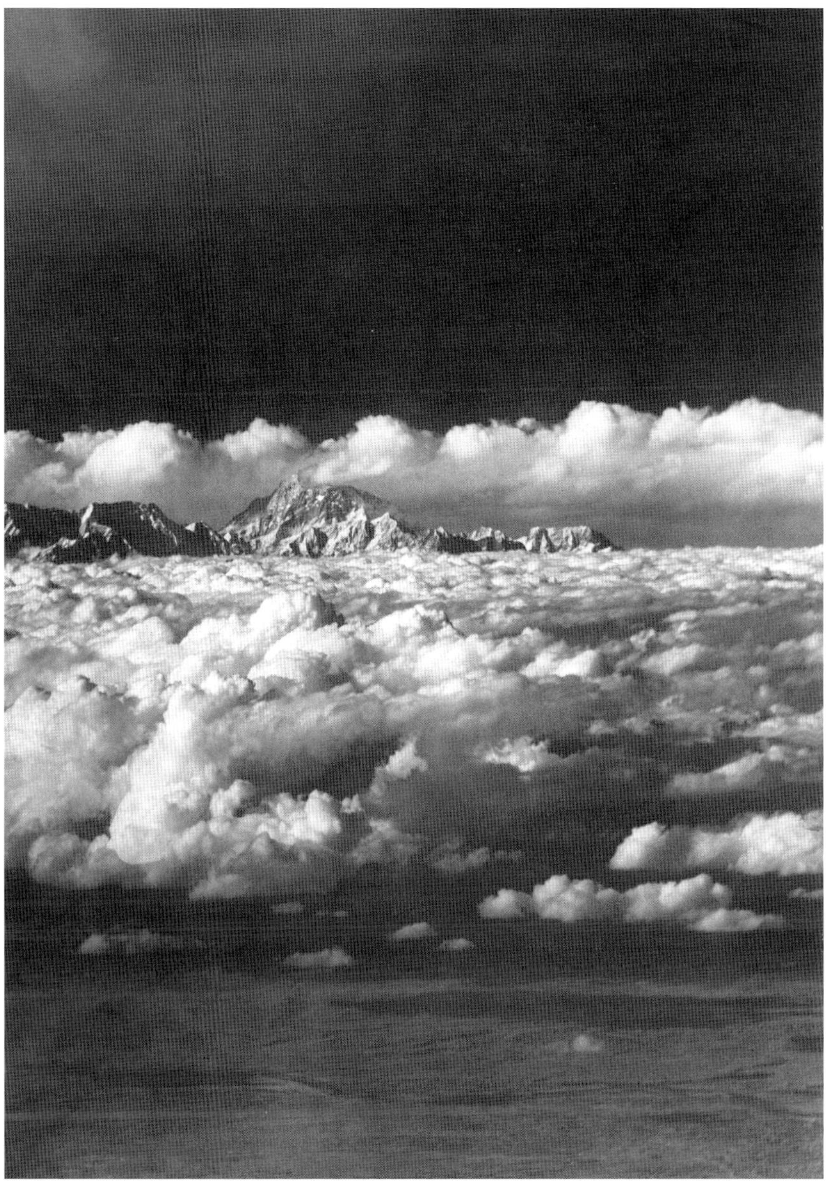

FIGURE 36. Infrared photograph of Makalu, taken by the Houston-Mount Everest Survey Expedition from over a hundred miles to the south. © Getty Images / Ullstein bild.

turned out to be most useful in promotion. They were widely disseminated, published, and exhibited under the sponsorship of Ilford and *The Times*, especially the picture of the "Everest range" that became the emblem both of Ilford and the expedition. When initially published, it was mistakenly captioned as Everest in both *The Times* and *The Illustrated London News*, though it actually depicts the peak of Makalu (fig. 36). Acknowledging the mistake, Blacker claimed, "Fortunately nearly all the photographs speak for themselves, and Mount Everest speaks for itself from the photographs." To confuse Makalu with Everest threatened the uniqueness of Everest and the value of the expedition. He was keen to stress that "there can be no possibility of mistake in the long run."[26]

This photograph nevertheless came to represent the survey and Ilford, and to stand as an aesthetic achievement as much as a technical one. When *The Times* published the picture, it described it as "a study in chromatic values."[27] In *First over Everest*, the authors noted, "The infra-red plates penetrated the mist admirably, but not the actual clouds themselves, nor the dust rising up from the plains. The result of infra-red photography on a cumulus cloud was to bring out the colour values remarkably well, and to emphasise the beauty of graduation of light and shade, not only on the clouds, but on the mountains themselves, in a way that was beyond the scope of other photographic art."[28] Following the opening of the Ilford Exhibition of Everest Photographs in May, the collaboration between Ilford and Blacker continued. The next day they held a joint event at the Royal Society, again showing the same photograph of the summit of Makalu.[29] Over the following year, the picture was circulated in photographic exhibitions around Britain, and in more illustrated lectures by Bloch, Blacker, and others.

The Times' art critic (unnamed but likely Charles Marriott), in keeping with the newspaper's investment in the whole enterprise, described the image as "easily the most impressive, and probably the most remarkable photograph ever taken." Emphasizing the extraordinary realism of the infrared process, he undid any claim to cartographic use: "The map-like effect entirely disappears, and everything, from the clouds to the distant peaks, is fully rounded and solid . . . the illusion of being there is perfect."[30] In a later review

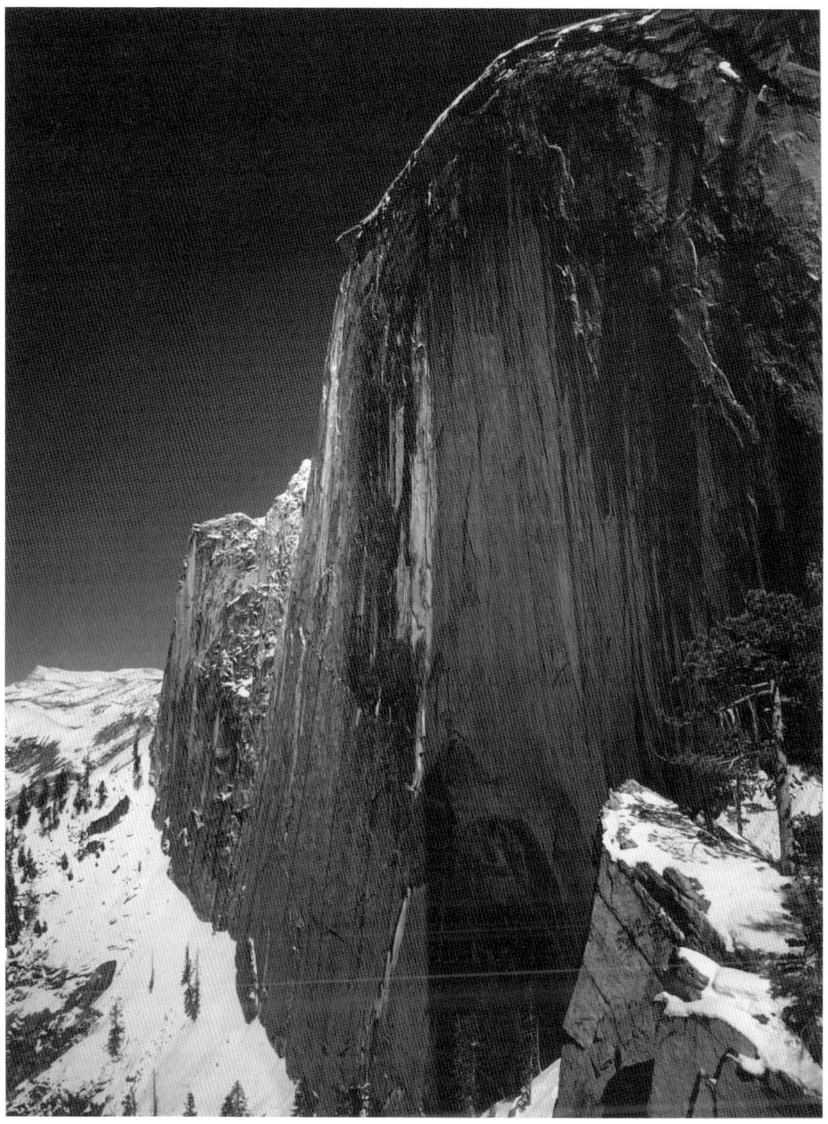

FIGURE 37. Ansel Adams, *Monolith: The Face of Half-Dome*, Yosemite National Park, California, 1927. Ansel Adams Archive, Center for Creative Photography, University of Arizona © The Ansel Adams Publishing Rights Trust.

he praised the infrared pictures for their "photographic purity." Marriott was a popularizer of modernism, and this review endorsed the modernist idea of medium-specificity, stating that the infrared photographs "bring home the truth that photography gets nearest to art . . . when it is used on its own lines."[31]

Today, these scenes have lost their novelty, and it is hard to recapture the sense of wonder such aerial photographs evoked. For comparison, consider *Monolith: The Face of Half-Dome*, a photograph by Ansel Adams taken six years previously on April 10, 1927, at Yosemite in California (fig. 37). Though the process was not infrared, Adams also used a dark red filter, this time over a Wratten panchromatic plate, and the visual effect is similar, darkening the sky to a deep black. If Adams's extraordinary photograph undermines some of the hyperbole surrounding the photograph of Makalu, it also reinforces Marriott's sense that the latter picture was modernist in style. However, in this case the photograph's modernity, both technical and aesthetic, worked to support far-right claims regarding Britain's imperial "vigor." Ilford's infrared plates cut through the atmospheric haze, stripping the mountains of the aura of "the last mystery" only to overlay them with another kind of aura, of imperial power and modernity.

17 PYRRHIC VICTORIES

► Giacomo Ciamician is the Italian scientist from whom I took the term "daughter of coal" (chapter 4), though he applied it not to photography but to the whole of civilization. In the early twentieth century he conducted research at the University of Bologna on the photosensitivity of plants. His work focused on how to harness the ability of plants to capture the energy of the sun and how plants could be converted into fuel or energy sources as a replacement for coal, which he anticipated would soon run out. Against the obsession with remaking the world from coal tar, which was at that moment facilitating the complete transformation of photography through improvements in sensitizing dyes, Ciamician wrote: "A battle is raging between chemical industry and nature, a battle which does honor to human genius. Up to now the products prepared from coal tar have almost always been triumphant. I do not need to remind you of the various victories but it is possible these may prove to be Pyrrhic victories."[1] It is easy now, more than a century later, to agree with Ciamician and see the triumphs of the coal tar industry as Pyrrhic victories. Pyrrhus was the Greek king of Epirus, who fought the Romans and won in two battles (circa 280 BCE). In winning, he lost many of his recruited soldiers and also "almost all his particular friends" who died in the battles, according to Plutarch's *Life of Pyrrhus*.[2] What we have gained from coal tar are countless technological advances but also countless losses.

That coal tar derivatives are deadly should hardly be a surprise; these substances are used in the manufacture of high explosives and poison gas, in warfare, and in the deliberate mass-extermination of populations, most notoriously in the form of Zyklon B, the hydrogen cyanide gas used for delousing troops and prisoners of war before it was turned on Jews as well as Roma, Poles, and Soviet prisoners of war in the gas chambers of Auschwitz. Hydrogen cyanide had been known for centuries before it began to be produced in large quantities in the 1890s for use in mining. It was manufactured using ammonia, which could be derived from coal gas works. The idea of killing people with hydrogen cyanide was pioneered in the United States, which used it to carry out death sentences from 1924. Zyklon B was initially developed to kill insects and other small animals. In his book *Terror from the Air*, Peter Sloterdijk argues that this shift from pesticide to a means of executing human beings reflected, on the one hand, the "pragmatic humanism" that claimed gas was a more "humane" method of killing (an argument that had been rehearsed by the apologists of poison gas warfare since the First World War) and, on the other, the "extreme pragmatism" of the "Nazi death industry," which literalized Nazi propaganda depicting Jews as "vermin" via the "psychotic acting out of a metaphor."[3]

If the use of deadly gases on humans was premised on first reducing them to vermin, the reverse was also true, in that the practice of exterminating insects was aggrandized as a kind of warfare. Around 1900, anti-malaria campaigns in British India declared "war" on the mosquito. This reflected existing British attitudes to insects in general, which divided them into good and bad species. As the historian Rohan Deb Roy writes, the British Indian government, from the 1860s, viewed insects as a threat to commerce, "most consistently as detrimental to the tea, cinchona, coffee, sugarcane and cotton plantations."[4] Pesticides were a product of the diversification of the coal tar dyestuffs industry, stimulated by intense international competition among the dye companies and high rewards for innovation: in Germany, for example, IG Farben targeted lice, while in Switzerland, the Geigy aniline dye company focused on the protection of fabric from moths, filing patents from the late 1930s.[5]

In 1940 Geigy patented the substance known as DDT, first used in Switzerland in 1941. During the 1940s, the company produced the pesticide, trademarked as Geserol, at its Swiss and German plants.[6] The US patent was filed in March 1941. DDT's disastrous environmental consequences are well-known thanks to marine biologist Rachel Carson's *Silent Spring* (1962). It was used against lice and malaria-carrying mosquitoes, dusted over human populations, and sprayed across whole areas in places like Greece, Italy, Saipan, and Burma (Myanmar) during wartime. By the mid-1940s, evidence was accumulating about the toxicity of DDT to humans and larger animals, and the ecological problems were beginning to be recognized, too. But despite these early warnings, by the 1950s it was being used on a massive scale, destroying whole ecosystems and causing health effects that last for generations.[7]

Coal and coal-derived compounds are carcinogenic, something noticed as early as 1775 by Percivall Pott, whose published account of cancers among London chimney sweeps may have been the first description of a specific occupational cancer.[8] By the 1890s, doctors were reporting cases of bladder cancer among workers in the German and Swiss dye factories. Ludwig Rehn, a Frankfurt surgeon who had previously been the factory doctor at the Chemische Fabrik Griesheim, made the connection between the coal tar dye industry and bladder tumors in 1895. Rehn used information from a doctor at the Hoechst on the Main aniline dye works, about cases in the fuchsine department at Hoechst (fuchsine was named after the fuchsia flower because it was a similar shade of pink).[9] Subsequently, more cases were discovered in other German aniline dyestuff companies, including the photographic manufacturer Agfa.

In Basel, Switzerland, a urologist named S. G. Leuenberger found cases at Geigy and CIBA over the first decade of the twentieth century. By 1923, Rehn had discovered ninety-four cases of bladder cancer in the German dye industries. In Britain, where the industry was much smaller, cases were identified from 1926 onwards. In the US, the synthetic dye industry only got going in the interwar period, and so cancer cases only began to be identified in the 1930s.[10] "Aniline cancer" was one of the first industrial cancers to be definitively identified, according to science historians Heiko Stoff and Anthony S. Travis.[11] The fact that this cancer has a latency period of ten to nearly thirty years is one reason why it was only noticed in the 1890s, by which time, tens of thousands of workers were employed in the coal tar industry and these dyes had been widely used for nearly forty years.[12] Looking at the history of the coal tar dye industries, it is hard to ignore what Kirsty Sinclair Dootson calls, in her book *The Rainbow's Gravity*, "the very close affinities between dyeing and death."[13]

Industrial scientists were aware of ongoing global warming caused by the mass production and consumption of fossil fuels as early as 1938, when the engineer Guy Stewart Callendar published an article titled "The Artificial Production of Carbon Dioxide and Its Influence on Temperature" in the *Quarterly Journal of the Royal Meteorological Society*.[14] Some writers have claimed that global warming resulting from industrial emissions was a topic of discussion prior to the Great War, with some tracing the discovery to the work of the Swedish scientist Svante Arrhenius in the 1890s.[15] The role of carbon dioxide in atmospheric warming was known even earlier, for example in the work of the Irish physicist John Tyndall in the 1850s. Even before Tyndall, the American scientist Eunice Foote had discovered how carbon dioxide in the air absorbed heat from sunlight, and she had linked increasing levels of CO_2 and water vapor in the atmosphere to climate change.[16]

Yet scientists like Callendar, who correlated an increase in global temperatures to fossil fuel use, did not necessarily see this change as a bad thing. They also underestimated the growth in the use of fossil fuels. It was already known that coal burning and coking (the process that produces gas and coal tar) caused atmospheric pollution. Silver, that sensitive substance, was among the first to register the change: photographic manufacturer W. J. Streeter writes in his polemic *The Silver Mania* that by the 1890s, "years of burning coal and other fossil fuels had added enough sulfur to the earth's atmosphere that silverware tarnished quite rapidly," a factor in the introduction of stainless steel cutlery and other silver replacements.[17] Sulfur also fogs silver halide photographic emulsions, and by the 1880s, the "vitiated" sulfurous atmosphere of cities was bemoaned in the photographic press as a principle cause of the deterioration of silver photographic prints.[18] As art historian Siobhan Angus has shown, platinum prints were less sensitive to contaminants in the atmosphere than silver, and this was a key selling-point for the platinotype process from the 1870s.[19] At the same time, platinum prints were popular among pictorialists because they lent themselves to "suggesting the idea of 'atmosphere.'"[20]

By the late nineteenth century, coal smoke was increasingly regarded as dangerous to the wider environment and human health. The 1880 book *London Fogs*, by the meteorologist Francis Albert Rollo Russell, described how "only lately has the sudden, palpable rise of the death-rate in an unusually dense and prolonged fog attracted much attention to the depredations of this quiet and despised destroyer."[21] London was not only a cold and foggy city with a vast number of coal fires and coal-fired gasworks, it was also a major center for the production of coal tar. In 1904, one London firm, the South Metropolitan Gas Company, claimed to produce ten million gallons of the stuff per year, "about 1/20th of the tar produced in all the European gas works."[22] The "ammoniacal water" or "liquor" from the coal tar works would be used to manufacture ammonium sulfate for fertilizer, emitting sulfur dioxide and toxic particulates into the air surrounding the plant. The same London company exported anthracene, another by-product of coal distillation, to Germany for the manufacture of alizarin dyes. Coke, the leftover after distillation, was sold by the gas companies as a remedy to the London fog and a means to realize the dream of a "smokeless London," but its much-touted purity resulted from "the objectionable matters in coal being carried off in gas and vapors," many of which made their way into the atmosphere during processing at the gasworks.[23]

We now know that coal mining, burning, and coking, for gas and electricity production but also for the production of new chemicals, produced an excess in the atmosphere of carbon dioxide, the largest contributor to global warming; methane, also a greenhouse gas; and nitrous oxide, which is also emitted in the production of nitric acid. The fossil fuel industries, combined with mass deforestation, are largely responsible for the rapid climate change that our planet is undergoing. Together with exponential population growth, they are also producing mass extinctions among both plant and animal species. As a species we are losing our "particular friends," our companions and codependents on this planet. Photographs and films gave us ways of knowing this, and ways of mourning it, and they are also, ironically, a product of the industries at the root of this tragedy. Photographic materials, as well as being manufactured using the by-products of coal gas, were often produced by the very same aniline dye companies responsible for the production of Zyklon B and DDT, and that tried to suppress or minimalize the high rates of cancer among their workers. Seen from this angle, the history of photography is a dirty history, which is simultaneously about the representation and the contamination of the Earth.

18 WAR IS BEAUTIFUL

One of the most canonical essays of photography theory, Walter Benjamin's "The Work of Art in the Age of Mechanical Reproduction" (1935–36) had particular impact in the English-speaking world when it was published in translation in the collection *Illuminations* in 1969. Describing photography and film as forms of technical reproduction, Benjamin situated them as technologies that had both transformed the experience of art and allowed people to objectify themselves to the point that humanity "can experience its own destruction as an aesthetic pleasure of the first order."[1] This analysis resonated with 1960s understandings of capitalist spectacle and with arguments about the politicization of aesthetics. Benjamin's conclusion to his essay is much cited, and often summarized as a general point about how fascism aestheticizes violence or about the opposing strategies of communism and fascism in relation to aesthetics.

I must have discussed this essay hundreds of times with successive groups of students, but now, reading it while thinking about the industrial and chemical connections between photography and poison gas, its closing paragraphs appear in a new light, as a set of reflections about imperialism as much as fascism, and as a very specific and idiosyncratic take on the ways in which a love of technology comes to substitute for an older, collective relationship with nature. Written in the mid-1930s, Benjamin's essay is also a reminder that the "interwar period" was nothing of the sort, and that in these decades, in the context of competing European imperialist drives, the Italian fascists pioneered new, brutal tactics of chemical warfare.

In his discussion of fascism, while also alluding to Nazism, Benjamin was referring directly to Italian fascism and the Italian Futurist poet Filippo Tommaso Marinetti's notorious claim that "war is beautiful." This phrase (more literally translated as "war has its own beauty") is a repeated mantra in Marinetti's "The Futurist Aesthetic of War" (*Estetica futurista della guerra*). This text, which Benjamin identifies in his essay as a "manifesto on the Ethiopian colonial war," had been published in several Italian newspapers on October 27, 1935, around the time that Benjamin was composing the "Work of Art" essay, and the precise date that (we now know) Benito Mussolini authorized the use of poison gas from the air on the battlefields of Ethiopia.[2]

Ethiopia (then also known as Abyssinia) had been one of the few African states free from European colonization. On October 3, 1935, Italian forces invaded the country from Eritrea and Italian Somaliland, using bombs, mustard gas, and heavy artillery to kill hundreds of thousands of Ethiopian civilians and soldiers over the next two years. Marinetti had been an early supporter of war in Ethiopia, calling for an invasion months before armed conflict first started (at the end of 1934) and promoting war not just in writing but in speeches and recruitment drives. He arrived in Ethiopia as one of the first volunteers to fight on the front lines, despite being sixty years old.[3] He would remain an apologist for fascism's brutal tactics in the war, later defending the use of poison gas.[4]

"The Futurist Aesthetic of War" declared "man's dominion over the subjugated machinery" at the same time as it eulogized "the dreamt-of metallization of the human body."[5] In Marinetti's prose, gas masks, "terrorizing megaphones," flamethrowers, and armored tanks were technological prosthetics that "perfected" the new "mechanical man."[6] A couple of years later, in his April 1937 manifesto "Poetry and Corporatist Art," Marinetti would declare that the Italian Futurist movement set out to "create a 'non-human' poetry and art which is to say a poetry and art extraneous to humanity thanks to its systematic extraction of new beauties and new music from the technicisms of machine civilization." In "The Non-Human Poem of Technicisms" (*Il Poema non umano dei technicismi*) (1939), he would again compare the production of poetry to mining and extraction.[7] For Marinetti, as cultural historian Jeffrey Schnapp writes, automation "promised to free culture from the burdens of the old humanism and its cult of reflective distance and interiority."[8] The mingled bodies of workers and machines, or soldiers and machines, promised new subjects for modern experience. Despite their extreme political differences and vastly different conclusions, Marinetti and Benjamin share this sense of technology as something that was transforming both embodied experience and modes of perception, and destroying an older high-cultural aesthetic experience that was premised on contemplation and critical distance.

His prose rendering violence visual, symphonic, and architectural in turn, Marinetti substitutes the nature symbolism of traditional poetry with the machinery of death and destruction. He celebrates the "fiery orchids of machine guns," the perfumes mingling with the smells of putrefaction, the silence that falls on the countryside but only due to a pause in the gunfire. He even evokes the pictorialist device of chimney-smoke rising from a picturesque cottage, twisting it into "the smoke spirals from burning villages." Benjamin reads this as self-alienation but also as "the consummation" of art-for-art's-sake—in which aesthetic pleasure overrides all moral or social values.[9]

Marinetti's paean to technological destruction aestheticizes not just war but specifically, in Benjamin's phrase, "imperialistic warfare." Marinetti was born into a colonial family in Egypt, and much of his work is characterized by a violently racist fascination with Africa and Africans, yet both he and Mussolini were characteristically indifferent to the specificity of Ethiopia, to the lives and cultures of Ethiopians, and to the politics of the nation, arguably even to its significance to Italy beyond being an expression of Italian power and an addition to its territories.[10] The war in Ethiopia was war for war's sake, a means for fascism to achieve a masculinist and militarist nation. According to historian Alexander De Grand, Mussolini's hopes were vested in "the future generation which might be infused with new martial values," and in the 1930s, "the entire propaganda apparatus, especially the schools, geared up to stress military values, the glories of war and Italy's destiny in the Mediterranean and the colonies."[11] Following the invasion of Ethiopia, fascism could take its gloves off—"Ethiopia was the place where fascism could shed any inhibitions, escape from the shadow of defeat and feelings of inferiority, and impose its core beliefs of hierarchy and absolute obedience." No weaponry was ruled out on moral grounds: mass executions and poison gas were used and bacterial warfare only avoided because unnecessary.[12]

Benjamin states that Marinetti's manifesto has the virtue of clarity. His Futurist rhetoric exposes what is implicit in the more conventional fascist aesthetics with its exaggerated classicism and adoration of the armored body. In his deliberately shocking enthusiasm for, and aestheticization of, violence and destruction, he justified imperial warfare in terms that Mussolini would not, although they are no less ideological. Contrast Mussolini's own radio speech of October 2, 1935, in which he described the invasion as a mass march against injustice, as the Italians taking their rightful "place in the sun," as revenge for the postwar settlement when "only the crumbs of a rich colonial booty were left for us to pick up." Mussolini described the conflict in conventional rhetoric familiar to Western European nations: in terms of the "heroic" dead of the Great War, the aversion to spilling blood, the conflict between civilization and barbarism (with Ethiopia cast as a "barbarian nation" deserving of invasion), and most of all in terms of a justifiable "colonial conflict" that did not deserve to become a European war.[13] If Mussolini's rhetoric rationalizes the invasion, Marinetti's justifies it on sensual, aesthetic grounds.

Marinetti's shocking aestheticization of war also has clarity because it expresses a convulsive ecstasy of destruction that Benjamin recognized as one he had written about seven years previously. In "One-Way Street" (1928), the concluding section titled "To the Planetarium" deals with precisely this—the way that technological warfare recreates, in corrupted and perverse form, an ancient collective and ritualistic relation with the cosmos. If the ancients had the ability to collectively immerse themselves in the cosmos through ecstatic trance, Benjamin argues, this had been broken with the advent of modern science and particularly astronomy, which produced a modern "optical connection with the universe." In the "Work of Art" essay, he suggests that photography takes this further, taking artistry away from the hand so that it "devolved only upon the eye looking into a lens." However, this new kind of perception is not necessarily distancing, rather it brings the object into "close range."[14] In both cases, though, a ritual relationship (with nature, with art) is destroyed.

In "One-Way Street," Benjamin argues that the older, ritualistic cosmic experience has been individualized, reduced merely to poetic and aesthetic experience, to "the poetic rapture of starry nights." But the drive for collective cosmic experience reasserted itself, in much more dangerous ways, in technological warfare. In Benjamin's description, the Great War was a twisted revival of that ecstatic commingling, a kind of collective ritual, a "wooing of the cosmos" on a "planetary scale." He writes: "Human multitudes, gases, electrical forces were hurtled into the open country, high-frequency currents coursed through the landscape, new constellations rose in the sky, aerial space and ocean depths thundered with propellers, and everywhere sacrificial shafts were dug into mother earth."[15] Reading this, it is easier to make sense of Benjamin's cryptic claim, in the "Work of Art" essay, that modern warfare is a kind of uprising or rebellion of technology. In "One-Way Street" he suggests that technology is, or ought to be, a means to mediate between "nature and man" but instead is employed by "the imperialists" as a means to master and subdue nature.[16] Technology produces both alienation and destruction not because of something innate to it, but because of something innate to imperialism, yet technology is also an agent in its own right, rather than simply an extension of human power.[17] Seeking a kind of ecstasy of union with the universe, but confusing it with mastery, imperialism creates the conditions in which "technology betrayed man and turned the bridal bed into a bloodbath."[18]

19 ABOLISHING THE AURA

► In *Terror from the Air*, Peter Sloterdijk writes that gas warfare is the first environmental warfare: it attacks not the individual but the environment on which they depend, the atmosphere they breathe and through which they see. Unable to resist breathing, the victim of a gas attack becomes "an unwilling accomplice in his own annihilation."[1] Marinetti implicitly recognized this when he enthusiastically reported that Ethiopians found themselves under "a new unbreathable sky," contrasting their predicament with that of the aerial Italian soldier, safely enveloped in the individualized atmosphere of his airplane, protective suit, and gas mask.[2] For Sloterdijk, gas warfare marks the transition from classical warfare to terrorism, which "exploits the fact that ordinary inhabitants have a user relationship to their environment."[3] And in his appeal to the League of Nations in 1936, Emperor Haile Selassie emphasized that the aerial gas attacks in Ethiopia were on civilian "populations far removed from hostilities, in order to terrorize and exterminate them."[4] Italo Brandimarte, writing about the atmospheric aspects of the Ethiopian war, emphasizes how the form this terror took was a kind of "lethal weather."[5]

Indeed, poison gas (which was not always technically a gas but a vapor) often took the form of artificial weather, appearing as rain, a rolling fog, hail, or, in the case of chlorine gas in the Great War, "A great yellow, greenish-yellow, cloud" about twenty feet above the ground.[6] Gas warfare, Sloterdijk claims, introduces "a whole black meteorology."[7] Brandimarte describes how this lethal atmosphere became normalized as weather: "While Ras Imru, facing gas for the first time in late December, hopelessly claimed that he was 'stunned, [he] did not know how to respond, [and he] did not know how to fight this rain that was burning and killing' . . . Ras Kassa—defeated in the Second Battle of Tembien, 2 months later—already stoically accepted its lethal pervasiveness: 'Our conscience was clear: you cannot kill the fog.'"[8] In his 1936 speech, the Ethiopian Emperor described how the Italians began with tear gas from the air, then barrels of mustard gas "hurled upon armed groups," and finally:

> Special sprayers were installed on board aircraft so that they could vaporize, over vast areas of territory, a fine, death-dealing rain. Groups of nine, fifteen, eighteen aircraft followed one another so that the fog issuing from them formed a continuous sheet. It was thus that, as from the end of January 1936, soldiers, women, children, cattle, rivers, lakes and pastures were drenched continually with this deadly rain. In order to kill off systematically all living creatures, in order more surely to poison waters and pastures, the Italian command made its aircraft pass over and over again. That was its chief method of warfare.[9]

Tens of thousands of Ethiopian citizens died as a result of these mustard gas attacks. Brandimarte's article is titled "Breathless War," and as he explains (and Marinetti's own writing reveals) gas warfare constructs two opposing and racialized subjects—the "Italian aerial soldier" invulnerable in his flying machine and the breathless Ethiopian (anticipating more recent understandings of racialized oppression as removing the right to breathe).[10] The gases affect more than breathing, since mustard gas has powerful blistering and burning effects, as does chlorine gas, used early in the Great War. Bodies disintegrate and lose their substance: survivors' accounts of the first chlorine gas attacks in the Great War describe how the bodies of the dead and wounded "were almost visibly blowing up—their bodies were going colored, but they were blowing up. You could put your finger and make a little hole, almost, in them."[11]

While the aerial and armored Italian soldier could transcend the earth, Ethiopians were condemned to their contaminated environment. Brandimarte argues that defining bodies and lives according to their access to breathable air produces a hierarchy dictating both what it means to be modern and what it means to be human, rendering the racialized Ethiopian as both outside modernity and less than human.[12] Or, to put this within the frame of my argument so far, the gendered and racialized opposition between aerial and earthbound bodies that pervades European visual culture of the 1920s and '30s (see chapter 12) plays out here in lethal ways.

When Walter Benjamin introduced the subject of gas warfare at the end of his "Work of Art" essay and in relation to Marinetti's "manifesto on the Ethiopian colonial war," he could not have known that Mussolini had already given the order to rain down poison on Ethiopia. Nevertheless, he connected Mussolini's and Marinetti's imperial war directly with gas attacks, asserting that through gas warfare society has "found a new means of abolishing the aura."[13] The claim is part of the culmination of an essay in twenty-one sections—and only the twenty-first addresses war. The rest of the essay is famously about technical reproduction, specifically about how filmic and photographic reproductions strip "the veil from the object" and thus contribute to the decay of the aura.[14]

How are poison gas and photography connected? The answer is found in atmosphere. "Aura" is an atmospheric term that came into several European languages (including French, English and German) via Latin; it was derived from the Greek term meaning "morning mist." By the time it made its way into late Middle English, it meant a gentle breeze or breath. Benjamin writes that "to follow with the eye—while resting on a summer afternoon—a mountain range on the horizon or a branch that casts its shadow on the beholder is to breathe the aura of those mountains, of that branch."[15] This sentence tells us several things about how Benjamin understood aura: as an experience of communion with nature, as an optical experience, and as something breathed in, that is, something in the air. One experiences aura as a body immersed in an atmosphere, endowed with senses able to access not only that which is proximate but also that which is remote. He describes aura as a phenomenon characterized by "a strange tissue of space and time: the unique apparition of a distance, however near it may be."[16] This echoes his description in "One-Way Street" of the ecstatic trance: "It is in this experience alone that we gain certain knowledge of what is nearest to us and what is remotest from us and never one without the other."[17] The experience of aura is a simultaneous experience of connectedness (after all, the air is within us as well as around us) but also of what is far away.

As a concept aura is doubly atmospheric, both in the sense of being in and of the atmosphere and in the sense of a mood or feeling that suffuses an image, scene, or space. One way of thinking about Benjamin's claim that both gas warfare and technological reproduction put an end to aura is to consider how they demolish an older experience of atmospheric immersion in the cosmos and replace it with something else. In the case of technological reproductions, the remote can be brought near, but in the process it loses its specific atmospheric and located relationship to the beholder. In the case of gas warfare, atmospheric immersion is ended through a technological violation of people's relationship with their environment, because it makes the act of breathing deadly.

But there is another, more well-known aspect of Benjamin's argument in the "Work of Art" essay, which is that "the artwork's auratic mode of existence is never entirely severed from its ritual function"—the respectful contemplation of an artwork is a secularized ritual. In this context aura can be imagined as a halo worn by artists and artworks, singling them out as special and transcendent (recall the French word for halo is *aureole*—see also chapter 11). When threatened by the rise of photography, Benjamin argues, art responded by entrenching itself further in its quasi-religious status, via what he terms "a theology of art," that is, art-for-art's-sake. And here we come full circle, since Marinetti's aestheticization of human annihilation is the "consummation" of art-for-art's-sake according to Benjamin.[18] How can these two things be the case at the same time: that the gas warfare Marinetti will end up defending abolishes aura, while his writing is saturated in this theology of art? Marinetti the modernist is aware, just as Benjamin is, of how the bourgeoisie and its technologies have (in the words of the *Communist Manifesto*), "stripped of its halo every venerable occupation. . . . All that is solid melts into air, all that is holy is profaned."[19] But Marinetti the fascist responds by transferring a ritualistic and worshipful contemplation usually reserved for religion or for art to the shock technologies of modernity. In "One-Way Street" Benjamin writes that in technological warfare, "annihilation" is accompanied by "a feeling that resembles the bliss of the epileptic."[20]

Marinetti can aestheticize warfare because he invests technology with a cosmic experience which has already been rendered impossible by the wartime misuse of technology. As Jeffrey Schnapp has argued, Marinetti's concern with the "conquest of matter" was part of a broader attempt in Futurism to "to relocate the spiritual within new technologies and the materials with which they were affiliated."[21] His deliberate substitution of violent technological objects (such as machine guns) for the traditional natural symbols of poetry (such as orchids) hangs not just on the desire to shock but on the desire for technology to provide something analogous to a cosmic ecstasy. Though Marinetti is extreme and idiosyncratic, his writing makes vivid something that is not peculiar just to him: a "sense perception altered by technology" that finds its greatest expression in war.[22] Reading the informative, empirical and chronological accounts of the developments in the European chemical industries in the late nineteenth and early twentieth centuries, I find it useful to hold Benjamin's perspective in mind: to see developments in science and warfare as not simply driven by rationalism and realpolitik, but also by a secularized theology with deep roots in European mythology and imperialism.

20 THE COAL TAR COMPLEX

BEAUTIFUL TAR, the outcome bright
Of the black coal and the white gas-light,
Of modern products most wondrous far,
Tar of the Gas-works, beautiful Tar!

Opening verse of "Beautiful Tar: Song of an Enthusiastic Scientist" (1888)[1]

► There is another, much more straightforward and material way in which photography and poison gas are connected. As I have already argued, the industrialization of photography and the production of toxic gases for warfare were both dependent on the coal tar industry. Indeed, through their shared reliance on coal tar, the dyestuffs, pharmaceutical, photographic, and pesticide industries were intricately related, with some companies involved in all of these areas. Coal tar was processed and supplied by intermediate fine chemical companies, but it was not simply a relationship of supply and demand. Rather, the very exploitation of coal tar required the prior development of photography and photographic chemistry. The dyestuffs companies first set up in the nineteenth century to exploit coal tar leaned on the photography complex of the time. Those chemical companies and their products, including poison gases and photographic materials, have left their traces in the present in very literal and material ways as their residues remain in our soil, water, and atmosphere. In picturing lifeworlds, they have also materially transformed them.

Coal tar is not a by-product of mining coal, but of "coking" it, which is to say, heating it at very high temperatures without oxygen. This process produced "town gas" or coal gas, the dominant form of commercial gas for heating and lighting in nineteenth-century Britain and until the 1960s, when it was replaced by natural gas. It also produced coke, essential to iron production and later marketed as a relatively smokeless alternative to coal for domestic consumption. Initially, coal tar seemed the least useful of these products, good only for shipbuilding and preserving wood (in the form of creosote), and often treated as waste and dumped in rivers.[2] But as chemists experimented with it, it began to yield more and more compounds with a huge variety of uses: artificial colors, soaps, disinfectants, medicines, scents, flavorings, sweeteners (saccharin), fertilizers, explosives, and, finally, photographic sensitizers.

In Britain, the interdependence of the dyestuff and photographic industries and materials is vivid from the start. Both depended on the same suppliers of fine chemicals. There are hints of connection, for example, in the fact that the student who produced the first well-known aniline dye, nineteen-year-old William Henry Perkin, had previously "dabbled in photography."[3] His photographic experiments were presumably carried out in the same house where he converted a room into a lab with a furnace in the fireplace and, in the evenings after college, attempted to use coal tar to synthesize quinine. Perkin's experiments with coal tar, under the tutorship of the German chemist August Wilhelm Hofmann at the Royal College of Chemistry in London, led in 1856 to the production of aniline purple or "Perkin's Mauve," made by nitrating coal tar hydrocarbons (in this case, benzene).[4]

The chemicals Perkin used included nitric acid. For the commercial exploitation of his new color, he used Chile saltpeter (sodium nitrate), in reaction with sulfuric acid, to produce the strong and volatile nitric acid.[5] The artificial dyes that came after Perkin's also required nitrating, and the dye industries, like the photographic industries, worked with suppliers of nitrates from South America. Though these nitrates—saltpeter and guano, deposits originating in sea fog and the droppings of bats and birds—are usually understood in terms of their major application as soil fertilizers, they were used by a wide range of industries and trades, including in the dyestuffs and photographic materials industries, and in great quantities for explosives manufacture (see chapter 22). In the early twentieth century, the larger photographic companies had their own nitrating facilities to produce silver nitrate, while smaller companies bought it ready-made, sometimes as a by-product of gold refining (as in South Africa).[6] Another way of producing nitric acid was from the ammonium by-products of the (coal) gasworks.[7]

In the late 1850s, Perkin's family set up a dye works to exploit the new aniline color. They needed a fixing agent and found that albumen worked, and fortunately a thriving albumen trade (from chicken eggs) already existed for photographic printing. When Perkin needed intermediate chemicals for his process, the problem of how to manufacture them was solved by the fine chemicals firm Simpson, Maule and Nicholson, until then a company that primarily served the photographic business.[8] Very quickly, by the early 1860s, this company shifted from being a photographic chemicals supplier to being the biggest British producer of synthetic dyes.[9]

In other examples, this development happened the other way round, with companies moving from dye production into photographic materials production as some of the new aniline dyes derived from coal tar proved to be photosensitive. By the late nineteenth century there was a growing demand from the photographic materials companies for dyes that could compensate for the strong blue-violet bias of silver halide emulsions and increase the speed of their response to different wavelengths. Dye sensitizers were first introduced into photographic emulsions by the German chemist Hermann Wilhelm Vogel, but the results were hard to replicate. Eventually, though, the techniques for using dyes as photographic sensitizers were commercialized. The most well-known example of a dyestuffs company that supplied the photography complex becoming a photographic materials manufacturer is the German giant Agfa. Agfa was initially an aniline dyestuff company—the name is an acronym for *Aktiengesellschaft für Anilinfabrikation*, but it set up a photographic department in 1888 and started making plates in 1893.[10] By the early twentieth century, led by Agfa and then by the chemical cartel IG Farben, Germany would come to dominate the production of photographic sensitizers.

Starting in 1856 with Perkin's experiments and continuing throughout the second half of the nineteenth century, the black, sticky, unpromising coal tar opened up a rainbow of colors—mauve, magenta, various blues and violets, alizarin red, and tartrazine yellow. In the 1860s this industry was mainly located in France and Britain, but from 1871 Germany began to overtake both. By the 1880s Germany was manufacturing half of the world's synthetic dye output, and by the end of the nineteenth century it produced more than 90 percent.[11] While a British dye company's employees would number in the hundreds, the German companies, such as BASF, Bayer, Hoechst, and Agfa, employed thousands.[12] Their coal tar was produced by the coal gas and steel industries of the North Rhine and Westphalia. If in Britain, the dye industry had emerged out of the coal tar industry, in Germany it seems to have been the reverse: Germany's coal tar industry followed the dye industry, expanding to meet its demands, and the photochemical industry followed it, too.[13]

During the First World War and immediately following it, there was an increasing monopolization of the European chemical industries, including the photographic industries. This was driven both by governmental ambitions to consolidate expertise and the fear among managers of an increasingly unionized and powerful workforce. This concern is expressed in the 1918 sales report of the Gem Dry Plate company (a subsidiary of the Imperial Dry Plate Company later absorbed by Ilford Limited):

> Prices will not come down to the old level in the near future because general costs of manufacture have risen and labour will exact its toll. Already one works (Ilford) has been dealt with by a trades union; another (Kodak) is being vigorously canvassed; a third (Illingworth) is marked out for attack & probably Imperial will be the fourth. Experience in the fine chemical trade proves that if the employers federate, trade unionism is easy to deal with & is beneficial in many ways. If manufacturers deal individually with the trade unions the whole trade suffers because any special arrangement made with one manufacturer is used as a lever to enforce a similar

> arrangement with all other manufacturers. Much time, trouble & friction is saved if the problem is dealt with en bloc. We might reasonably expect to get along without any disputes as our workers are now well treated but it is as well to bear in mind that the newly emancipated are sometimes the most violent of revolutionaries.[14]

This last phrase would have had particular resonance following the Russian Revolution.[15] In fact, corporate federating, in the form of "combines" (or mergers) and takeovers, was already beginning. British Photographic Industries Limited had been formed in 1915 out of six companies, and Amalgamated Photographic Manufacturers (APM or Apem) would be formed in 1920–21 out of a further seven.[16] The photographic equipment companies had been consolidating since the turn of the century. In the Great War, the new Department of Scientific and Industrial Research (established in 1915) funded schemes to boost research in British industries by forming cross-company research associations, including the British Photographic Research Association.

Ilford Limited had been acquiring its competitors from 1895, enabling the firm to expand into new products and absorb technical expertise from other companies.[17] Its gradual absorption of other British firms (a "horizontal" expansion) was barely visible to consumers, since the smaller family firms they took over retained their original names and branding (see plate 4). APM also did this; according to their sales manager, A. J. Catford, they "worked on the assumption that the more brands they could offer the more customers they could get."[18] Ilford's stealth approach allowed the company to have fingers in several pies, able to capitalize on successful brands and test out the viability of products without staking its reputation in the face of a British suspicion of monopoly and combines, which were associated with price fixing.[19] Overseas, every brand had its own agent, so the persistence of the different brand names long after Ilford had taken over the companies was partly to do with their dependence on these agencies.

Around 1920 Ilford (and Imperial and Gem, both now effectively subsidiaries of Ilford) joined with APM to form Selo Limited as a sensitizing company for roll film. In Britain, the post-Armistice boom of 1919–20, characterized by a flurry of corporate amalgamations and reorganizations, was followed by postwar recession in 1920–21 and then relative stagnation. Nevertheless, Ilford did continue to acquire other manufacturers. By 1930, APM and Selo Limited had been absorbed by Ilford, which had rationalized production across five factories and now owned nearly all the British photographic materials manufacturers, excluding a few small firms and its major competitor, Kodak Ltd, the British subsidiary of Eastman Kodak.[20]

In Germany, worker demonstrations against poor conditions and low pay in the chemical industries in the context of the postwar slump drove the employers' decisions to form cartels. IG Farben (full title I.G. Farbenindustrie AG) was formed in 1925, unifying BASF, Bayer, Agfa, and four other firms.[21] As Esther Leslie writes, it became "the largest chemical concern in the world, producing . . . colors and dyes in all varieties, pharmaceuticals, photographic materials, artificial silk, nitrogen and cellulose products. The whole German color industry was one vast monopoly, rationalised and concentrated."[22]

Writers on the chemical and photographic industries tend to treat the militarization of these industries in wartime as exceptions or aberrations to the normal run of things.[23] IG Farben underwent official "Nazification" in 1936–37, including the removal of Jewish scientists and board directors, although the purge of Jewish chemists had begun as soon as the Nazis came to power, with the influential chemist Fritz Haber, by then professor at the University of Berlin, forced to resign in 1933.[24] IG Farben's products directly facilitated the Holocaust and underpinned Germany's armaments industry. But the militarization of the chemical industries had started long before. In the First World War, coal tar had been used to make high explosives, and the coal tar dye industry made possible the manufacture of chlorine gas, first used in battle on April 22, 1915, when the German army launched a chlorine attack on French, Algerian and Canadian troops north of Ypres in Belgium. Haber himself had been head of the chemistry section in the German Ministry of War, and it was he who led the German poison gas research, mobilizing dye factories to produce these chemicals (mustard gas, for instance, was produced at

Bayer). In Britain, too, military, governmental, and industrial interests were already tightly intertwined. Even in the depression Britain had, as Eric Hobsbawm puts it, a growing and increasingly modern "state-sponsored field of armaments."[25] Certainly Ilford profited from the growth of the picture press, advertising, and tourism, of the entertainment industry and the amateur photography market, but it was also part of an expanding military-industrial complex.

21 THE CHEMISTS' WAR

► The new dyes were not just intended to increase the sensitivity and speed of photographic emulsions. They were also, crucially, about targeting and specializing photographic sensitivities for different purposes. As Kelley Wilder has shown, it is not enough to explain these changes in the chemistry of emulsions as improvements to make photography faster and more accurate or more like human vision. New emulsions were also more specialized. Wilder explains that "designer emulsions" were made for different scientific uses, which "were not about 'speed' but about specificity," tailored to register only specific wavelengths of light or to absorb or block certain kinds of radiation.[1]

It is tempting to describe the panchromatic emulsions that were perfected in the early years of the twentieth century as principally enabling films and plates to record colors in monochrome shades that corresponded to the spectral range of human vision, and in terms of increasing speed. But Wilder's point applies here, too, in that panchromatic emulsions served specific, specialist purposes. The new photosensitizer dyes enabled these emulsions to pierce through blue haze and cloud cover. Panchromatic emulsions were characteristically fine-grained, lending themselves to enlargement and the recording of fine detail. Long before panchromatic film and plates became widely available to amateur photographers in the 1930s, these emulsions were being used for military and scientific purposes, especially for aerial photography.

The photosensitizing dyes used in panchromatic emulsions were made in Germany to secret formulas and sold under trade names, though it was known that they were "quinoline dyestuffs of the cyanine type."[2] Quinoline had been one of the earliest substances derived from coal tar and it is also a component of the natural substance quinine, which explains why Hofmann and his students thought that coal tar–derived quinine might be possible. Like quinine itself, quinoline is light sensitive.[3] One dye, Pinacyanol, discovered by the Czech-German chemist Benno Homolka in 1904, was crucial in the manufacture of panchromatic emulsion. British firms bought the dye from Germany and used it to manufacture panchromatic glass plates, especially for aerial photography. The pioneer in Britain was the small firm Wratten and Wainwright, where the joint managing director C. E. Kenneth Mees produced the first plate in 1909. Ilford Limited's version of the panchromatic plate was launched in 1911. Like the Wratten plates, the Ilford plates used dyes manufactured in Germany. With the outbreak of the Great War and the blockade of Germany, the German photosensitizers became unavailable to both British and American companies. And since these dye sensitizers were required for aerial photography, to cut through atmospheric haze, this had serious implications for military reconnaissance.[4]

Both Ilford Limited and Eastman Kodak were successful in reproducing Pinacyanol, also known as Sensitol red, though in Britain this was only possible because of state investment. Government funding via the new Department of Scientific and Industrial Research allowed Ilford's chief chemist at the time, Frank Foster Renwick, to begin research in dye production. He collaborated with William Jackson Pope, professor of organic chemistry at Cambridge University, and two other chemists, J. E. G. Harris and Frances Hamer (one of very few women chemists in the period).[5] In 1918, in his presidential address to the Chemical Society of Great Britain, Pope used the plates as a key example of how Britain could excel in organic chemistry despite assertions to the contrary, claiming that since the photosensitizing dyes produced in Britain had overtaken the earlier German ones, "It is safe to assert that the manufacture of panchromatic plates has now attained a degree of perfection in this country such as will long defy competition."[6] By the end of the war, boosted by government investment, Ilford's plates began to outstrip the Wratten plates, resulting by 1920 in Ilford's much-praised Special Rapid Panchromatic plates for aerial photography.[7]

After the war (and presumably because the German patents were made available to the Allies) there was very rapid development. The chemical constitution of the quinoline cyanines was now known, the first infrared sensitizer (called "neocyanine") was synthesized, and new and better infrared sensitizers rapidly followed.[8] Ilford's infrared plates were introduced in 1932, allowing photographs to be taken in the absence of visible light or through atmospheric haze. Following their use by the Houston-Mount Everest survey expedition (see chapter 16), these plates would go on to play a crucial role in aerial reconnaissance and bombing during the Second World War, as panchromatic plates had done in the First. Plates for recording nuclear particles were also produced by Ilford in the early 1930s.[9]

The photographic firms' supplies to the military were specialized, not only in terms of the tailored emulsions used, but also with distinct product formats designed for different uses and various kinds of cameras (including motion picture cameras). In the interwar period, Ilford Limited and Kodak Ltd. had a growing trade with the military, supplying an ever-widening range of materials. In the 1930s Ilford alone was manufacturing at least five types of aerial film, including three types of panchromatic film, an orthochromatic film, and infrared film.[10] The firm produced aerial film on Selo no. 20 stock (equivalent in size to 120 film) to be used in a Hythe gun camera. This was a precursor of contemporary simulation technology, designed to train aerial gunners and enhance their accuracy with a real gun. The organizations being supplied film by Ilford Limited include the Royal Air Force, the Admiralty, Nobel's Explosives, and GEC Birmingham (General Electric Company, which was also a defense contractor).

Their trade reveals how photographic supplies were now weapons of war, interbred with other kinds of war materials, from explosives to poison gases, and how the new sensory economy was inseparable from a war economy. The First World War has been described as "the chemists' war," and in 1918 Pope claimed that, despite Britain's flagging dye industries, "the whole Empire is now one vast chemical and engineering laboratory."[11] The war also gave a boost to the coal tar producers whose gases and explosives were in such high demand. According to Ilford's general sales manager, A. J. Catford, it was assumed that the photography trade would disappear in the war, but within a year it became clear that the industry's profits were actually increasing, "initiating for it a period of prosperity such as it had never anticipated and from being a purely luxury trade it assumed, for the first time in its history, the dignity and importance of an essential industry."[12]

In *Synthetic Worlds*, Esther Leslie writes: "Film and photography mediated the rape of nature in war. The chemical industry had it all covered: it produced the bombs and the gases and the photographic materials."[13] In the period between the two World Wars (in which other imperial wars continued), this did not stop. As the historian David Edgerton has argued, militarization gathered pace in Great Britain, and, against the usual histories of Britain as a declining power in the interwar period, he claims it became a "warfare state."[14] Edgerton disputes existing arguments about Britain disarming in the interwar period (including claims made at the time) by looking at evidence of battleship production and refitting, the capacity and production figures of various armaments businesses, employment figures for the aircraft industry, and export figures for arms.[15] He could equally have given the example of photographic suppliers to the military.

Edgerton points out that during and following the Great War, British scientists and engineers were prone to complaining about lack of scientific expertise and antitechnological orientation among the civil service. He writes that a "longstanding and influential" view was the "technocratic critique of the higher civil service, which is seen as being made up of men trained in the classics and at best, history, rather than the scientists and engineers that a modern society was felt to need as state administrators."[16] Pope's presidential address of 1918 (mentioned above) furnishes an example. He argued that the coal tar dye industry in Britain remained undercapitalized, with only a tenth of the capital of the German dyestuffs firms, and that if nothing was done to improve the situation "we shall be again in the hands of the foreign producer."[17] He attributed this to the British government's failure to understand and invest in organic chemistry, and to the narrow class of people from whom the ruling elite were drawn, with their overdependence on a stultifying classical education: "opulent, indolent Great Britain has for the past century permitted all its educational interests to pass into the hands of a particular caste which despises all knowledge difficult to attain."[18] However, Edgerton argues that the ubiquity of such claims points to the *success* of the military and the technocracy, and that in the interwar period "Britain was no longer as liberal—economically, politically, intellectually—as many supposed it remained."[19] In other words, the influential image of Britain's liberalism and nonaggression (before and after the Second World War) was a piece of propaganda rather than an accurate representation.

Nevertheless, British chemists viewed the German firms as much more well-funded than the British. Britain and France had fewer research laboratories and academically trained chemists within the major firms prior to the First World War. One reason was that German patent law encouraged the development of new processes rather than just new products, so new ways of synthesizing existing compounds were supported, and since patents formed a large part of a company's assets, German firms were motivated to employ academic consultants and research laboratories. They were also encouraged to diversify, actively looking for compounds that could work as pharmaceuticals, pesticides, and photographic chemicals.[20]

Although the sensitizing dyes for panchromatic emulsions were being produced in Germany prior to the Great War, it took decades for panchromatic roll film to be widely used. In 1936, Mees argued for the huge historical significance of the introduction of these sensitizing dyes into film manufacture, claiming that the years 1928–30 should be viewed as hailing a new era in photography through the "introduction of panchromatic materials into everyday use."[21] Later he would argue that the dyes brought about "an almost revolutionary change in the art of photography" in the decade after 1925, and that sensitizers developed after 1930 made it possible to realize "a new system of photography in color" because they did not "wander" from the silver bromide to which they were attached, as earlier dyes

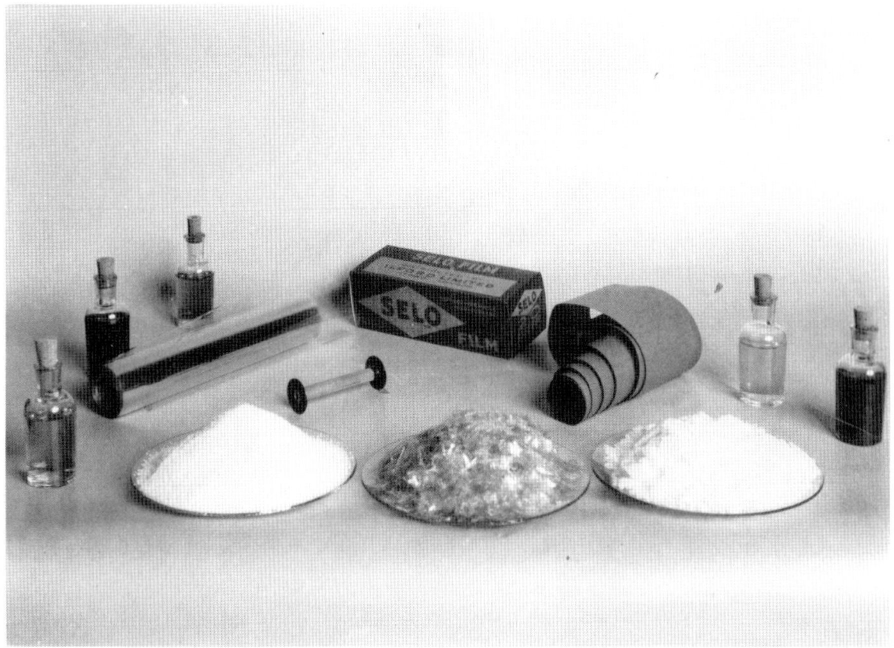

FIGURE 38. "The Ingredients of a Roll Film," 1938. Ilford Limited collections, Redbridge Museum and Heritage Centre. Courtesy of Redbridge Museum and Heritage Centre 2025.

had done.[22] This facilitated the color separation of the three layers of emulsion in the new Kodachrome film introduced in 1935. Certainly, these dyes had transformed Mees's own career: Eastman Kodak took over Wratten and Wainwright in January 1912, to make it possible for Mees to run the new Kodak laboratory in Rochester, New York, and produce the Wratten plates there.[23] He would go on to become vice president of the Eastman Kodak Company.

It is in these two years, 1928–30, that panchromatic emulsion first made its way into amateur practice, facilitating low-light and high-speed photography and resolving problems of spectrum insensitivity (in which blues would be rendered as whites and reds as black on a positive print). In the 1930s, panchromatic versions of Ilford's Selo film were introduced. These were fine-grained and relatively fast films that recorded a spectrum of light close to that perceived by the human eye. The *Imperial Handbook* of 1933 announced, "There is no doubt whatever that the future of the panchromatic plate is a rosy one."[24] But the introduction of panchromatic emulsion did not always go smoothly. It caused problems in cinema when combined with warm-colored incandescent lights and necessitated strangely colored "panchromatic make-up" for the actors.[25] The sensitivity of the emulsion also meant that a red safety light could no longer be used in development as it would fog the film or plates. In a speech given in the 1940s, Ilford factory manager Cecil N. Potter remarked about how much darker the Ilford plants had become when panchromatic and "super speed orthochromatic" plates arrived. He claimed that the women who packed the plates "subjected their eyes to considerable strain." leading to what he patronizingly termed "an outbreak of hysteria in the Packing Department among the girls." He added, "It was only after they were advised to work by touch and not by sight that they settled down." He also mentioned a trial to establish whether the job could be "successfully performed by blind persons," but unfortunately "the results were not up to expectations."[26]

It took a while for panchromatic roll films to become ubiquitous. The publisher and poet John Lehmann describes in his autobiography how, circa 1936–38, he took a photograph while in Batumi, Georgia, with Soviet marines in shot: "I was placed under arrest in an instant."[27] He was forced to have his film developed by a photographer in the town, and speaking very little Russian he was unable to explain that no red light should be used in the developing room because the film was Panatomic (Kodak's panchromatic roll film). The film came out entirely black, confounding the Soviet servicemen, as well as the local photographer who had not yet encountered panchromatic roll film and presumably did not use a lightproof developing tank. Lehmann comments drily that this was "evidently some diabolical trick of the British Intelligence Service, to manage to obliterate all pictures at the moment of unmasking."[28]

22 OUT OF THIN AIR

► In August 2022, with my daughter Honor, I visited the National Trust property Tyntesfield House, a nineteenth-century estate built in the style of the Gothic revival, close to the city of Bristol. We were mainly interested in spending time in the extensive and beautiful woods and grounds surrounding the house, and we marveled at the wealth that must have been needed to run such a large estate. National Trust signage explained that the owner, William Gibbs, was a member of the Gibbs family who controlled the guano trade from Peru (guano is nitrate-rich seabird or bat excrement). From 1842 until his retirement in 1858, he was the sole partner in the Liverpool agents Anthony Gibbs & Sons, securing a complete monopoly on the guano trade and in the process amassing huge profits.

In 2022, the National Trust had been caught up in the so-called culture wars being inflamed by right-wing politicians and media who attacked National Trust projects concerned with decolonization and raising awareness of the history of slavery. By August a right-wing organization called Restore Trust, supported by Conservative members of Parliament, was attempting to gain control of the National Trust's governing council (it did not succeed).[1] Despite the National Trust's perceived "wokeness," Honor and I did not spot any references to slaving history on the signs and information panels around the estate, but curious about the guano trade, I looked online. The

Trust's website does include the following short paragraph: "Guano extraction was dangerous and living conditions were poor. Most extraction was performed by indentured Chinese labourers who worked in slavery-like conditions, and who were economically bound to earn back their transport, board and lodging costs before making a profit. Before slavery was outlawed in Peru in 1854, guano was also extracted by some enslaved people, as well as convicts, conscripts and army deserters."[2] The National Trust information, though it is probably sufficient to invoke the ire of the culture warriors, goes on to give the misleading impression that Gibbs ameliorated the working conditions of indentured laborers.[3] In fact, conditions seemed to have worsened significantly in the period of Gibbs's monopoly. In 1882 the London *Times* assessed that the trade in Chinese laborers (known as "Coolies") was "worse than the worst excesses of American slavery," reporting that "In 1860 it was believed that not one of . . . the Chinese Coolies who had been shipped to those islands since the trade began, in 1844, had survived, all those who had not died of exhaustion had put themselves voluntarily to death."[4]

This trade commenced in the eighteenth century when, rather than continue to strip their own soil of nitrogen, the European imperial powers started to import nitrates from the colonies, depleting their soil nutrients instead.[5] The start of war in 1792 hugely increased the demand for gunpowder and thus for nitrates. In France, the Gunpowder and Saltpeter Agency built a massive gunpowder factory in Grenelle, a district in the southwest of Paris, which exploded two years later. Six hundred workers were dead and another thousand injured.[6]

From 1841 Anthony Gibbs & Sons promoted the use of South American guano as a nitrate-rich fertilizer across Britain, and by the end of 1842, the year William Gibbs became sole partner, "imports of South American guano rose by over 700 percent: 2,881 tons in 1841 and 20,398 in 1842."[7] Gibbs & Sons acted as the main financial agents for the trade in guano from South America to Europe until 1862.[8] That the scale of their profit was immense can be seen by looking at the sheer volume of guano exported from Peru: in the forty years between 1840 and 1880, it was over fourteen million tons.[9] Though Peru was not part of the British Empire, the nitrate trade was a colonial trade—as writer and activist Louise Purbrick puts it, "British nitrate companies, such as those run by John Thomas North or Antony Gibbs and Sons, planted the machinery and created industrial colonies."[10]

Guano was the most important trade from Peru until the early 1870s, derived from deposits in the Chincha Islands off the Peruvian coast, and incentivizing Chile to expand north into Peruvian territory.[11] After Gibbs's retirement, and with the guano beds becoming exhausted, the company shifted from guano to caliche, sometimes called Chile saltpeter, nitrous mineral deposits found in the Atacama desert. Gibbs & Sons became a partner in the Tarapacá Nitrate Company by 1865.[12] While caliche came from nitrate fields mostly located in Peru, the trade was controlled by Gibbs and by Chile through the port city of Valparaiso. The trade inevitably affected the economies of these countries and also of Bolivia, leading to war between the three countries. Known at the time as the "Nitrate War," this conflict, which lasted from 1879 to 1883, resulted in Chile acquiring all of the nitrate fields.[13]

Eventually though, even the trade in caliche was supplanted by a new source of nitrates that seemed to be inexhaustible. War and European colonial interests also drove this new method of nitrate production. Before the Great War, Germany was looking to reduce its dependence on the British-controlled nitrate trade from South America, especially since nitrates are the basis for explosives. War accelerated research, and the solution was found by none other than Fritz Haber, pioneer of gas warfare, working at BASF (see chapter 20). Together with another chemist, Carl Bosch, Haber came up with new way of producing ammonia by extracting nitrogen from the air. It was named the Haber-Bosch process and manufacture began in 1913. This did not remove the reliance on coal, since it is very energy consuming, but Germany was rich in coal mines.[14] When Germany lost the war, the Allies were able to copy the German techniques, and in Britain and France, synthetic ammonia factories were soon under construction.

The Haber-Bosch process would become the method by which nitrates were produced for photography, too, though not on the scale that they were deployed for nitrate fertilizers. Today 160 million tons of ammonia are produced, mainly through Haber's method.[15] Haber's poison gas research has devastating consequences that are known, but the long-term effects of the Haber-Bosch process are not known. The process is disrupting the nitrogen cycle of the earth and risks crossing a boundary that might result in "irreversible degradation of the earth system."[16] Before the introduction of the Haber-Bosch process, no nitrogen at all was removed from the earth's atmosphere, but by 2010, 121 million tons were being removed annually.[17]

The United States could still rely on Chilean supplies of saltpeter. Nevertheless, the American chemical company DuPont adopted the Haber-Bosch process to manufacture ammonia. It specialized in cellulose and nitric acid–based products. DuPont had started out as a gunpowder manufacturer in 1802, and then expanded into a range of explosives, including dynamite, until it had a virtual monopoly in the United States. In the early 1900s it took over a number of paint, solvent, synthetic leather, and rubber companies to make new products from cellulose and nitric acid. As production capacity increased worldwide, the price of ammonia dropped, and in the interwar depression, prices for both ammonia and nitric acid collapsed. DuPont sought more uses for ammonia and more products to manufacture at their specialized plant—introducing new plastics, antifreeze, and alcohol derivatives.[18] The example of DuPont shows how the Haber-Bosch process increased not only the supply of ammonia, but also the range of ammonia-based products, driven by the capitalist imperative to exploit excess supplies. Inevitably, DuPont ventured into photography. The firm was already manufacturing gun cotton, which was made from cotton or wood pulp soaked in nitric and sulfuric acid and then dried. In 1922, Ilford's chief chemist Frank Foster Renwick was invited to work for DuPont, to set up a photographic film plant in a "former gun-cotton plant." DuPont had "the know-how on making the cellulose nitrate from years of experience with nitrated cotton and its solvents, but the emulsion making, and coating and drying, were still problems."[19] By the 1940s, Ilford was using DuPont-manufactured film base.[20]

As a child I read a terrifying book, about a boy who declares he does not like or need things such as animals and plants. One by one the objects around him disappear—his woolen blanket when he declares his dislike of sheep, his wooden bed when he rails against trees. Eventually he is left unanchored, floating in space. The Haber-Bosch process does not make something out of nothing (there's no such thing as a free lunch) and it was known from the start that its energy consumption was enormous, but it seemed to fulfill an old dream (or nightmare) of breaking human dependence on the environment. It also quelled more recent anxieties about Europe's reliance on its colonies. Now materials could be manufactured from thin air! As Esther Leslie suggests in *Synthetic Nature*, it seemed as if science, and especially chemistry, could "outbid nature," and the world of nature could be replaced by a synthetic utopia (or dystopia).[21] Substances like rayon ("artificial silk") and cellophane seemed almost miraculous. Images could be made of sunlight, and now stockings could be conjured from the atmosphere thanks to nylon, the first entirely synthetic fiber, brought to the market in 1939. Indeed, the ads for nylon boasted that it was made from "coal, water and air."[22]

► Just as nothing is manufactured out of thin air, meaning without cost, so nothing is produced without waste. Industrial plants, including photographic materials manufacturers and processors, are producers of potentially toxic waste which, left unchecked, leaches into air, soil, and water. Questions about waste and pollution are not mere technical or managerial questions, though they are often framed like that, but political questions—about land, rights, ownership, and ecological violence.

The philosopher Michel Serres describes pollution as a way of marking one's territory, like a cat spraying its urine: "Whoever spits in the soup keeps it." A factory takes possession of a place as it "empties its effluents into a nearby river, diffuses them in the atmosphere, or transports them to a remote mangrove swamp."[1] Pouring chemicals into a river suggests a sense of right to that place, but also a deep disregard for it. Pollution is ownership at arm's length, a seizing of territory, in which the thing possessed is not cared for or cultivated by its owner. It is predicated on a sense of environment as "out there," as something surrounding but not penetrating bodies, as if lungs did not need air to breathe and stomachs food to eat. The form of ownership that polluting practices enact is the territorial claim of colonial worlding, which generates new materialities, reconstituting the world physically and chemically, as if it were a tabula rasa, mere blank space, simply raw material. Polluters have no regard for local peoples or for ecosystems, no sense of the way that the apparently docile, inanimate, or inert might suddenly bite back.

In *Pollution Is Colonialism*, Max Liboiron describes pollution as "an enactment of ongoing colonial relations to Land"—where Land (capitalized) signifies "the combined living spirit of plants, animals, air, water, human, histories and events recognized by many Indigenous communities."[2] Pollution is colonial because it assumes access and rights to Land, and while not all pollution has anything to do with capitalism, the capitalist imperative of economic growth requires more and more "sinks" for pollution. "Waste management" relies on "thresholds of harm," on the basis that a given "sink" has the capacity to assimilate a certain amount of pollution. As Liboiron shows, this notion of assimilative capacity, first developed in the 1930s, legitimates environmental violence and still allows for the dumping of waste—it allows polluting industries "the ability to waste, even the right to waste."[3] It is also premised on isolating individual substances, assessing them singly against a given measure.

Silver is a case in point. Photography companies and photoprocessors were major consumers and recyclers of silver, with Kodak by 1920 the largest US consumer of it "outside of the U.S. Mint."[4] In 2006, the *Handbook of Industrial and Hazardous Waste Treatment* claimed that "extensive studies" in the 1990s had proved that the silver from photographic plants was not "an environmental concern."[5] Some studies claim that photographic silver leaves photo processing plants as silver thiosulfate (thiosulfate is "hypo" or fixer), degrading into compounds that are harmless, stable, and do not accumulate in organisms.[6] Yet other studies have found evidence that silver waste from processing plants and manufacturers *does* accumulate. It collects in plankton and in the bodies of clams, mud snails, oysters, and fish as well as in the ducks and gulls that feed on them. While some animals seem to adapt to and tolerate this excess of silver, others, especially as embryos, cannot.[7] Geologists and marine scientists have long argued that it should be considered toxic to organisms in water and soil.[8] Silver is a well-known antimicrobial, killing fungi and microorganisms, which is why recovered silver from photographic manufacture can be used, for example, to produce antibacterial soles for sneakers (Harman Technology now offer "bespoke antimicrobial products" using recovered silver).[9]

Research on silver concentrations in the sediments at the bottom of Lake Nahuel Huapi in Patagonia has found that the peaks and troughs of photographic processing in the city of Bariloche could be tracked "through a rapid increase of Ag [silver] concentrations after 1964 and the subsequent rapid declines."[10] In the UK, it has been claimed that in the 1990s, as consumers shifted from film-based to digital photography, the silver compounds in the Thames and the rivers and canals that feed into it (including the River Roding in Ilford) reduced, allowing the Thames to come back to life after having been declared "biologically dead."[11]

Waste treatment plants and silver recovery processes (which became more common and profitable with the widespread use of color film) reduced the pollution from photography processing and manufacturing to below the supposed "threshold of harm." But such measures ignored the cumulative and combined effects of silver-containing effluents. As the authors of a 1992 study of pollutants in British estuaries suggest, silver is likely to contribute to "stress to organisms" when combined with other pollutants. If other "metal-binding" substances are present in river or lake sediment, the take-up of silver by organisms is increased: "some organic materials (e.g. humics) in sediments may suppress the availability of Ag [silver] whereas others enhance it."[12] Think of the deadly Meuse Valley fog or the London fogs, in which the causes of death and sickness were not reducible to one contaminant but a combination that produced an air heavy with moist, greasy, and noxious particles. One 1943 study of Cincinnati's cold winter fogs argued that even the apparently harmless steam emitted from power plants was extremely hazardous to health because it combined with the "flue products of the city fires," holding them suspended in the air, with the consequence that steam trains and industrial power plants provided "one of man's worst winter hazards in urban areas."[13] Photographic pollution is also "never discrete."[14] The agency of these substances is the agency of an assemblage that is far more than the sum of its parts.[15]

Chemical residues accumulate. They are hard to trace—they "disobey boundaries, appear where they shouldn't appear, alter environments, and enter communities and bodies without permission."[16] Residues have effects that vary over time; they mingle and spread. They also come back to haunt us. A 2020 study of the Chiswick Ait, a mud island in the Thames, about five kilometers downstream of the Grand Junction Canal, found extraordinarily high levels of silver. The researchers were able to broadly date when the silver had been deposited and suggested that the residues of twentieth-century industrial activities, buried only a foot (thirty centimeters) below the surface of the mud, might rise again, carried by floods or dredgers. They call this "buried legacy contamination."[17]

Where do photographic residues start and stop when the production chain of chemical photography is so long and complicated, with all sorts of effluents discharged along the way? These include the toxic gases produced in the process of extracting nitric acid from saltpeter, which filled nineteenth-century suburbs with a thick and choking smoke. They encompass the cyanide used at gold-mines to extract silver from the gold ore and the effluent from the dye factories that was dumped into waterways such as the Grand Junction Canal in London (where Perkin's dye factory was located) or the Saône in Lyon, into which magenta producer La Fuchsine poured its arsenic waste, killing several citizens.[18] The photography factories leached ammonia, silver, thiosulfate (hypo), dioxin, and various other dissolved salts, sulfates and metals (including mercury), acids and alkalis. In the 1880s, at Ilford, if a plate or batch was spoiled the glass would be recovered for reuse, but the emulsion would be washed off into the nearby River Roding, the silver nitrate cumulatively destroying the river's ecosystem.[19] Celluloid film was manufactured from cellulose in a chemical process that required large amounts of water extracted from nearby lakes and rivers and chemicals including sodium hydroxide (lye or caustic soda), nitric and sulfuric acids.[20]

Nor can we separate photographic pollution from the wider polluting practices of the chemical industries. The River Roding was lined with chemical plants, including the Britannia Works (Ilford Limited's photography plant); the gasworks; Howard and Sons, manufacturers of pharmaceutical chemicals and later of solvents; and the Euplypton works, a celluloid factory that made not film but collars and cuffs supplied to the white-collar commuters or "City men" of Ilford.[21] Today the river appears to have recovered from some of this contamination, but it faces new challenges. It has the highest levels of "forever chemicals" (PFAs—per- and poly-fluoroalkyl substances manufactured since 1949) of one hundred and five rivers tested in England, despite the hard work of local campaigners to clean up the river and guard it from further pollution.[22]

Eastman Kodak provides the best example of how photographic materials plants became essentially chemical plants. Siobhan Angus writes that in the mid-twentieth century, Kodak Park (in Rochester, New York) "produced over 350 different chemicals for use in photography—a single day's production would fill eighteen tractor trailers—and more than four thousand research chemicals."[23] Alison Feser vividly describes the "fantastical pink clouds" and "strange ash" emitted from the smokestacks at Kodak Park later that century.[24] And as Rowan Lear points out, as a chemical company, Kodak's reach went far beyond photography, manufacturing "a range of synthetic substances for the clothing, fertilizer, pharmaceutical and military industries" at sites across the United States and globally.[25] Its polluting activities caught up with it in the early 1990s, as a series of lawsuits revealed its effects on the health of its workers and the surrounding population in Rochester. Kodak admitted to violating environmental regulations and eventually reduced its toxic chemical waste, just as the film market was collapsing.[26] Richard Maxwell argues that Eastman Kodak's "environmental debts and social liabilities" contributed to its 2012 declaration of bankruptcy (usually attributed to its failure to adequately anticipate digital developments). In the early 1990s it spent hundreds of millions of dollars on fines, lawsuits, and new waste management facilities while complaining it was a "victim of regulatory overreach."[27]

We live with the residues of our media, and many of these residues are irreversible. Even so, there are plenty of examples of nostalgic affection for a polluted lifeworld in which, in other ways, life was good. Feser's ethnographic fieldwork with former workers at Kodak Park reveals nostalgia for a good life beneath the factory smokestacks, even as these emitted "thick white plumes" into the sky above.[28] Jennifer Tucker's historical research on the late nineteenth-century

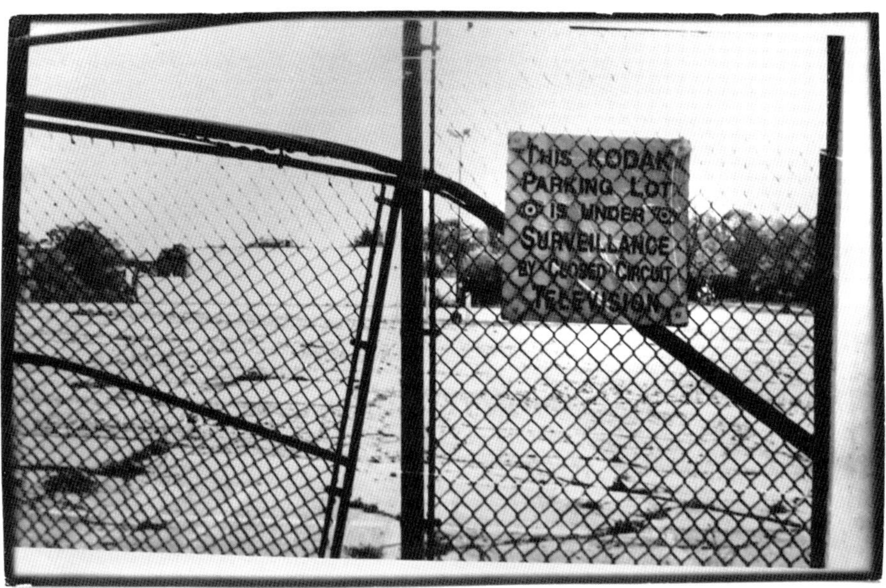

FIGURE 39. Alice Cazenave, *Legacy: The City That Kodak Built IV*, 2024. © Alice Cazenave. One of a series of photographs using expired Kodak gelatin silver paper, negatives, and prints developed with plant-based photochemistry foraged from Kodak Park, George Eastman House, Kodak contaminated land, and Kodak landfill areas. The contamination is too great to be remediated by these plants, but Cazenave uses them as vehicles to think about Kodak's chemical legacy and the disparities of wealth in the city of Rochester.

chemical industries in the north of England notes that the signs of pollution (smoke and alkali waste deposits) were associated with "prosperity and jobs" while "clean air was associated with trade depressions and strikes."[29] And, as I have argued, the London fog, too, inspired affection as well as opprobrium, and was sometimes seen as a sign of the power and modernity of the imperial city (see chapters 9 and 13).[30]

As Soraya Boudia and her coauthors point out, the residues of the past "can't simply be undone, undermining the idea that there is a 'pre-' era to which we can return, even if politically and economically powerful actors want it so."[31] We have been exposed, and exposures cannot always be stopped. Think of Humphry Davy and

Tom Wedgwood, who soaked white leather in light-sensitive chemicals circa 1800. They could make a photographic image using a long exposure but were unable to stop there: each lamplight or candlelight viewing added to the exposure until the image went entirely black. Just as exposure to light produces an irrevocable but sometimes invisible change in photographic materials, so exposure to the toxic residues of photographic manufacture continues. There is no returning, no undoing of the environmental exposure, and it is difficult to disentangle the chemical dangers of photographic image making from the pleasures of it. As the anthropologist and filmmaker Elizabeth Povinelli writes, talking of her childhood practice of running with the "fogging trucks" which sprayed DDT across her Louisiana neighborhood:

> We can ask, what were we thinking? What was anyone thinking? We can try to calculate the effects of this widespread pastime. But I will tell you something true: when I think of our races, their wild, uncontaminated intoxication is not diminished by the knowledge that they were toxic—any less than our flying ahead of raging fire is diminished by the knowledge that we could well have died, that we might be imperceptibly burning inside out from the long-lasting effects of DDT. These fires, fogs, and winds were a part of us. They were elemental to what we were because they were the elements that composed us.[32]

Acknowledging the toxicity of photographic chemistry and the polluting history of the chemical industry, as well as the environmental toll of contemporary digital manufacture), might seem difficult to square with a love of photography and the thrill of making photographs, let alone their ubiquity. But resignation is not a solution. Povinelli is not recommending that we resign ourselves to the hazards of toxic environments. Her point is closer to what philosopher Michelle Murphy calls *alterlife*, the messy condition of "existing in worlds that demand chemical exposures as the conditions for eating, drinking, breathing."[33] This concept requires acknowledging the staying power of residues and paying attention to their spread. It means following how, as Boudia and her colleagues express it, "chemicals continue to act, to exert their agency, in unforeseen ways, creating new realities after they have left the chemists' laboratories and manufacturing plants."[34] It means looking at the unevenness of this spread, how pollution accumulates around and in certain bodies, treats some places and not others as "sinks."

24 HOTWELLS

For Europeans of the late eighteenth and early nineteenth centuries, light and air were not separate entities as they are for us today. The early experiments that Geoffrey Batchen calls "proto-photographic" were happening at broadly the same time as the isolation of the different wavelengths of the visible spectrum, the discovery by William Herschel of invisible light in the form of infrared, and the division of air into its component gases.[1] Even so, for both scientists and artists, light often seemed to be something atmospheric and chemical, optical phenomena an issue of air as much as light. The teenage Humphry Davy, in work he soon repudiated, theorized that oxygen carried light with it, and should therefore be called "phosoxygen."[2]

When Davy collaborated with Tom Wedgwood, son of the ceramics magnate Josiah Wedgwood, on some of the earliest photographic experiments, Wedgwood had already been involved in Davy's experiments with nitrous oxide (laughing gas).[3] Davy would write up Wedgwood's experiments with silver nitrate two years later, in the first issue of the *Journals of the Royal Institution of Great Britain* (1802), as "An Account of a Method of Copying Paintings upon Glass, and of Making Profiles, by the Agency of Light upon Nitrate of Silver."[4] Though Davy would become far more famous, Wedgwood was a prodigy who had already developed his own theories on the nature of light and heat. He had had two papers read before the Royal Society in London in 1792, at which time he seems to have first observed that silver nitrate darkened through exposure to light.[5]

The pneumatic experiments that Davy conducted with laughing gas and the work on photography led by Wedgwood (using silver nitrate) are rarely discussed together. Yet these experiments are materially connected, and both dependent on the use of the acrid, choking, and volatile nitric acid. They are also likely to have been linked in the minds of both men, who were influenced by Romantic ways of understanding atmosphere and weather, and for whom these were explorations in aesthetics, driven by curiosity and natural philosophy as much as by the possible commercial or medical applications. In the development of nitrous oxide, now largely thought of as an anesthetic, Davy was deeply interested in sensual, aesthetic effects because, to him, laughing gas did not inhibit sensation but enlarged it.[6] Evidence of this came from the first-person accounts of friends and volunteers who inhaled the gas, and from his own experiences (he quickly became an addict). He recorded the hallucinations of his volunteers (the "bibbers") carefully: "My teeth are thrilling!" exclaimed one participant; "I felt as if I were lighter than the atmosphere, and as if I was going to mount to the top of the room," said Wedgwood.[7] Davy's initial reason for investigating the gas, as a possible cure for consumption (tuberculosis), was overtaken by his curiosity about its effects on sensation and perception.

One of the few historians of photography who does pay attention to the connections between the experiments with nitrous oxide and silver nitrate is Jordan Bear. Bear points out that the categories we now use to understand photographic history—"fixity, experiment and evidence"—were not yet secured. He sees both sets of experiments as examples of Davy's struggle to give form to sensation, "to transform these subjective sensations [the laughing gas experiments, photos viewed by candlelight] into useful, shareable data."[8] How could breath yield knowledge, how could hallucinations and impressions be something more material than the wild imaginings of individuals? Bear talks about how the experiments were "vulnerable to a rapid decay" and plagued by their "ephemerality," though he does not connect this with the tendency of nitrates to decompose and the volatility of the chemicals on which the processes depended. He suggests that the tendency to view the Davy and Wedgwood experiments as failures was due to the "inability to arrest the ephemerality of the impressions" and their "non-artifactuality," noting that until Batchen brought "proto-photography" into its purview, the history of photography depended on artefacts—on "the necessarily unrepresentative, extant subset of photographs."[9] Photography, for Wedgwood and Davy, was an unfixed, airy, hallucinatory affair.

What was the atmosphere of these experiments? What was the air in which photography begins to emerge? One answer is that it was a radical, revolutionary air: Davy's experiments with gas took place at Dr. Thomas Beddoes's Pneumatic Institution for Relieving Diseases by Medical Airs, located in the Hotwells area of Bristol, England.[10] Beddoes and his circle, including the poets Robert Southey and Samuel Taylor Coleridge, had a reputation for radical politics. Radicalism and Romanticism went hand in hand. The group wrote political pamphlets and gave speeches in support of the French Revolution. The laughing gas experiments became an object of ridicule, a political target for critics of Enlightenment thought, who used them to link science with atheism, revolution, orgies, and intoxication. Opponents of Beddoes's apparent "Jacobinism" made parallels between the "wild gas" used by Davy and "dangerously infectious" threats to the moral and social order.[11] The experiments may have damaged Beddoes's reputation irrevocably, but as Davy's biographer Jan Golinski suggests, they may have actually enhanced Davy's later self-construction as a charismatic genius at the Royal Society.[12]

While scientists at the Royal Society in London moved away from the kinds of experiments that might imply subjective and fallible judgment, in favor of a trust in instruments (a kind of "mechanical objectivity"), the Bristol experimenters moved the other way, foregrounding subjective experience even as they edged toward the view that thought itself was material and chemical.[13] Photography emerged alongside these "wild" airs that blew apart ideas of selfhood and tugged at the seams of language.[14] In 1799, Southey wrote to his brother, "Davy has actually invented a new pleasure, for which language has no name."[15] In Davy's notebooks on the nitrous oxide experiments, philosopher of science Birgit Griesecke locates "the dreamy, ecstatic prehistory of anesthesia."[16] Perhaps here too is a "dreamy, ecstatic prehistory" of photography.

Until the arrival of Beddoes's Pneumatic Institution, people had come to Hotwells to take its waters, not its airs. It was a health spa, the waters from its wells touted as a cure for illnesses such as diabetes and consumption. At its height in the eighteenth century, it was lively and fashionable. During the summer season, there were public breakfasts, evening balls, promenades, musical performances, and river excursions. Across the river Avon, visitors could visit Long Ashton village for strawberries and cream, and even the "keepers of fashionable shops at London and Bath" opened branches nearby.[17] While the "real Invalids" stayed near the wells, the wealthy and healthy rented splendid accommodations on the hills in Clifton village.[18]

Yet even at the peak of its elegance, while it had the spectacular gorge and forests to the north and west, Hotwells was close to the crowded and industrialized city center. Some visitors were critical of how dirty and industrialized the spa was.[19] Some objected to the tidal and naturally muddy river, others to the sight of the bustling port and the poverty surrounding it, the shipbuilders and dockworkers, quarrymen and souvenir sellers.[20] But the biggest problem was the air. Situated in the river basin, Hotwells was only a short distance from the coal-fired furnaces of the soap and glass factories and the brass works, copper works, lead foundries, and other industries that kept the city in a perpetual fog of "almost impenetrable obscurity."[21] Across the river were the coalfields of Ashton Vale and Bedminster, already producing large amounts of coal by the 1780s.

While Hotwells was marketed as a picturesque and "scenic enclave," wealthy visitors could hardly ignore the increasingly desperate and gaunt consumptives accumulating at the spa. When Beddoes opened his first pneumatic laboratory in 1793, seeking gaseous cures for illnesses such as tuberculosis, asthma, paralysis, and syphilis, the spa was already becoming a "last-chance saloon" for hopeless cases. Cultural historian Mike Jay writes, "The promenade was crowded with cadaverous patients in the final stages of consumption, and the Strangers' Burial Ground up the hill in Clifton was crammed with the bodies of those for whom the last miracle cure had failed."[22]

It was not just the depressing sights and polluted atmosphere that put off more affluent visitors, but also the greed of the Society of Merchant Venturers (Bristol's "mercantile oligarchy"), which had decided to make the spa more profitable, forcing the leaseholder to charge for drinking the waters.[23] Tourism declined as the end of the Napoleonic Wars opened up the European spas to the wealthy. Laughing gas and photography, sibling children of Romanticism and radicalism, were born, or at least conceived, in this place, in this filthy industrial air. Yet by 1816 the spa had fallen quiet, having "the silence of the grave, to which it appears to be the inlet."[24]

Hotwells is no longer a spa. Instead, it acts as an entry point for traffic into the city center and an exit toward the industrial port Avonmouth further north and to the Mendip hills to the south. The center of Hotwells, where I used to live, is like a large traffic island, its tall, terraced houses divided into flats and circled by vehicles entering and leaving the city. Situated in the river basin, below the grand Georgian terraces of Clifton village, Hotwells was heavily polluted when I lived there in the 1990s. Our local pub, The Rose of Denmark, on the island next to the slip road, had an atmosphere thick with cigarette smoke and a clientele of elderly Londoners, who had presumably moved to Bristol decades before to work in the docks or the nearby tobacco factories, and whose voices were hoarse from the combined effects of smoking and traffic pollution.

Today, Hotwells is within Bristol's Clean Air Zone, along with the docksides where the glassworks, the soap manufacturers, tanneries, brickworks, and brass works were once located. The zone, which aims at reducing traffic pollution, is placed where the pollutants heavier than air are most likely to sink—that is, along the river basin. Urban river basins are historically the places where the poorest live and work, while the wealthy situate themselves on the hills, though in Bristol the river basin is now largely gentrified. Numerous writers have shown how bad atmosphere is distributed according to wealth—how the poorest people occupy the industrial areas, or are located near flyovers and motorways, or (in the past) in the city center docks. Since the 1930s, studies have tracked and correlated sootfall, cancers, and respiratory diseases, which follow not just the river but the "smoke stream"—the route that smoke tended to take across the city.[25]

The terraced townhouse that was Beddoes's Pneumatic Institution still stands in the corner of Dowry Square. It is here that the nineteen-year-old Cornishman Humphry Davy came to live and work, living in quarters on the top floor of the building and acting as its superintendent; and here that Tom Wedgwood volunteered to imbibe nitrous oxide, desperate for a cure for the ailments that would kill him in 1805, aged only thirty-four. Nothing here recalls Wedgwood, but Davy's presence is now commemorated, rather unromantically, in the name of the slip road that joins the river basin road to the elevated A370 heading south. Nitrous oxide also lingers here—in the NOS canisters that litter the parks and roadsides, left by present-day "bibbers," and in the atmosphere. It is a major contributor to ozone depletion and the third most significant greenhouse gas.[26]

25 ENCAPSULATION

► This story begins, like the story of the Industrial Revolution, in a cottage: not a weaver's cottage but the photography studio of Alfred H. Harman at no. 3 Albert Cottages, Hill Street, Peckham, south London. Harman, the founder of the company which became Ilford Limited, was going up in the world. He moved from cottage to villa by 1864, when his new home and business address became Gunnersbury Villas, Harders Road. Four years later he was able to separate home and business, with a larger studio and darkroom at 79 Peckham High Street and a suburban home in Surrey, to the west of London. He was one of many photographers profiting from the period of "cartomania," producing cartes de visites and cabinet cards for a public with an insatiable appetite for portraits (fig. 40). He was not a notable or innovative photographer, judging by portraits he made in the 1860s and 1870s, but he was a successful businessman. He may have been one of the first to offer copying, enlargement, and retouching services for a clientele of amateur photographers. He also took orders by mail, using the highly efficient Victorian railways to dispatch prints across the country. In 1879, Harman gave up the studio business to focus on producing dry plates, establishing his new factory, the Britannia Works, in the village of Ilford, to the northeast of London.[1]

I am telling you this brief history of Harman's business not to focus on Ilford's founder, but to draw attention to the darkroom as a space, as it moves from

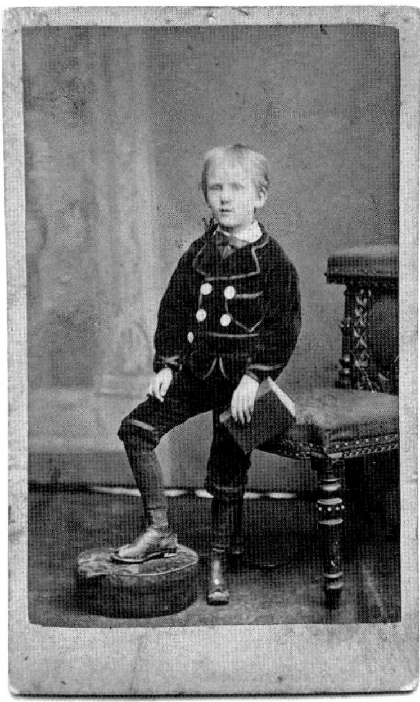

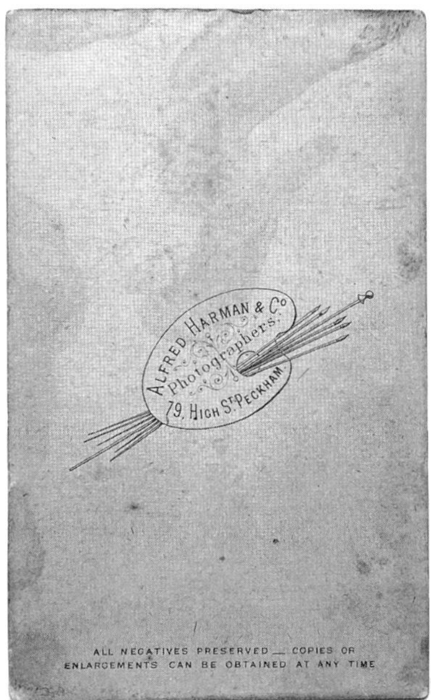

FIGURE 40. Alfred H. Harman carte de visite (front and reverse) taken after 1868. Private collection.

cottage to villa, separates from domestic life, and is eventually incorporated into the factories producing photographic materials. One concept which is useful for thinking about the darkroom as a partitioned-off interior space is "encapsulation"—a term used by the historian Chris Otter to describe a process that begins with people becoming more accustomed and acclimatized to life inside buildings, and with the division of domestic spaces or vernacular buildings into discrete rooms given over to distinct functions (dining rooms, bedrooms, kitchens, and so on), and that ends with a "technosphere." This latter term appears to have been coined by the geologist Peter Haff in 2014, who used it to describe the sum of the global technical infrastructures on which humans depend.[2] Otter connects this notion of the technosphere to the idea of sealed interiors or capsules and artificial atmospheres. His technosphere is "a giant apparatus within which encapsulated beings are fed, watered, mobilized,

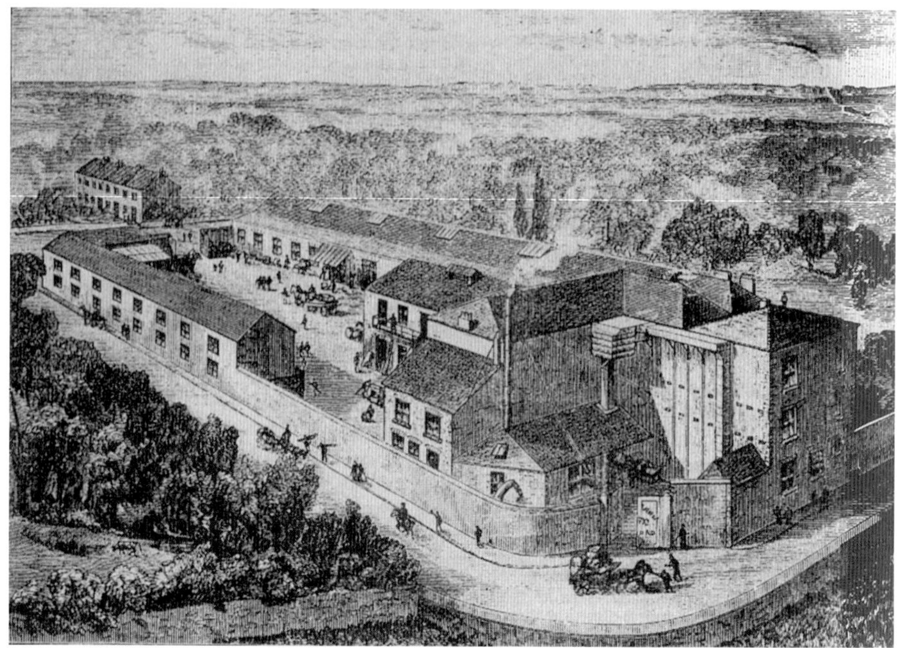

FIGURE 41. The Britannia Works Ilford. Illustration from the *British Journal of Photography* 35, no. 1469 (June 29, 1888). The building in the foreground with the chimneys is the engine room, washing and drying of glass plates took place in the buildings on the left and across the courtyard from them were the coating rooms.

entertained, and maintained in states of historically-unprecedented bodily comfort."[3]

The home darkroom is a capsule within a capsule. It is one of the first spaces in domestic settings (along with the larder) to take the partitioning of space to the point where temperature, atmosphere, and illumination are distinct from the rest of the dwelling. During use, the door cannot be opened, all windows must be covered, even the keyhole may need to be stuffed with tissue to prevent daylight invading. Better still, the darkroom is situated in a windowless cellar, shed, or loft (though often it was a temporary bathroom conversion). Such precautions produce their own problems: in protecting photographic materials from light, they also prevent a flow of air, and since the darkroom is full of chemical fumes and vapors, lack of adequate ventilation can become dangerous. Situated below ground, a cellar is also likely to be damp, and the moisture and cold detrimental to photographic materials.

Ventilating darkrooms while keeping out light and dust was a challenge from the beginning. It was especially important because of the wide range of toxic chemicals used in photographic darkrooms and factories. Bill Jay has listed the numerous ways in which photographers and photographic workers in the nineteenth century were at risk from poisoning when dealing with often deadly darkroom chemicals. The risk was not just from accidentally drinking these chemicals, although this did happen, nor from cut or grazed hands coming into contact with them, but from the atmosphere of the darkroom itself. Daguerreotypists breathed in mercury vapor, iodine, and bromine fumes. Those using wet plate collodion also encountered mercury and ether fumes. Potassium bichromate, used in many photographic printing processes, had nasty effects when inhaled, while the fumes of potassium cyanide, a fixative often employed instead of the more harmless hypo (sodium thiosulfate), could kill, or in smaller quantities produce serious symptoms lasting months.[4]

Despite newspapers and photographic journals being full of reports of poisonings in the darkroom and advice on how to mitigate risks, limited attention was given in *The Ilford Manual of Photography* to health and safety in the darkroom. Early editions devoted only one page to darkroom ventilation, to guard against the dangers not of chemical fumes but of human emissions:

> A point of the greatest importance is the proper ventilation of the workroom and neglect in this matter leads to unpleasant consequences, which are often erroneously attributed to other causes. It is indispensable that there should be an exit for foul air and entrance for fresh air, a small room, if not ventilated speedily becomes so filled with products of respiration as to be distinctly injurious to health.[5]

The professional photographic business required that darkrooms become further partitioned, as a series of different operations must be conducted in varying degrees of darkness, and at different temperatures and humidity. Even when Harman became a dry plate manufacturer, he did not fully divorce the domestic from the business: the Britannia Works began in a large house in Ilford.[6] Here, the partition of spaces commenced: the basement was devoted to emulsion making (using a secret formula) and coating, and different work took place on the other floors.

In keeping with the domestic setting, the secret emulsion was, according to A. J. Catford, "prepared by Mr. and Mrs. Harman assisted at times, by their housekeeper" (Mrs. Harman and the housekeeper disappear from later accounts).[7] They poured the emulsion onto the glass plates with a teapot and placed the plates on drying racks inside a cupboard supplied with warm air.[8] It is at this point that the plates are most vulnerable to spoiling, so the drying plates acquired their own distinct space and unique climate. As the business expanded, the different steps of the process were allocated different buildings: a cottage for the manufacture of emulsion, a house for plate coating, and so on. This arrangement meant that emulsion had to be carried in lightproof containers from one building to the next, always with the risk of spillage or light leaks. Eventually, in 1883, Harman had a factory purpose-built, and then a second in the 1890s. The inner workings of these factories were secret, few visitors were allowed, but Hercock and Jones write that in similar factories plates were moved on conveyor belts through coating and drying rooms and into packing rooms—all of which, of course, were also darkrooms.[9] We also know that from 1895, the company was operating an air-cooling system.[10] As the Britannia Works expanded in the 1920s and '30s, it absorbed surrounding cottages, displacing their occupants and either demolishing them or absorbing them into the factory structure.[11]

In the early twentieth century, homes became more like factories as Frederick Winslow Taylor's "scientific management" principles were applied to domestic space, most obviously to kitchens, in order to rationalize housework—Margarete Schütte-Lihotzky's famous Taylorized Frankfurt kitchen was built in prototype in 1926, and in Britain, ideas of labor-saving homes were circulated via the popular press and the Ideal Homes Exhibition.[12] By this time, the Ilford Limited company's factories did not resemble domestic spaces as much as they once had, though the company did retain some of the original

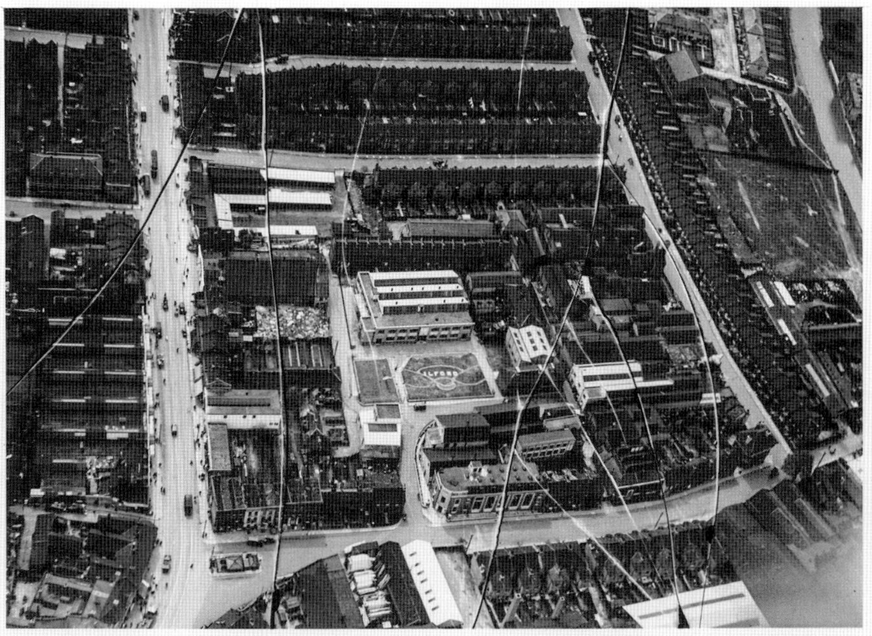

FIGURE 42. Cracked glass plate depicting aerial view of the Ilford Limited factory in Ilford, with the river on the right side of the image, 1932. Ilford Limited collections, Redbridge Museum and Heritage Centre. Courtesy of Redbridge Museum and Heritage Centre 2025.

cottages. Now they shared some of the characteristics of the Fordist-Taylorist factory system, such as conveyor belts moving products to workers performing distinct tasks (see plate 6).

Ilford's factories now employed large numbers of workers. Describing the plate factory in wartime, staffed mainly by women and old men, factory manager Cecil N. Potter named the different employees by the distinct, repetitive tasks they were assigned, among them rackers, coaters, slippers, packers, and breakers. Potter's description of the work of a racker reveals how specialized and rationalized the factory labor had become: "Our best racker can pick up and put into racks 3,600 plates an hour. She could not do this unless there was perfect co-ordination between eyes and hands and strict economy of movement. The body is not moved and practically all

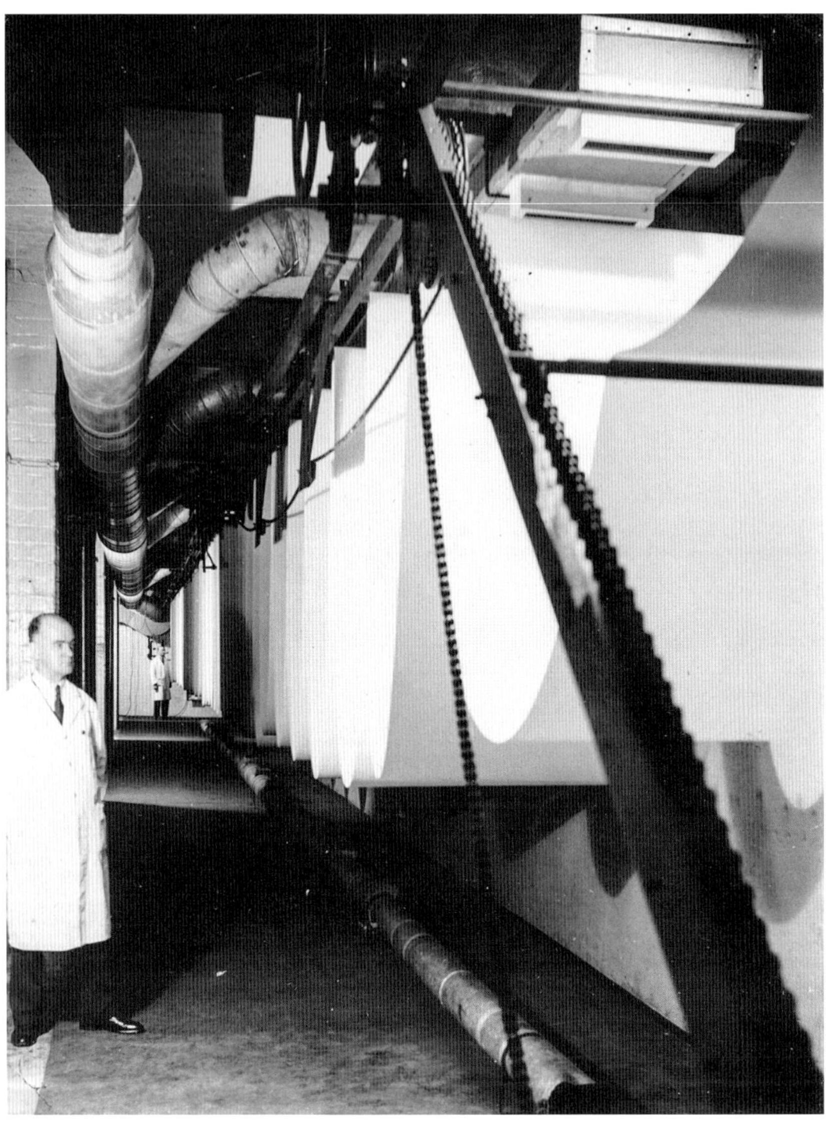

FIGURE 43. Paper coating M/C Drying Track at the factory of Thomas Illingworth & Co Ltd, Park Royal. Ilford Limited amalgamated with Illingworth in the 1920s. Photo undated but likely to be between 1912 and the mid-1930s when paper coating operations were transferred to the Rajar factory in Mobberley. Ilford Limited collections, Redbridge Museum and Heritage Centre. Courtesy of Redbridge Museum and Heritage Centre 2025.

movement is from the elbows to the wrists. You will not be surprised to learn that it takes two years to train a racker."[13] Artisanal skills are replaced by a new kind of skill, the skill involved in repressing normal bodily movement and moving instead like a well-designed machine. Her pace matched exactly the pace of the new coating machines, which by 1930 allowed 3,600 glass plates to be coated in an hour. The machinery became more like that of an assembly line, with moving beds and new spiral dryers for celluloid film installed in the 1920s.[14]

Sensitized materials were moved through a series of specialist rooms, each of which was designed to protect the coated glass plates and films from light and other kinds of contamination. Dust was a big concern: the cutting of plates and films produced it, electric fans spread it, workers carried it into the factory. At various stages in the process, the coated plates and films required warming or cooling. Step by step, factories were redesigned to incorporate different lighting conditions and different climates within a series of differentiated spaces, each partitioned from the last. Gradually, the darkened spaces of photographic materials factories resembled the domestic darkroom less and less. Even so, a Thomas Illingworth experiment book begun in the 1920s and held in the Ilford archives at Redbridge contains, clipped to the front of the book, a drawing of "The Bathroom Dark-room," as if the company laboratory should never forget its humble origins.[15]

26 KEEPING OUT THE FOG

► The photographic industry depended on the wider chemical industries for its materials, but it was also manufacturing a product that was highly sensitive to contamination and transformations in the atmosphere. To isolate photographic materials from atmospheric exposure, and turn factory darkrooms into purified spaces, new systems of air-conditioning were introduced, themselves making use of damaging chemicals —including coal derivatives, but also, from the 1920s, fluorinated gases (F-gases or HFCs), which today make up around 2 percent of greenhouse gas emissions. The term "air-conditioning" was first used in a patent by Stuart Cramer in 1906, to cover "humidifying and air cleaning and heating and ventilation," though today we might associate it more with cooling systems.[1]

Generally, and despite the hazards to humans of the vapors given off by chemicals such as potassium cyanide and potassium bichromate, atmospheric control in the darkroom and the factory was principally about ensuring a conducive environment for the various photographic materials. Climate control centered on commodities, for example in the cooling systems for food preservation used in European colonies and the mechanical systems for ventilating and heating glasshouses for tropical plants. Both of these examples show how artificial atmospheres are the products of both capitalism and colonialism.[2] Heating, ventilation, and air-conditioning systems were first introduced in factories to cool beer, dry out tea, control the humidity of fibers in weaving, and prevent mold growth on celluloid.[3] Like these products and like tropical plants, photographic materials demanded their own microatmosphere in order to thrive.

While writers who have studied mechanical air-conditioning tend to see it in terms of human comfort, sometimes air-conditioning would actually increase the discomfort of people, especially in factory settings where the protection of sensitive materials and goods was at least as much of a priority as the protection of workers. Sometimes the factory environment was made hotter and more humid to make raw materials easier to manipulate.[4] In the industrial darkroom, workers had to adapt to the needs of the materials, most obviously by working in low light or total darkness.

In his 1969 book *The Architecture of the Well-Tempered Environment*, the architectural historian Reyner Banham argues that air-conditioning was initiated in factory environments because of the economic advantages, and the risk of damage to materials and products that occurred without it. This can be seen in the case of Ilford, though Banham does not mention that rises in industrial emissions, and specifically coal burning, are among the factors that spur its development. Ilford Limited's venture into air-conditioning was a response to a specific pollution event and to the ongoing negative effects of the polluted London fog on their product.

Despite its picturesque qualities, fog had long been the bane of photographers—or as an 1870 article in *The British Journal of Photography* described it, "the great winter enemy of London photographers."[5] During a fog, it was almost impossible to take photographs, not only outdoors but even within photography studios. The monstrous fog penetrated the studio and the darkroom, even the camera. It was far easier to keep out light. In 1881 an article in the same journal, titled "Dust and Fog," described how the fog entered the studios of London photographers, noting that the studios contributed to the problem since the rooms were heated by coal.[6] Another *BJP* article published in November 1898 talks of the fog as a problem in the studios of commercial portrait photographers, stating that "however brightly the sitter may be illumined, the fifteen or twenty feet of smoky fog intervening between the sitter and the lens necessarily mars the brilliancy of the picture."[7] For the nineteenth-century photographer, the distinctive yellow color of the pea-souper made exposures especially difficult for the yellow-insensitive photographic emulsion of the time. Hydrogen sulfide in the fog affected emulsions and developers. By the 1890s, the coal fires in surrounding houses and the fumes from the gasworks, as well as the London fogs, were affecting photographic plates in the Britannia Works' drying rooms.[8]

The London fogs had been worsening throughout the second half of the nineteenth century due to a combination of misty river basin weather conditions in winter and vast amounts of coal smoke from both industrial and domestic sources. Gas was promoted as a cleaner fuel, and indeed it did help to clear the air in the city center, but the gas industry was coal fired; while it was relatively clean at the point of use, and air quality improved as a result, it was filthy at the point of production. Polluting the air with smoke and particles and the soil and waterways with "ammoniacal liquors" and surplus tar, the early gasworks were the cause of the hydrogen sulfide in the atmosphere. This substance, with its familiar "rotten eggs" smell, was a by-product of purifying the gas stream using lime.[9] By 1900 London's gas companies were consuming about four million tons of coal a year.[10] The gasworks proliferated and expanded at the margins of the city and in working-class communities. They displaced the fog and pollution from the city center to the suburbs and to the dockland areas of the East End. Their acrid fumes also drove the middle classes to flee from these areas, which became increasingly impoverished. As one article in an 1864 London newspaper commented, "Wherever a gas-factory—and there are many such—is situated within the metropolis, there is established a center whence radiates a whole neighborhood of squalor, poverty, and disease."[11]

The town of Ilford had thus far escaped the worst of the manufacturing industry and chemical works—the "unfragrant" and "offensive" trades proliferating in other Essex towns on the periphery of London—and it had a relatively affluent population.[12] It already had the Ilford Gas Company, which dated back to 1839, the same year as the arrival of the railway in Ilford and the announcement of the new invention of photography to the world; but by the standards of the late nineteenth century, the air was fresh and sweet in Ilford.

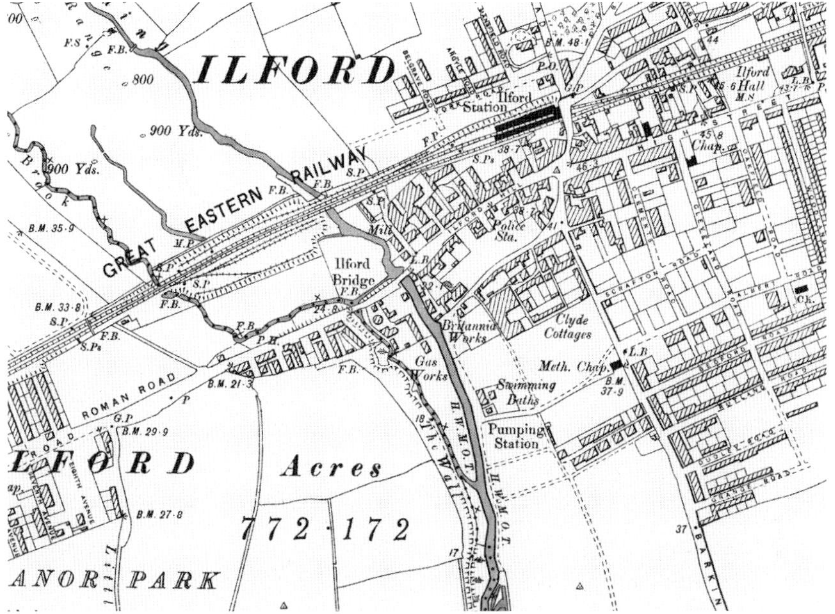

FIGURE 44. Detail from Ordnance Survey Map of Ilford, LXXIII.NE published 1898. Map of Ilford in 1893–94 showing the Britannia Works and the Gas Works on opposite sides of the River Roding. Reproduced with the permission of the National Library of Scotland.

Photographic manufacturers, and later the film studios, all wanted to avoid the pollution of the "big smoke" but needed to benefit from London's trade links. So they chose locations at the edges of the metropolis, which also had the advantage of lower land prices. The attraction of Ilford to the photographic manufacturers was evident: clean air, clean water in the River Roding, and easy access by rail and road to the city. When Harman moved his business there, the first factory had green fields on three sides and the population was about seven thousand. But as London rapidly grew, and as it expanded into Ilford, it brought with it new industries and a larger population, all heating their homes with coal fires. By 1900 there were over forty thousand people, and the Britannia Works was enclosed on all sides by houses.

As London came, the London fogs came, too, especially to east-

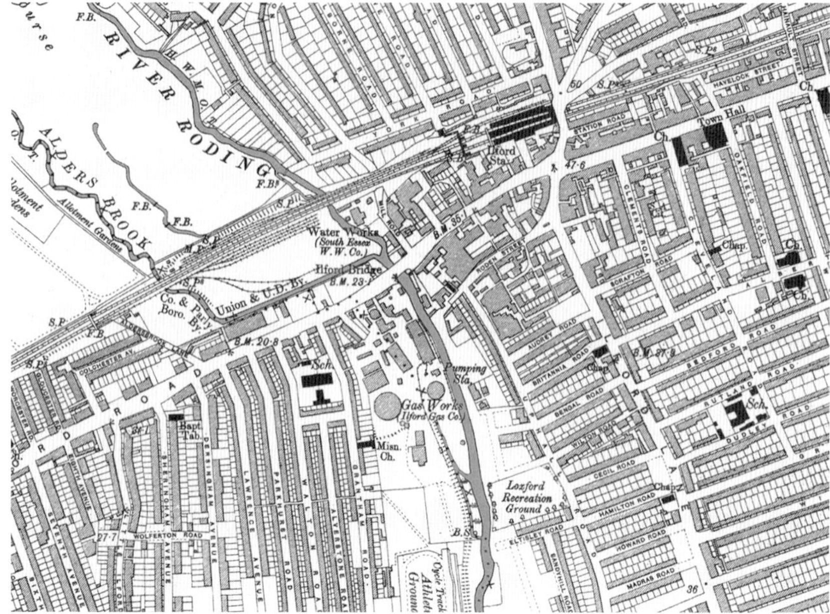

FIGURE 45. Detail from Ordnance Survey Map of Ilford, London Sheet H, published 1921. Map of Ilford in 1914 showing the Britannia Works and the Gas Works on opposite sides of the River Roding (river highlighted). Reproduced with the permission of the National Library of Scotland.

ern districts like Ilford, since the prevailing wind was from the west. Later, in the early twentieth century, the movie studios would try to outwit the fog by situating themselves on higher ground to the west, yet even they were beset by fog, to the point that they either had to embrace the low-contrast, soft-focus aesthetic that resulted or shut entirely during winter.[13] In the 1920s, promotion for the Selo company included the following claim: "It was not mere capriciousness that made the firm build their Factory on the hill-top, but cold logic, for here on the heights the air is pure and clear and free from many of those things which would mar the good quality of Selo sensitive products."[14] But in the 1890s, Harman's factory was not on a hill-top but on the banks of the River Roding in Ilford. As the population grew, so did the gasworks just across the river. A photograph from 1945, after a rocket attack on the Britannia Works, reveals the prox-

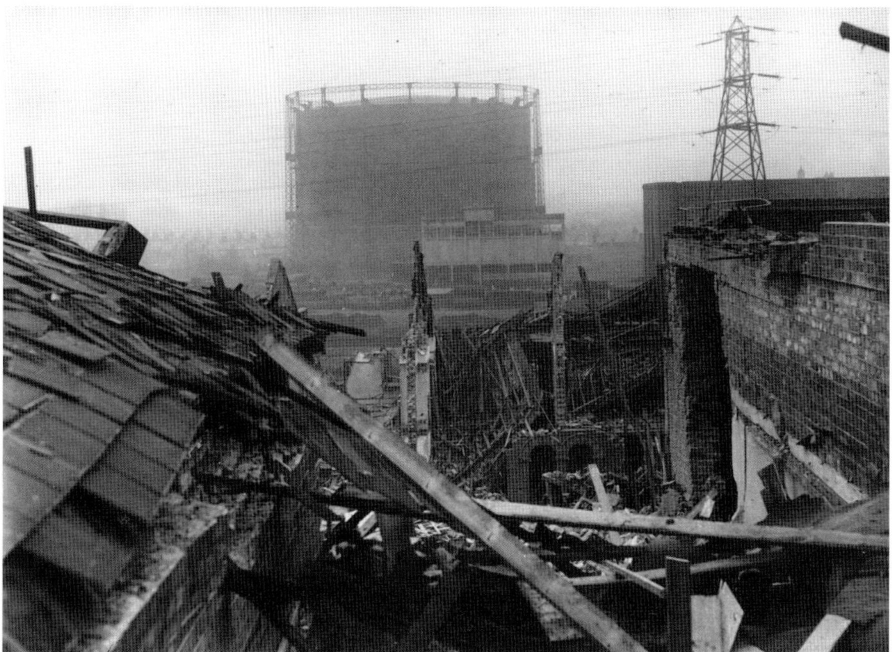

FIGURE 46. Ilford Factory facing the gasworks, after rocket incident, showing roof of drying room (left). February 20, 1945. Ilford Limited collections, Redbridge Museum and Heritage Centre. Courtesy of Redbridge Museum and Heritage Centre 2025.

imity of the factory to the gasworks (fig. 46). In June 1899, the Ilford Gas Company gained statutory powers and resisted an attempt by the local authority to turn it into a municipal supplier.[15] That year the gasworks began to manufacture sulfate of ammonia from their ammonia by-products to sell as fertilizer. The resulting fumes were highly damaging to wet photographic emulsion.

In April 1899, in just one day, 25,000 photographic plates were ruined through fogging in the drying rooms at Ilford's factory. Company historian and manager A. J. Catford places the blame for the incident firmly on the hydrogen sulfide emissions from the Ilford Gas Company's new works.[16] The loss of these plates, and further thousands of pounds worth of damage caused by the gasworks' fumes entering the photographic paper factory, meant that the firm was faced

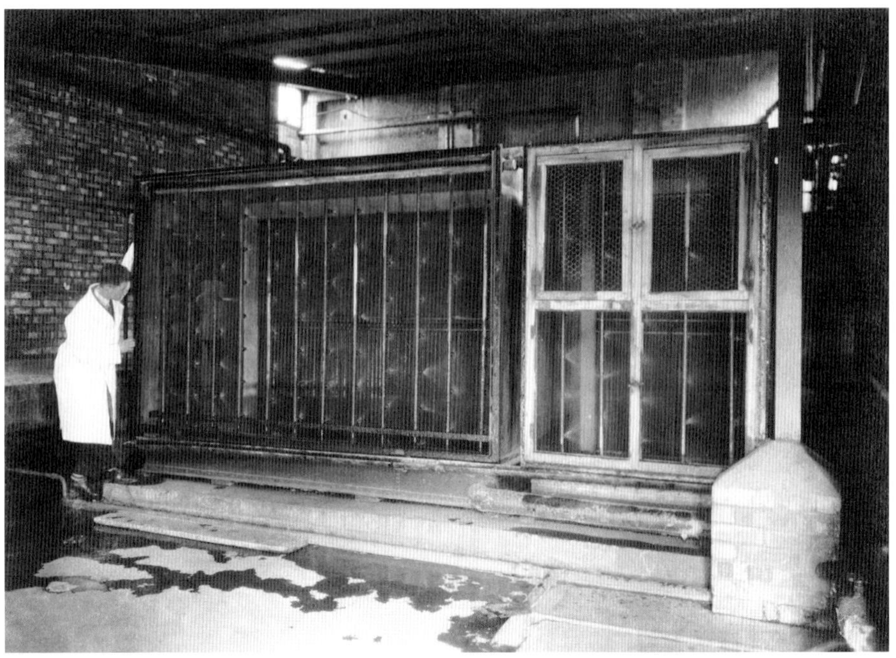

FIGURE 47. The Air-Washers at the Selo Factory, 1938. Ilford Limited collections, Redbridge Museum and Heritage Centre. Courtesy of Redbridge Museum and Heritage Centre 2025.

with three options: first, to try to limit the pollution from the gasworks; second, to escape the polluted air by physically moving away from it; and third, to create an isolated pocket of purified air.

They chose all three. In May 1900, they issued a writ against the gas company. They designed and built a new factory further out of London, in Brentwood; and they made changes to the existing factories in 1902 and 1905. These changes included a novel solution: an early form of air-conditioning. They brought in a consultant engineer to invent a means to purify the air in the glass plate drying rooms. This involved filtering the air, washing it with water, then cooling it to remove the water vapor, reheating it and then circulating it into the rooms.[17] Of the three options, this was the most successful. The writ against the gas company was withdrawn, and although an elaborate new factory was built in 1901, it closed only nine years later (reopening after the war as the Selo roll-film factory) (fig. 50).[18] These two strategies had been premised on the notion that the firm could outrun or combat the spread of pollution. But the air-conditioning solution was different, because it was premised on the idea of creating a space of exception within the larger polluted environment. It meant accepting that pollution of the wider environment was, if not inevitable, then unstoppable. Instead of supporting the regulation of pollution, the firm chose to focus on creating a regulated factory interior.

27 POCKETS OF EXCEPTION

► Writing about the difficulty that eighteenth-century scientists had in isolating air as an object of study, Steven Connor says they had to learn to enclose the air, to create "new pockets of exception in the immersive totality."[1] Ilford Limited set out to do something similar: to isolate a purified atmosphere from the contaminated one beyond the factory walls or even just beyond the drying rooms.

By separating off the atmosphere of the factory from the external atmosphere, air-conditioning was not only a practical solution but also embodied a new concept of mechanically controlled atmosphere that went beyond earlier heating and ventilation systems. The year that Ilford introduced air-conditioning was 1902, the same year that the self-proclaimed "father of air-conditioning," American engineer Willis Haviland Carrier, installed his first climate-control system in the Sacklett-Wilhelms printing plant in Brooklyn, and the same year that the Royal Victoria Hospital in Belfast became "the first major building to be air-conditioned for human comfort" according to Reyner Banham.[2]

As A. J. Catford put it, "Our company was a pioneer not only in the manufacture of photographic material but also in the manufacture of weather, or as it later became known, air conditioning."[3] Another term for air-conditioning, used well into the 1930s, was "man-made weather" (fig. 48).[4] The irony is that the weather outside was (as we now know too well) also increas-

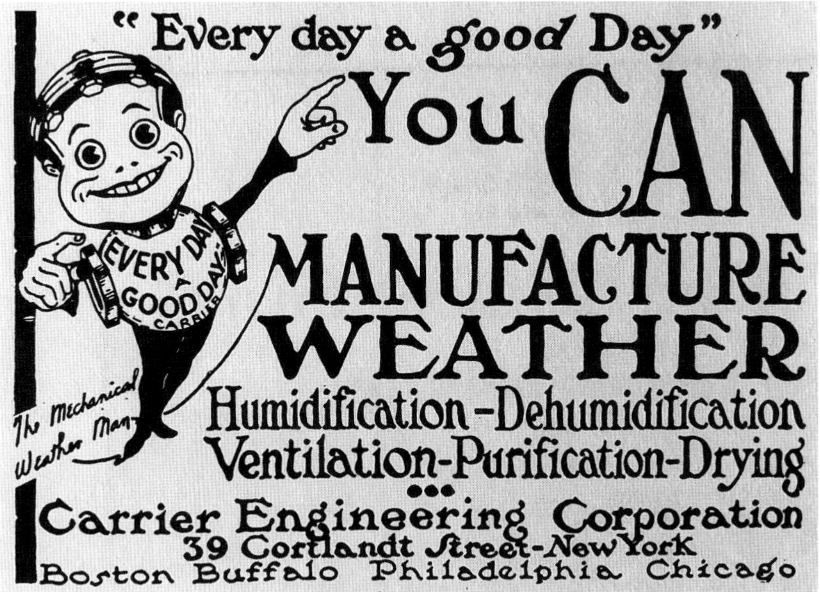

FIGURE 48. Advertisement for Carrier's air-conditioning system.

ingly "man-made." It's this external weather (the polluted fog) that drives the production of the purified interior.

Even in the earliest of Ilford's factories, there were microenvironments for coated and uncoated plates: washing rooms, coating rooms, chilling tunnels used to set the emulsion, and heated drying rooms or tunnels.[5] The sequence was as follows: the glass plates were washed prior to coating by glass-washing machines which passed them through rollers and brushes. Initially these were placed by workers onto racks and dried in warm rooms or even outside in the open air, but the later machines incorporated a heated tunnel through which the plates were passed to dry.[6] The plates were then coated and proceeded through a chilling tunnel (at first cooled with ice, later with the water from the refrigeration unit) to set the emulsion and finally to the heated drying rooms.[7] In the early days, all of these operations would have taken place in controlled lighting under a safelight, but with the introduction of panchromatic emulsions the

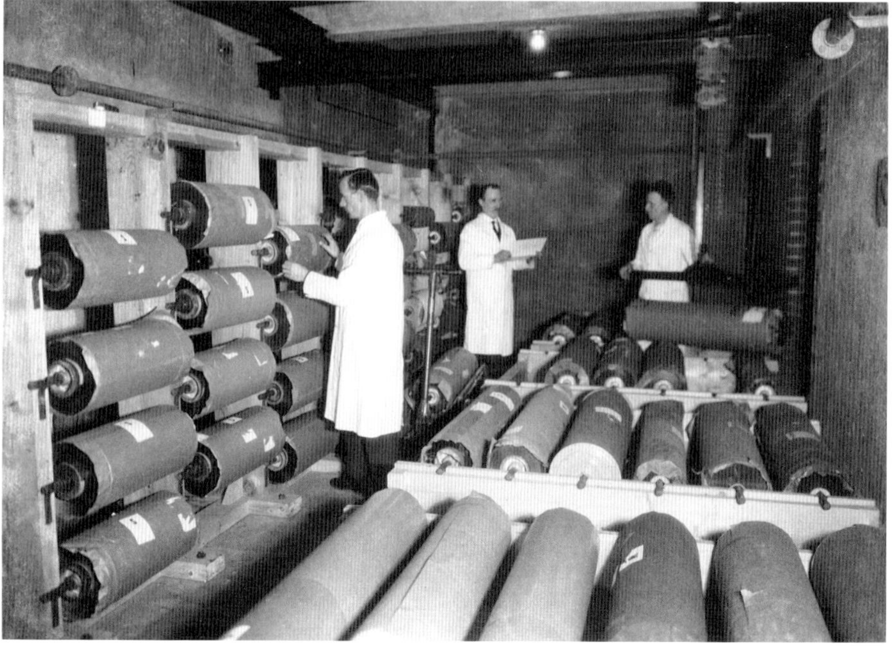

FIGURE 49. Celluloid stores, Selo factory, Brentwood, 1938. Ilford Limited collections, Redbridge Museum and Heritage Centre. Courtesy of Redbridge Museum and Heritage Centre 2025.

workers had to operate these machines and move the delicate glass plates and films from coating machine to drying room in almost total darkness. Thus, the capsules proliferate as spaces are subdivided, like cells, each with their own distinct environments. Encapsulation involves the spread of climate-controlled spaces; as Chris Otter writes, it "generates innumerable climate bubbles proliferating across the earth . . . everything—humidity, temperature, dust, smell, sound and light—is meticulously regulated."[8]

Such systems need to be driven by an energy source, and generally, in Britain, this would be coal. A drawing of the first of Harman's factories, which appeared in *The British Journal of Photography* on June 29, 1888, shows a large chimney emitting either steam or smoke (fig. 41).[9] The new factory built in the 1890s had a boiler and steam engine driving the machinery, both coal fired. It also had an

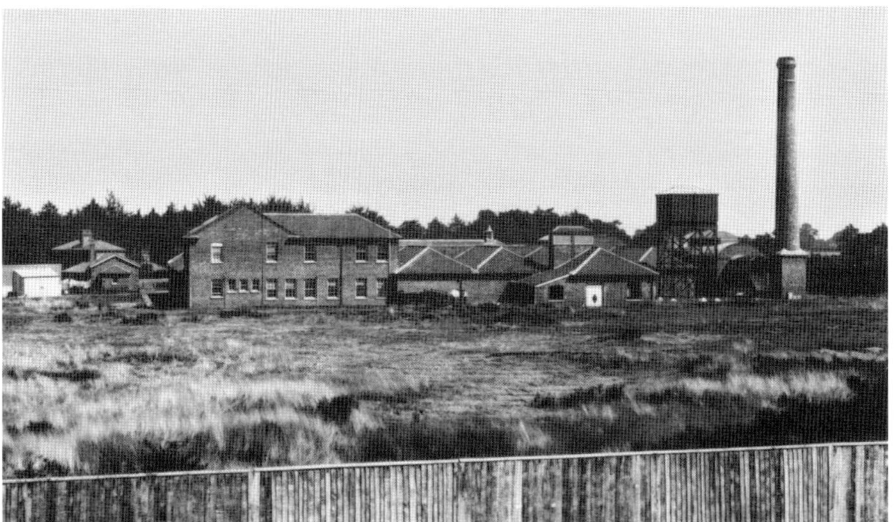

FIGURE 50. The first factory built at Warley, Brentwood, in the early 1900s. It was closed in 1910 following the success of the air-conditioning systems at the original Ilford factory but then turned into a roll film factory a few years later. Ilford Limited collections, Redbridge Museum and Heritage Centre. Courtesy of Redbridge Museum and Heritage Centre 2025.

engine and dynamo to drive the electric lighting.[10] The Selo factory at Brentwood included an electricity generating plant (coal fired) and a refrigeration plant. In the 1920s, prior to the introduction of the National Grid, the factory could not rely on a public electricity supply, so each factory had generators with a gas one on standby. Even after the public supply was available at the Selo plant, in 1928, these older generators were used in cases of power failure. As late as 1931, Geoffrey B. Harrison of Ilford Limited's Rodenside Laboratory authored a text on "Automatic Control of Mains Voltage" discussing the difficulties of ensuring a consistent electricity supply to the factories.[11] The Ilford factories moved from manually cranked machinery to machines driven by steam and later by electricity, but all, ultimately by coal.[12]

Otter writes: "Climate engineering, then, aspires to create spaces which are atmospherically sealed from a dirty, noisy, turbulent outside."[13] He points out that these spaces are always connected to the

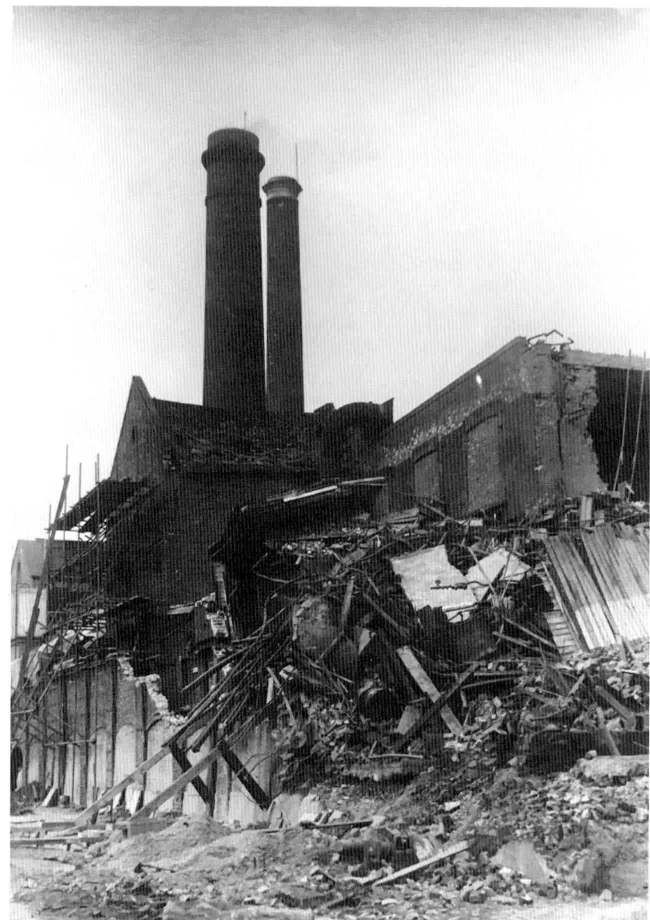

FIGURE 51. The Ilford Limited factory in Ilford, following the WWII rocket attack, Feb. 20, 1945. Buildings shown are the battery house, filter and fan chamber. Ilford Limited collections, Redbridge Museum and Heritage Centre. Courtesy of Redbridge Museum and Heritage Centre 2025.

outside, by infrastructure, by the raw materials they depend on, and by treating that outside as a "sink" into which waste is put. We can see this in the case of the Ilford climate engineering. The clean air of the drying rooms requires an infrastructure (cooling and heating systems and fans) which pulls the outside in, and that depends on a supply of steam and coal. Dirty air—expelled outward—is produced at the same time as purified air is drawn inwards. The irony is that the system ends up purifying the air that it has itself polluted.[14]

Climate control appears in the Ilford darkrooms at the beginning of a century in which fossil-fueled HVAC systems would become ubiquitous in domestic, industrial, and public buildings in wealthy countries. Today cheap air-cooling systems use hydrofluorocarbons (HFCs) to trap heat, despite concerns about the role of these gases in increasing global warming. Dipesh Chakrabarty writes about the growing use of these in India's densely populated cities, where there is an urgent need to combat the high temperatures caused by global warming.[15] Chakrabarty cites the economist Michael Greenstone: "The very technology that can help protect people from climate change also accelerates climate change."[16] Or, to quote the philosopher Roberto Eposito from his book *Immunitas*: "The risk from which the protection is meant to defend is actually created by the protection itself."[17]

For Chakrabarty, the example serves to show how political questions of freedom and equality do not always mesh easily with the "ecologically desirable."[18] In the context of my discussion, however, the irony of contemporary air-conditioning contributing to the rise in temperatures is a continuation of the earlier irony, of coal-fired artificial atmospheres designed to protect against the smoke and fumes produced by coal and coal derivatives. Their own fuel produces the contaminated or unbreathable atmosphere that the system then works so hard to wash and filter.

However, air-conditioning's role in the production of an ever more degraded world, from which it is intended to create safe haven, is not just about the exclusion of chemical and particle contaminants. It is also, simultaneously, a symbolic detachment of interior from exterior, which in the search for an ideal interior environment brackets off the exterior altogether. The process of encapsulation relies on an opposition of inside and outside that makes the outside easier to neglect, less visible. So, air-conditioning contributes to the degradation of that which it excludes as impure. We can see this in the proliferation of what the architect Rem Koolhaas calls "junkspace," which he describes as "the product of an encounter between escalator and air-conditioning, conceived in an incubator of Sheetrock." The degraded junkspace environment is what is ejected and rejected by technologies of interiorization and encapsulation: "Junkspace is what remains after modernization has run its course, or more precisely, what coagulates while modernization is in progress, its fallout."[19]

The privileging of interior over exterior has roots in a paranoid, imperialist mentality. In a vivid description that would inform Otter's concept of encapsulation, Walter Benjamin described a nineteenth-century bourgeois domestic interior as a kind of enclosed and suffocating "spider's web."[20] Benjamin was reworking an earlier passage he had written under the influence of hashish (in 1928), translating his hashish-induced paranoia, in which "thoughts of the 'outside' become almost agonizing," into the generalized drugged state of the imperial bourgeoisie, whose stuffy and cluttered rooms express not just an "aversion to the open air" but a fear of an external world that they themselves have helped to bring into being. The imperialist bourgeoisie brought the empire home in the form of hothouse plants, but, especially in the colonies, they treated themselves like the plants, isolating themselves within environments increasingly designed to accommodate white "thermal comfort" and to exclude a tropical nature and colonized peoples viewed as suspect, potentially hostile, and definitely contaminating.[21]

Architectural historians trace some of the earliest mechanical systems for ventilating and heating buildings to the hothouses or glasshouses designed to accommodate the Victorian fashion for tropical plants.[22] Histories of air-conditioning tend to distinguish between mechanically driven systems and the broader work of regulating atmospheric conditions in interior spaces, which begins much earlier and can be traced back through centuries of architectural design. They point to the changes in greenhouses because, like photographic materials, tropical plants demanded their own microatmosphere in order to thrive. In the mid-nineteenth century, horticulturalists and gardeners went beyond the architectural innovations of glasshouses, with iron structures heated by stoves, to more complex systems using flues and pipes to move steam around, simulating the humidity of tropical climates as well as the temperature.[23] For Peter Sloterdijk, the palm house is an imperialist attempt to stabilize environment, the flip side of the attack on environment represented by imperialist, terroristic gas warfare.[24]

Air-conditioning evolved partly from the cooling systems for food preservation that were used in the hotter parts of the United States and for colonial settlements across the European empires—as On Barak writes, “Standardizing the temperatures of empire was one of the most exciting potentials of artificial cold production.”[25] Climate-control systems were introduced very early on European expedition ships, around the same time that photography was introduced in the late 1830s and ’40s. For example, architectural historian Dustin Valen describes the atmosphere-purifying experiments of Scottish engineer David Boswell Reid on Captain Henry Dundas Trotter’s 1841 expedition to the Niger. Reid’s air-purification system was installed on the ships to protect the white sailors, in particular, from tropical disease. He fitted the ships with “medicators,” apparatuses “designed to purify the atmosphere below deck by filtering out mechanical impurities, regulating its humidity, and infusing the air with chemicals to neutralize airborne poisons.”[26] Reid used nitrous oxide in his purification systems, deriving the idea that the gas might have curative properties from the work of Humphry Davy at the Pneumatic Institution in Bristol. The system was a disaster that did not prevent widespread illness and death among the sailors on the Niger ships, though Reid was unwilling to accept the failure. As Reid’s approach suggests, practices of encapsulation and the production of microclimates helped to intensify the oppositions of inside versus outside, and inclusion versus exclusion, which would continue to structure colonial space.

28 REGULATION AND EXPOSURE

► Practices of encapsulation and the production of microclimates emerged partly from colonial anxieties about the dangerous, uninhabitable tropics.[1] Tropical heat was associated not only with the risk of catching tropical diseases, but with unproductivity and racial degeneration. From the late nineteenth to the mid-twentieth century, the concept of "thermal comfort" used to promote the new air-conditioning systems linked coldness with masculinity and vigor, heat and humidity with laxity and dissolution.[2] It built on older notions of stuffy interiors and an overprotected, luxurious life as negative to health, combined with the miasma theory that associated humid climates and stale atmosphere with disease. Writing on the history of environmental medicine, Vladimir Jancović argues that as early as the eighteenth century, innovators promoted ventilation systems by constructing their absence as the problem: "Foul air ceased to emanate from the unwashed multitudes; it was created by bad ventilation." Engineers were effectively social reformers peddling solutions to modern life, and the fact that it was now technologically possible to prevent bad air "made discomfort still more intolerable."[3]

Increasingly, technological solutions to environmental hazard were seen as both morally desirable and essential. Ilford Limited's decision to install new air-washing systems in its existing factories, build a well-equipped new factory, and not to sue the gasworks reflects the perception that a technological solution is a modernizing solution. Integral to such ideas of modernization are concepts of regulation and the avoidance of risk. The air-conditioning systems at Ilford Limited's factories were systems designed to regulate and control the exposure of photographic materials to contaminating substances and atmospheric conditions.

Exposure has a meaning for the photographic materials factory that it does not have for other manufacturers, since every photograph is an exposure. Photography begins with the controlled admission of light into a darkened room or a camera, onto a surface hitherto entirely protected from light. The camera itself is a dark room that continues the protection begun in the factory. In a camera, controlling the exposure of the light-sensitive material to the external world is done by adjusting the size of an aperture and the timing of its opening. Controlled exposure also means preventing accidental light leaks, and manipulating the direction and angle at which photons fall on the surface of a film, a glass plate, or a sensor. It means limiting contaminants in the air and regulating temperatures and humidity, particularly when the emulsion is not yet dry. In the early twentieth century, one of the central tasks of the photography industry was to control that atmospheric exposure, to do what it could to protect the sensitive surface from moisture, pollutants, sudden changes in temperature, dust carried in the air, and so on.

Jancović points to the relationship between the concept of exposure and ideas of risk. He shows how in the context of human health, risk might be measured according to vulnerabilities specific to a person (for example, pale, light-sensitive skin), the level of exposure (amount of time spent in strong sunlight), or environmental conditions (the strength of the sun's radiation and absence of shade).[4] We might think about photographic exposure in this way, too, as exposure to risks assessed according to the fragility of the materials themselves (badly coated plates perhaps, but also heightened sensitivity to particular kinds of radiation), levels of exposure (to light, to contaminants, to temperature fluctuations), and environmental conditions (the interior atmosphere of the darkroom and factory, the polluted fog, the weather).

Even before Ilford installed air-conditioning in 1902, photography studios had begun installing new ventilation and heating systems. These were also described in terms of control and regulation. For example, in 1897, *Scientific American* reported on the technical improvements made to Lafayette's photography studio to exclude "fog—one of the most deadly enemies of the camera." The article describes how the technical system ensured that "both in winter and in summer the warming, cooling, and ventilation of the building is under entire control."[5]

By the 1930s, Ilford Limited was keen to emphasize the well-disciplined and orderly nature of its factories, in order to guarantee the quality of the product. For example, *The Ilford-Selo Record* of March 1935 quoted the Ilford plate factory manager, Cecil N. Potter: "If you could see us and our factory, if you could see the painstaking and meticulous care observed in every operation, you would marvel that a defective plate could get out and if one did you would be inclined to preserve it as a souvenir of something unusual." Potter emphasized "dust prevention": "To start with all outside yards and passages are kept wet, being watered many times a day; all rooms are supplied with washed and conditioned air under pressure so that no unconditioned air from outside can get into the rooms; floors are oiled, and every department has a vacuum cleaner and is cleaned daily."[6]

In 1939, *The Ilford Courier*, another company magazine, made a very similar point with a little more emphasis on the technological, mechanical aspects of the cleaning: "There is nothing haphazard in the manufacture of Ilford and Selo materials. Everything works with precision. Air is purified and washed before it enters the factory. Dust which might be carried into the factories by employees is eliminated by vacuum cleaning the staff before they enter the coating, drying and packing rooms, and floors and passages are cleansed thoroughly every day with special electrically driven machines."[7] Not only the rooms but the staff themselves are vacuumed! Here technology serves both a practical and symbolic function in relation to risk. Practically, it helps to remove or lower the chances of the photographic materials being exposed to damaging chemicals or particles, or excessive moisture. Symbolically, it communicates to customers, investors, and the wider world that the company is competent and able to manage any risks it faces. Industries respond to attempts to regulate and control them with technologies that give the *appearance* of regulation, control, and cleanliness.

Historically, ideas about routines, regimens, and regulation had been connected to moral ideas of discipline and orderly living.[8] By the twentieth century methodical routines and regimens were embodied in machines. The development of mechanical climate-control systems was motivated by ideas about the well-regulated modern life.[9] Such machines operate as "cultural techniques" in the sense Bernhard Siegert uses the term (in his book of the same name), making it possible to "filter out" the symbolic from the real, signal from noise, sense from nonsense, culture from nature.[10] Cultural techniques create "order by introducing distinctions."[11] They include the methods used in laboratories and offices to routinize and standardize operations and produce reliably comparable results.[12]

The washing and filtering of the air in the factory is similarly a cultural technique, standardizing the quality of the product to ensure uniformity and consistency and to create a sense of orderliness. This extends to the workers themselves, as *The Ilford-Selo Record* explained: "Employees have special dressing rooms and have to be vacuum cleaned daily before entering their work rooms. In the coating rooms the need for care is so great that the girls remove their frocks in their special dressing rooms and vacuum clean each other's underclothing before putting on overalls designed to prevent fibers or dust from their underclothing escaping into the rooms. Even the hair is concealed in a cap which comes down to the ears. The illustration will show how the girls are enclosed in their overalls."[13] While there are clearly practical reasons for this, the communication of it is also about constructing appearances. The choice to emphasize the women workers here, rather than the men (who also entered the coating rooms), may be a way to add "interest" for a presumed male reader, but it also aimed to highlight the modernity and rigor of the company. Of course, rules about dress also have the added function of disciplining workers.

Altogether, the regulation of the factory environment and manufacturing processes—expressed in terms of cleanliness, rigorous procedures, and technical equipment—seemed to speak of a responsible, modern, and efficient firm. Why else repeatedly advertise the factory's air-conditioning system or the procedures for removing dust from workers, their "meticulous care" and the "scrupulous cleanliness" of the coating rooms?[14] What we don't see are the difficulties, the inconsistencies of the process, the challenges of making reliable and identical materials. These remained hidden in the confidential experiment and formula books that were originally held in Ilford Limited and its subsidiaries' various labs, and which challenge the outward impression of regulation and control. They reveal a world riddled with "troubles": the trouble with fog; the trouble in obtaining hardness; trouble with black spots (and comets); trouble with slow speed and with fast speeds; with grease and froth; with emulsion running and setting; with mealiness, matting, and blistering.[15]

So one of the purposes of air-conditioning was symbolic. There are precedents for this—for example, Michael Osman has argued that speculative futures markets were enabled by novel technologies, specifically cold storage systems for perishable goods. These made it possible to exchange contracts in which buyer and seller agreed to sell perishable commodities at a set price on a future date.[16] In fact, the futures markets frequently operated without the physical goods ever exchanging hands, but refrigeration units acted as a guarantor, making it possible to believe that delivery of a commodity was at least contemplated between the traders. I am suggesting that air-conditioning at Ilford Limited, while clearly also serving a practical function, worked much like cold storage—as a guarantee, an insurance against risk even while risk is an inevitable part of the process (both in the speculative futures markets and the business of making sensitized materials).

These kinds of risk-managing or regulatory technologies substituted themselves for another kind of regulation, which is that of the atmosphere and environment beyond the factory. Their use grew as environmental regulation became more ineffectual and more compromised. For example, in France, the existing regulatory system to manage pollution was beginning to be dismantled as early as the 1770s, under pressure from capitalists who regarded it as "holding back competition and stifling innovation."[17] Over the course of the nineteenth century, according to environmental historians François Jarrige and Thomas Le Roux, while old regulatory systems were attacked as out of date and more appropriate to artisanal production, new factories were presented as well-regulated, with the division of labor and streamlining of production suggesting a moral and modern sense of control and efficiency. By the early twentieth century, new institutions and inspectorates oversaw pollution, but unlike the older regimes they prioritized economic interests, and "resulted in a greater tolerance for pollution, justifying its ubiquity in the name of economic liberalization and expanding wealth."[18]

In 1935, the Royal Photographic Society held the *Exhibition of Kinematography* at the society's headquarters at 35 Russell Square, in Bloomsbury, London. The exhibition did not show actual cinema films but photographs, including film stills, and pieces of cinematic equipment (such as cameras). Ilford's chemists were highly active in the society and Ilford Limited, along with other "leading firms," had provided much of the exhibition material. The very first exhibit—in the first section entitled "the life of a film"—showed a photographic factory's air-conditioning plant. As a contemporary reviewer described it: "The series of illustrations starts off with an air conditioning plant. Extraordinary pains are taken to avoid the presence of dust in a coating room, where the emulsion surface, still moist, positively invites the attention of floating particles. The whole ensemble of such a place recalls, in its scrupulous cleanliness and the white-shrouded workers, the operating theatre of a hospital. The air admitted here has not only been filtered, but frozen to remove excess of moisture which would otherwise have hindered the film from drying in scheduled time."[19] This rhetoric was used more widely. A description of the Brentwood Selo factory produced by the company similarly described the coating rooms as spaces in which "technical and mechanical efficiency" combined with "extreme simplicity." It described how the air passed through an air-washing system and how the workers were clothed in white "to prevent contamination of the film," and it used the same hospital analogy, together with a military comparison: "The efficiency of the battleship is combined here with the hygienic conditions of the hospital . . . ward-room and

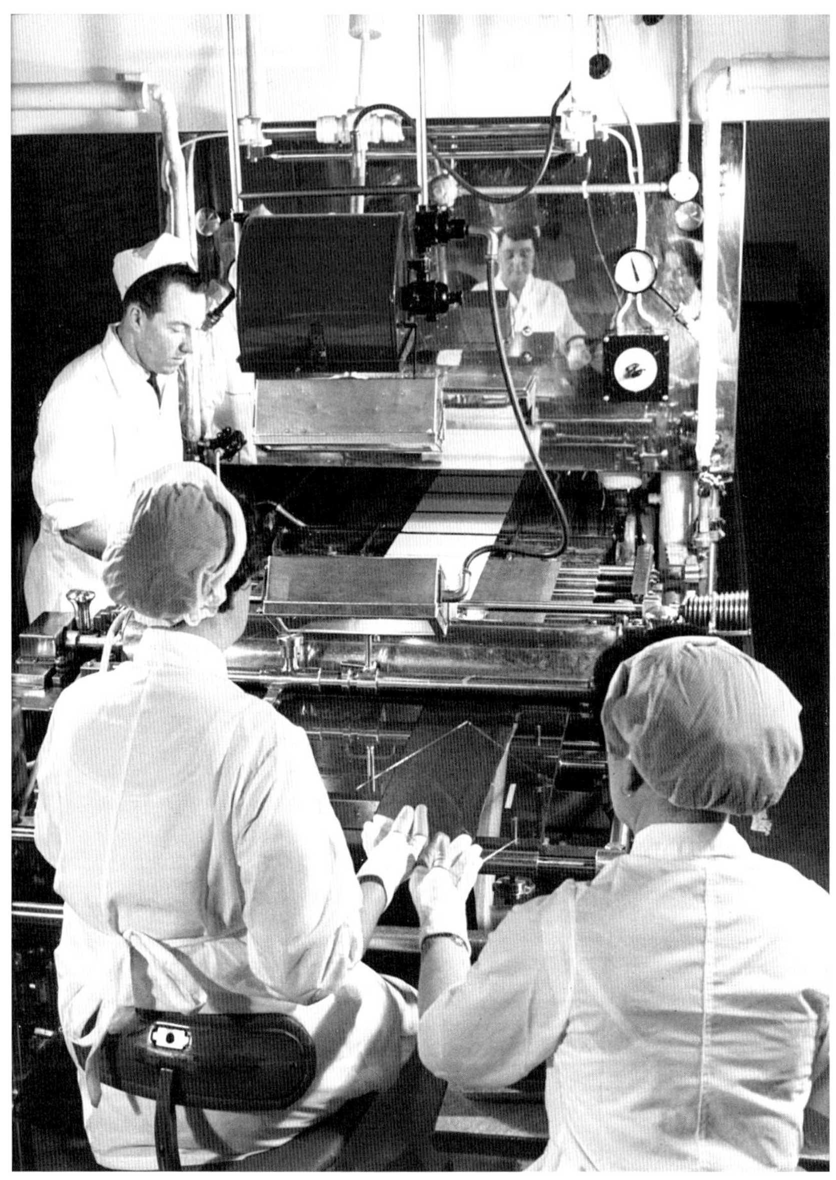

FIGURE 52. Maurice Broomfield, The Plate-Coating Section at Ilford, ca. 1960–1965 © The Estate of Maurice Broomfield. Original in color. Courtesy of Redbridge Museum and Heritage Centre 2025.

ward come together."[20] In the 1940s, Potter referred to "white-clad figures" in the coating room at the Selo works, "as clean as an operating room."[21]

If the coating room is an immaculately clean hospital, it is a maternity ward in which two things are being brought into the world: one is photography, breathing the newly purified air and protected by the velvety darkness around; the second is something monstrous, expelled, and disavowed. An exemplary capsule, the dark rooms of the photographic materials factory produce an outside that becomes ever more abject and degraded, even as they produce the sensitive material capable of documenting it.

29 YEAR WITHOUT A SUMMER

► The relationship between photography and fog—or rather the struggle between them—points to the difficulty in achieving an image in conditions of modernity. The nineteenth-century discourse of photography as the "pencil of nature" or the writing of the sun (heliography) seems to suggest that once the chemistry is in place, light simply imprints itself, as if photography is in harmony with nature. As art historian Niharika Dinkar has shown, talk about photography was rich with "solar metaphors," even while photographic practice increasingly made use of artificial light.[1] In *The Making of English Photography*, Steve Edwards suggests this emphasis on solar agency is akin to the emphasis on the agency of steam among scientists in the mid-nineteenth century—that is, it was used to displace and downplay human agency, and specifically artisanal skill, in manufacturing.[2] Others have argued that this heliocentrism has deeper roots, not least because we are ourselves light-dependent.[3] As media philosopher Joanna Zylinska writes, the conditions of existence of the photographic medium are "*also the conditions of our existence: light, energy, the sun.*"[4] But in any case, from its beginnings, photography wrestled with atmospheric and light conditions. It was always an attempt to make the most of limited sunlight, impure air, contaminated water, and difficult weather.

It was not until well into the twentieth century that artificial light was "on tap" for most of the population, including in Europe and the United States. Prior to the introduction of electric lighting, artificial light was a flickering, tenuous thing, a living flame of oil, wax, paraffin, coal, or gas—which could be heightened using minerals such as quicklime, saltpeter, and magnesium.[5] Light was atmospheric, sunlight and fire were its sources. In Western philosophy, sunlight is often identified with divine light and light with truth (Jacques Derrida wrote "the entire history of our philosophy is a photology").[6]

As Greg Lynall has argued in *Imagining Solar Energy*, Romantic poets and writers were preoccupied with the sun, drawing on classical figures and stories "associated with sun, light, heat and flame, including Prometheus."[7] Mary Shelley's *Frankenstein* was subtitled "A Modern Prometheus"; Percy Shelley published his play *Prometheus Unbound* in 1820; Lord Byron's poem "Prometheus" was published in 1816. In all of these, Prometheus is a figure with political resonance, associated with rebellion and protest, but also with life-giving sun and fire. They were also fascinated by the absence of sun, particularly in 1816, the year known as the "year without summer." Byron's poem "Darkness," written that year, describes a world in which the sun, the moon, and the stars are extinguished. The population sets about burning buildings, and then whole cities, just so they can see one another's faces. Eventually they burn all the forests "and all was black." A kind of madness consumes people, who die of war and famine, and finally the whole earth, sea, and land, is lifeless and silent. The poem opposes two kinds of light: from the life-giving sun and from the destructive, all-consuming earthly fires.

To me there is something proto-photographic about the poem, as if it anticipates the period after 1839, in which a "photomania" descends driven by people's desire to see their own and one another's faces, and in which the world suddenly seems replaceable (tear down the pyramids of Egypt, we have their pictures now!).[8] Byron's is also a vision of a world in which combustion takes over. By the early nineteenth century, combustion was associated with the furnaces of heavy industry, and with images like Philippe Jacques de Loutherbourg's fiery 1801 painting *Coalbrookdale by Night* (plate 10).[9] But it was also linked to photography, as is reflected in one of the earliest pamphlets on a process we would now term photographic, Elizabeth Fulhame's report on her experiments with "A New Art of Dying and Painting," which is titled "An Essay on Combustion." In it, she frames her experiments with a discussion of different theories regarding the causes of combustion.

Although literature and poetry are usually explained in terms of cultural rather than natural phenomena, "Darkness" is now widely understood as at least partly inspired by the dark, gloomy, and stormy weather of 1816, at the end of twenty years of war in Europe. Crop failures compounded the famine and suffering brought about by the Napoleonic Wars, and the unseasonably chilly weather came in what was already an unusually cold decade.[10] "The sun on its few appearances was a pale disk," and the rain incessant, causing widespread flooding.[11] Sunspots, large enough to be viewed through smoked glass, suggested to some that "a chunk of the sun would break off and destroy the world." In France, newspapers tried to reassure their readers that these spots did not augur the end of the world.[12]

The unusual weather of that bitterly cold and dark year is now thought to have been caused by the eruption in 1815 of the Indonesian volcano Mount Tambora in Sumbawa, and the resulting "globe-girdling veil of volcanic dust."[13] In his book *Tambora*, Gillen D'Arcy Wood describes the dramatic impact of volcano-induced cold weather across the globe: "Starving rural legions from Indonesia to Ireland swarmed out of the countryside to market towns to beg for alms or sell their children in exchange for food. Famine-friendly diseases, cholera and typhus, stalked the globe from India to Italy, while the price of bread and rice, the world's staple foods, skyrocketed with no relief in sight."[14]

Byron wrote "Darkness" at the Villa Diodati near Lake Geneva (Lac Léman), where he was staying with his physician, John Polidori, and Percy and Mary Shelley. Driven indoors by the bad weather, the group read ghost stories to entertain themselves and challenged one another to write them. Mary Shelley's *Frankenstein*, Percy Bysshe Shelley's "Mont Blanc," and John Polidori's *The Vampyre* were also either written during or inspired by their time spent at Lake Geneva in the year without a summer. The Byron-Shelley party were very receptive to their environment and the weather, and the eighteen-year-old Mary described it in diaries and letters. Around May 12 or 13, the Shelleys had traveled through Dijon and headed toward Geneva. As they passed through the mountainous Haut-Jura region, Mary Shelley wrote that "The spring, as the inhabitants informed us, was unusually late, and indeed the cold was excessive."[15] Nevertheless, they arrived in Geneva to warm sunshine, although this did not last long—by the start of June, she wrote: "An almost perpetual rain confines us principally to the house; but when the sun bursts forth it is with a splendor and a heat unknown in England. The thunderstorms that visit us are grander and more terrific than I have ever seen before."[16]

That same May, less than two hundred kilometers from Geneva at Chalon-sur-Saône near Dijon, Joseph Nicéphore Niépce was embarking upon a series of experiments in making images using sunlight.[17] It was in this year that Niépce first produced his heliographs (sun-inscriptions) on paper, a decade before he made the object that is now sometimes credited as "the world's first photograph."[18] What a year to embark on *heliographie*! Like Frankenstein's Creature, the art of light inscription arrives in darkness.

On May 28, 1816, Niépce sent his brother Claude four images on paper made with his new process. The pictures have long since faded, but the letter accompanying the photographs reveals they were negatives—since the roof of the barn is "to the left instead of the right" and "an opening between the branches of the trees" appears as a black spot. The next challenge was, as Niépce expressed it in another letter, "to fix the image in a solid way and to put the shadows and the lights in their natural order."[19] His paper and silver chloride–based exposures relied on natural light, although he explained in an earlier letter that "the sun does not have to be shining in order to do it."[20] An early commentator, Victor Foque, clarifies that this meant "that one can do it in all weather: it's enough that it is daytime."[21] Niépce made a point of adding this statement since he had forgotten to mention it previously. His comment reflects the fact that the spring weather was already proving unreliable and the sunlight inconsistent. Later, he would write to Claude, "I believe that in living memory we have not seen a spring and a beginning of summer like these."[22]

In June, Niépce tried various substances to produce a positive image rather than a negative, but some interacted badly with the humidity of the air while others needed more light than was available.[23] From the tenth of June, the dark and dismal weather descended fully and the "almost perpetual rain" made him fearful of floods and worried about damage to the family farm's crops.[24] He turned to other inventions and put his experiments with light on hold, most likely because of the impossible weather.

Other writers have noted how Niépce's photographic experiments took place following work with hydrocarbons and combustion, notably the boat engine named the Pyréolophore which he had devised with Claude circa 1807 and trialed against the currents of the river Saône. Writing about "the history of photography's uneven emergence within the chemical medium of air," Emily Doucet, Matthew Hunter, and Nicholas Robbins point out that the Pyréolophore translates as "Fire-wind-carrier," and that the engine drew on atmospheric air rather than relying on heating water (as the steam engines of James Watt and others did). The engine needed clean air and was uniquely "vulnerable to aerial pollution" even as it polluted the surrounding air itself. The Niépce brothers turned to the asphalt mines at Seyssel, near Geneva, to try to find the ideal fuel. By 1817 Nicéphore Niépce began to use this "light-sensitive and highly explosive" asphalt, called Bitumen of Judea, for his photographic experiments, pulverizing it, mixing it with lavender oil, and applying it to silver plates.[25]

In his book *The Song of the Earth*, Jonathan Bate reads "Darkness" as a response to the disastrous summer of 1816 and also as a future-oriented poem, with Byron as a "prophet of . . . ecocide."[26] I am slightly skeptical, insofar as that would make a prophecy of almost every apocalyptic tale. His reading of *Frankenstein* excites me more. Frankenstein's crime is usually regarded as a Promethean one: as Prometheus stole fire from the gods, so this "Modern Prometheus" stole the capacity to give life. But Bate argues there is a further crime. By rejecting and casting out his Creature, "Frankenstein's crime, committed in the isolation of his laboratory, has been to deny the principle of community. Like Coleridge's Mariner, he breaks the contract of mutual dependency, which binds species in a network of reciprocal relations."[27] The poor Creature is left to wander the earth, eventually wreaking revenge, like "the repressed nature which returns and threatens to destroy the society that represses it."[28]

This reading of *Frankenstein* (as with Bate's reading of "Darkness") points to its continuing relevance for present environmental concerns. But it also shows how much *Frankenstein* (and "Darkness") are products of a specifically Romantic sensitivity to the connections between humanity and nature. Byron and the Shelleys, but also Davy and Coleridge, cultivated their aesthetic and environmental sensibilities. The early German Romantics, among them Novalis and Friedrich Froebel, emphasized the relationship of human beings to crystals and stars, and insisted that inorganic matter spoke to them. The English Romantics—Byron and the Shelleys, Davy and Coleridge and Wordsworth, among others—turned themselves into weathervanes, sensitive to every tiny meteorological change. They marveled at optical illusions and atmospheric phenomena, and they experimented with drugs and recorded their hallucinations with great seriousness. They sensed the disastrous potential of industrialization and of modern science even as they were fascinated by them.

In *Burning with Desire*, Geoffrey Batchen situates the emergence of "proto-photography" in the context of Romanticism and German idealism, in concepts of nature, time, subjectivity, and the picturesque that are specific to the culture of the late eighteenth and early nineteenth centuries.[29] Batchen's analysis is influential and convincing, although some have read it as underplaying the agency of "free-willed individuals" and the social conditions of Romantic science.[30] But why stop there? What about the agency of volcanoes and tempests and cholera bacteria? What brings the Creature to life? Frankenstein's toils, the "sudden light" of his inspiration, or the unnamed instruments that "infuse a spark of being into the lifeless thing" (perhaps with an electrical jolt or, as interwar versions of the Frankenstein story have it, invisible rays)? What brings photography to life? Niépce's intentions and the support of the Napoleonic state, a cultural desire to fix a fleeting image, or asphalt mines and sunshine?

Writing about the eruption of Mount Tambora as a possible explanation for the visual style of J. M. W. Turner's paintings, Michel Serres revises his earlier reading of the paintings that had situated them in the scientific practices and understandings of the time, asking what made the ashen skies of Turner's London: volcanic dust or "fire-driven machines and the factory proletariat"? The bigger question he raises is about the intersection of the natural and the social, and the disciplines of geophysics and history: "How, then, are we to evaluate the supposedly brilliant strategies of Wellington and Napoleon? What wins or loses battles and overturns or builds empires? Infection? Meteorology?"[31]

The answer is all of these: human intentionality must reckon with the agency of bacteria and storms and volcanoes. All are lively, all exposing and exposed. The sun and the rain and the dirty air have all helped to shape our concepts and our images.

30 THE GALL

► A tiny wasp lays her eggs in the leaf bud of an oak tree somewhere in Syria, or possibly in Greece, Bosnia, or Turkey. As she does so, she injects the tree with something that causes it to develop a protective shell, a sphere known as a gall, around the egg. As it grows on the tree, the gall, which attains the size of a Ping-Pong ball, resembles a crab apple. Inside this gallnut the wasp lava develops and changes, until it is ready to bore a hole out and fly away. This is the first of a series of metamorphoses involving the gall, and it is the result of a symbiotic relationship between tree and wasp. This relationship exists across many countries and between many species of wasps or flies and trees, but it is on the *Quercus infectoria Olivier* that the biggest galls, and those thought to be the best, are found. In these dwell the larvae of a specific insect—*Cynips gallae-tinctoria Olivier.*

Humans have long prized galls for their medicinal uses, especially in the treatment of hemorrhoids and bowel complaints. People have used them in fortune-telling and to fatten pigs, and to detect iron and other metals in substances such as mineral water and blood. Most importantly, combined ("chelated") with iron oxide, the oak gall can produce rich black dyes and inks which were valued for their fastness and the depth of the black. The insect's name contains reference to this use, *tinctoria*, as does one of the common names for the oak—the "dyer's oak." In Asia, the dyes were used to produce black silks, and in Europe the ink was

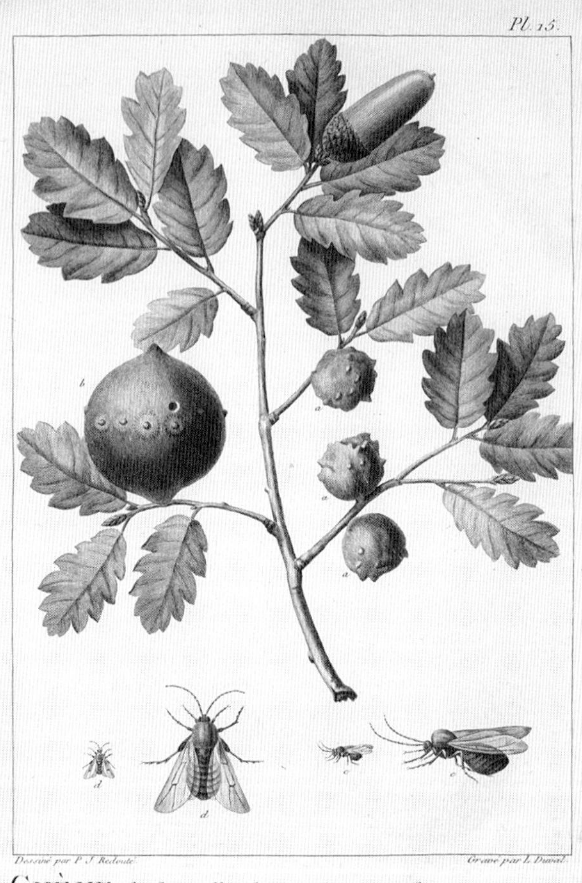

FIGURE 53. Gall Oak *Quercus infectoria* with gall wasp. From Guillaume Antoine Olivier, *Voyage dans l'Empire Othoman, l'Egypte et la Perse* (Paris: H. Agasse, 1801). Dumbarton Oaks Research Library and Collection, Trustees for Harvard University, Washington, DC.

used for legal documents because it bound to the paper and was impossible to erase, although it also risked corroding the paper because of its acidity. Another major use was in leather tanning, a process which requires the tannins that are found in many plants, but in exceptionally large quantities in galls. For centuries, then, galls were a prized commodity, and a common cargo of the ships moving between the Levant, India, and Europe. The most valued gallnuts still contained the wasp larva within them—"they should be heavy, and of a fresh green color. If the insect has escaped, they are yellow, and are not of nearly so good quality."[1] They were sold under the names of the ports from which they were shipped: Aleppo, Morea, Smyrna.

Galls could also be fermented to produce a substance called gallic acid. The result of decomposing and oxidizing the tannic acid in the galls, it was sometimes accidentally generated in leather tanneries. There, fermentation was undesirable as it caused "ropiness" and prevented tanning.[2] The Swedish scientist Carl Wilhelm Scheele first studied gallic acid in 1770 and published a paper on the subject in 1786. Scheele also heated the "white silky needles" of gallic acid, transforming them into the more active (and less acidic) pyrogallic acid or pyrogallol.[3] In these forms, as gallic acid and pyrogallol, the gallnut becomes part of the story of photography.

It goes like this: in March 1839, William Henry Fox Talbot purchased a supply of gallic acid from a chemist shop in Oxford Street, London.[4] It had been recommended to him by John Herschel and J. B. Reade, and initially he tried to add it to his photographic emulsion as a means to quicken the appearance of the image. It has been suggested that the idea may have come from observations of the effects of silver chloride solution on leather, which could have contained traces of gallic acid from the tannery; or that Talbot may have been inspired by the fact that oak gall ink appears pale in the bottle and quickly darkens with use, as a photograph darkens after exposure to light.[5] In any case, Talbot and others recognized that gallic acid worked as a reducer, able to liberate silver from silver salts. He soon employed it as a bath for the exposed sensitized paper, enabling the negative to darken more quickly, and in 1841 he patented the calotype technique using gallic acid as a developer.[6] Gallic acid was thus the first photographic developer, the result of symbiosis between a wasp and an oak tree.

This transformation of galls into gallic acid and then pyrogallic acid (a much faster and more efficient developer) has immense significance for photography. Before the use of developer, Talbot's "photogenic drawing" or salt print process involved paper saturated with a weak solution of sodium chloride salts, which was then treated with silver nitrate to produce light-sensitive silver chloride. After the exposure, the image would immediately start to appear on the paper, and a wash in a stronger silver chloride solution would enhance its stability, followed by a wash in water to stop and fix the image. This process, later called "print-out," needs enormously long exposures to allow the silver salts to gradually decompose to silver. The introduction of gallic acid changed this process, allowing a much shorter exposure, and introducing a delay between the moment of exposure and the moment of appearance of the image. Eventually other substances come to serve the same function, but gallic acid introduced something that continues to remain central to photography: the latent image.[7]

Considering latency in photography opens different ways of thinking about its temporality. Since the 1850s, writers have described photographs as "instantaneous" and as "frozen" moments. Yet the existence of a hiatus between exposure and development, which could be very brief or last years or even decades, ought to affect our perception of photographic immediacy. One of the most striking things about photographs, one of the things that photography theorists return to repeatedly, is what they do to our sense of time and of the past, distorting or replacing our own memories and seeming to preserve an instant for eternity. But after the introduction of developer, images could also be latent, unseen, ready to appear at an unspecified moment in the future—this quality, though it seems to me to be equally curious, has been much less discussed.

That the emergence of the latent image depended on a relationship between an insect and a tree, on gall pickers and on shipping routes from the Middle East, draws attention to the complexity of photographic agency, and the relationship of chemical photography to a much longer history of materials and their properties, linking photography to ink manufacture, to tanneries, and (even prior to the invention of coal tar dyes) to the dyestuff industries. Nor was the wasp the only nonhuman animal involved in the production of gallic acid: camels carried the heavy sacks of oak galls from the interior of the Levant to Aleppo and Beirut. Each sack could weigh around two hundred pounds (or around ninety kilos), and a side effect of this labor was that the animals developed their own "galls," sores from the chafing of saddles and harnesses.[8] Etymologically, the two kinds of gall are linked through the bitterness of the tannin-rich gall. When something is "galling" it is hard to swallow or to bear, when someone has "gall" they have the nerve to do or say something outrageous, cheeky, or unwarranted. It relates to chafing—saddles chafe and horses "chafe against the bit," though the latter has connotations of resistance and dogged refusal.

The bitterness of substances such as tannins (contained in large quantities in galls) is experienced by humans as a feeling through the sense of taste, and this taste connects to the emotion of bitterness (consider the feeling of revulsion, the wince, even the sense of betrayal that a very bitter substance gives). Bitterness often acts as a signal of toxicity in plants; as artist and researcher Hannah Drayson suggests, plants "speak" to us through our tongues, our tastebuds.[9] Bitterness defends animals, too: the black-and-yellow-striped caterpillar of the cinnabar moth that feeds on ragwort, or the monarch butterfly lava that lives off milkweed, become as bitter as their foodstuffs and highly toxic to predators.[10] The gall's high level of tannins also serves to deter predators.

All the meanings of the word "gall" circle around the concept of bitterness—galls are bitter because of their high tannin content; a gallbladder stores bitter bile (associated with ill-temper in the ancient theory of the humors); and things that are galling produce feelings of bitterness and resentment. From bitter experience we emerge scarred. We have ourselves soured: bitterness manifests in outward expressions of sarcasm, bleak humor, carping, cynicism, distrust. Bitterness is like the ugly feelings discussed by cultural theorist Sianne Ngai: "explicitly amoral and noncathartic, offering no satisfactions of virtue, however oblique, nor any therapeutic or purifying release."[11]

Tastes, physical sensations, and emotions are interwoven in this language, and so too are animals, plants, and humans. Drayson suggests that "plant agency and plant 'talk' might manifest in human cultural and affective tropes," and I would take this further, suggesting that our concepts and ways of seeing depend on wasps and oak trees, camels and humans—gall pickers, camel drivers, ships stokers among them—and the economic relations that link them.[12] As Bruno Latour, Jane Bennett, Donna Haraway, and others have argued, humans are caught up in networks of reciprocal relations with nonhuman and more-than-human entities. The parasitic symbiosis of the wasp and the oak does not just enable the production of developer (along with many other dyes and inks) but also raises the tantalizing possibility that our ideas of time and of latency depend on these plant and animal interactions.

31 THE LATENT IMAGE

► In the nineteenth century, no one could describe the exact process by which a brief, invisible exposure could be brought to visibility using developer; they could only observe that it happened. It was in the 1930s that the theoretical side of photographic research arrived at the point where this could be thoroughly understood.[1] The full explanation is complex, but put simply: after a relatively short exposure, the film, plate, or paper appears unchanged, but at a microscopic level there is now a speck of silver on each minuscule crystal. These then react with the developer, entirely blackening each crystal that contains a speck, while leaving the other crystals unaffected. In this way, the image appears. Even the physicists who have described this process at the level of photons and electrons and quantum mechanics struggle to explain the nature of the undeveloped image. For example, L. M. Slifkin, writing in the 1970s, resorted to an analogy between the speck of silver and a living creature with the capacity to remember: "It is this speck which retains the memory of the exposure to light; because at this stage it is so small as to be invisible, we call it the 'latent image.'"[2] Slifkin is not alone. The phenomenon of the latent photographic image proves remarkably difficult to describe. It is practically impossible to talk about those tiny specks of silver without using metaphor.[3]

In early photographic handbooks and dictionaries,

the latent image is often described as an "invisible image."[4] By 1945, the edition of *The Ilford Manual of Photography* authored by James Mitchell navigated the problem by distinguishing between a latent image and a "real, visible image," though this leaves open the question of what the latent image is, if not real.[5] This idea of an image that is invisible is peculiar—surely the definition of an image is to do with its visibility—or, in the case of a metaphor or thought-image, with our ability to visualize it in the "mind's eye"? The phrase "invisible image" rarely appears in discussions of subjects other than photography. As far as I can tell, before 1839, it would have only been used occasionally in relation to religion (the invisibility of the deity). The paradox of the phrase itself is indicative of how challenging it was to describe something that is present but not yet an image, something that is suspended in time, caught in the pause between exposure and development. In fact, if we could visualize what Slifkin describes (which these early writers could not, as they did not yet know about these microscopic specks of silver), each black dot would be floating free from the next, a galaxy of black stars against white in constellations we can barely map, disassociated and atomized, dispersed rather than cohering into an image.

To clarify the distinctiveness of the latent image, consider what its introduction meant for photography. The introduction of a development stage changed exposure times and the need for strong light: much less light is needed to produce a negative or print that will then be developed, such as a calotype, than in a "print-out" process like a salt print or a cyanotype which slowly appears directly after exposure and only has to be fixed and washed. "Certainly a much better picture can now be obtained in a minute than by the former process in an hour," wrote Talbot.[6] Obviously there are many other factors affecting the amount of light needed to make a photograph, from the quality of lenses, to the size of the camera or enlarger's aperture, to the sensitivity of the emulsion—but the introduction of developer and of latency was a step in this general direction. Latency also means that an exposed, sensitized surface can be kept in the dark for an indefinite period before the image is realized via development, depending on the stability of the emulsion and its substrate and the environmental conditions. There is also always the possibility that the film (or plate or paper) is never developed at all. I have three undeveloped films on my shelf as I write, and at least two more within cameras. Sometimes I have opened a camera, forgetting a film was inside, and immediately ruined any exposures that were on there. Latency describes a gap in time between the exposure and the appearance of the image, so that the image can be suspended, unrealized, for hours, days, weeks, or even years.

Furthermore, the process of development might produce very different images, depending, for example, on procedures followed and decisions made by the photographer doing the processing, or on the settings of the processing machine, the precise mix of the chemicals used, or the atmospheric and environmental conditions in which the film is developed and printed. Handbooks on darkroom photography suggest there is an ideal negative, developed correctly, at the perfect temperature and in the appropriate solutions, which is of the correct density, neither too contrasty nor too soft, which is not fogged, which has grain of a desired size. Yet the fact that they devote so many pages to development shows how variable the process can be. For example, in the revised 1956 edition of Sowerby's *Dictionary of Photography*, the development of negatives is described across twenty-eight pages, which cover not only different processes and developers but also the use of "development by inspection," in which the photographic worker (as they are described) does not rely on set timings but removes the negative from the developer when they judge the contrast to be correct (by eye, under a red light).[7] This had been a common procedure until the advent of panchromatic film, which the red safety light would spoil. In the case of "dry" glass plate photography or individual sheets of film (as opposed to roll films), it was possible for guides such as the *Imperial Handbook* of 1914 to give very specific guidance according to the subject matter of each photograph, recommending adjustments in the dilution of the developer as well as in the time the plate spent in the developing bath.

In the absence of reliable light-meters for taking photographs, these kinds of adjustments were essential. With roll film, however, it was more difficult since different subjects on different days and times and different lighting situations could be caught on the same roll. Before the introduction of standardized measures of emulsion "speed," by the firm Hurter and Driffield, the only measure was against the old wet plates. As the *Imperial Handbook* claims: "In the old days, say till 1890, the amateur knew his plates by a certain 'number of times.' A plate called, for instance, '30 times' needed twice the exposure of one '60 times' and so on. The 'times' meant the number of times the plate was quicker than a wet plate."[8]

By 1956, the grading of emulsions and the metering of light were both much more refined, and roll film and panchromatic emulsions were the norm. So as Sowerby's *Dictionary of Photography* explains, the most common practice was the use of timed developments, in which workers used tables and data published by the photographic manufacturers to figure out the best time and temperature for a given film. Developing negatives, as distinct from producing positives, no longer involved any degree of adjustment according to the specific subject matter and composition of each exposure. The lightproof development tank would be loaded in total darkness; positive prints allowed for more tailored adjustments under the red light. Even this more standardized and "blind" process relied on the person's judgment about the levels of contrast and grain desired. Development had become a standardized practice, and it barely seemed necessary to ask where or what the image had been before its sudden appearance on the negative.

The variability of the development process is a reminder that every photograph is a rendition of an event, one version of many potential versions. This idea is usually discussed in relation to Talbot's introduction of a negative-positive process, since many versions may be made from a single negative. Yet even the negative itself is only one version of many possible versions, which only becomes manifest in development—at which point these other potentialities cease to exist.

The arrival of digital or electronic photography changed the pause between the moment of exposure and the moment of viewing, most obviously in the case of live-streamed video but also when an image is viewed live on-screen prior to the shutter being pressed and the image being captured. Even so, the digital has its own kind of latency as images exist only as data until rendered visible on a screen. The impression of immediacy depends on the possibilities allowed by latency. The digital image's ability to be viewed so quickly after an event and then to be viewed again at another time, when the moment is long past, is made possible by a play of appearances and disappearances. Mobile phone photography enables one to send an image immediately to other computers or phones in other places and thus to one or many people, almost anywhere in the world. Though this capacity is distinctive, it also extends something already there in chemical photography after the introduction of gallic acid, which is that an image is seen, "captured" and then stored, invisibly, only to reappear somewhere else.

Another type of latency is described by the artist Hito Steyerl, who points to how mobile phone cameras are optically so limited that they depend on algorithms to distinguish image from noise, creating the image by scanning past and existing photos and constructing a picture from that data. This means that "you don't really photograph the present, as the past is woven into it."[9] As the art historian Kate Palmer Albers explains, this produces a "new form of latency" in which the "interlude between existing and not existing" is filled with automated association, assembling and reassembling data based on vast archives of other images.[10]

Artist and writer Yanai Toister suggests that the latency of electronic (digital) images is also to do with "the fact that without the occurrence of a prescribed event they exist solely as a data file, or worse, as voltage differences and nothing more."[11] The "prescribed event" is part of a program conducted automatically, the result of human interactions with computers and cameras, and of decisions embedded in software—save, delete, open. He concludes that in both chemical and electronic photography, "the tendency to think of photography as being 'a realm of realities' is misguided"; instead photography is "a realm of possibilities and potentialities wherein images are nothing more than a looming latency awaiting realization." So even today, latency is part of digital photography.[12] Yet I find myself agreeing with the artist Moyra Davey, who sees latency as something diminished by the digital image. For her, latency gives rise to a certain emotional and aesthetic experience—the wait, the anticipation, the not knowing, the accidents, and the finding out characteristic of chemical photography.[13]

► It is clear that latency does not depend on oak galls but was initiated using them, at least in paper photographic processes. In nineteenth- and twentieth-century technical accounts of photography, gallic acid became the emblem for latency because it was generally understood that Talbot's calotype technique, patented in 1841, had introduced the latent image.[1] Actually, latency in photography predated Talbot's calotype and was present in Daguerre's process, since in this, too, the image did not appear immediately but was conjured into being through the application of mercury vapor to the metal plate. As the photography historian R. Derek Wood points out, this is not seen as having such "historical status" as the introduction of liquid developers such as pyrogallol, since the daguerreotype eventually died without offspring.[2] Even though photographic latency arrived first as a form of vaporous magic with the daguerreotype, the humble oak gall had, by the 1850s, already become entirely bound up with this concept of the latent image. Wood wrote extensively on the use of gallic acid in early photography and was interested in the question of whether Talbot was aware of the significance of this change to the process. But I myself am struck by how a tree and a wasp produce the gall, from which is derived the latent image, which in turn shapes human thinking about the visibility and invisibility of images, and about what is present and not present.

That the latent image was a new concept and difficult to explain is evident from a French newspaper article published in *Le Constitutionnel* two days after François Arago's lecture at the Académie des Sciences in Paris, giving the details of Daguerre's process. Published on August 21, 1839, the article drew on an interview with Daguerre as well as Arago's account of the full technical details of the process. The author of the article, identified by Wood as Isidore Bricheleau, tried to explain it in layman's terms: he described the image as "latent" and in the same "indistinct" state as a chrysalis (a butterfly pupa), until the mercury vapor takes it out of the cocoon and brings it to light.[3] This insect metaphor seemed promising, especially given that the final stage of the butterfly metamorphosis is known as the *imago*, but it didn't satisfy Bricheleau. He wanted to emphasize that the image already existed on the daguerreotype, was already "traced" forever on the iodide-coated plate, and only required a catalyst to make it visible. So he turned to another analogy: letters written in invisible ink (made from onions) which become visible when exposed to heat. He even drew a comparison between the newly visible ink and the permanent oak gall ink, though the brown oak gall ink darkens to black immediately with use.

Bricheleau says that heat reveals the onion-writing "very legibly, the characters becoming as visible as if ink was formed there from nutgalls."[4] Wood is struck by this appearance of the gall in a discussion of the latent image that predates the patenting of Talbot's calotype technique. But for me, the interest of this article lies elsewhere—in the difficulty Bricheleau had in describing the concept of the latent image. Reaching first for one analogy and then another, Bricheleau found the latent image hard to explain in writing. Other writers also scrabbled for novel analogies. For example, in his book *The Silver Sunbeam*, published in 1864, John Towler wrote: "It is like the impression of the finger on a plate of copper, or of a warm piece of metal on a glass mirror; after the removal of the finger, or of the metal, the eye can not distinguish the spot where the impression was made; but . . . breathing upon the glass will make the impression manifest, will show that the image was there in a latent or invisible condition."[5]

This struggle to define the latent image suggests that before photography there was no clear precedent. Yet latency already existed as a concept: it meant something hidden, or dormant, something not immediately available to human senses. According to the etymological dictionaries, the English term "latent" comes from the Latin term *latentem*, meaning hidden, secret, concealed, or unknown, and also from the Greek *lēthē*, meaning forgetfulness or oblivion. Sometimes "latent" meant the opposite of "sensible," in the old meaning of something perceptible to the human senses. The Scottish moral philosopher Thomas Reid, in his book *Essays on the Intellectual Powers of the Human Mind* (published circa 1785), distinguished between "the sensible qualities of objects and their more latent qualities."[6] These latent qualities may include causes of which we can only perceive the effects—he gives the examples of gravity, magnetism, and electricity. The powers ascribed to objects by people are not powers of those objects in themselves, but signs of less perceptible forces: "Thus we say that a ship sails, when every man of common sense knows that she has no inherent power of motion, and is only driven by wind and tide."[7] Reid described latent qualities as those "not immediately discovered by our senses; but discovered sometimes by accident, sometimes by experiment or observation."[8] We can learn the latent qualities of objects, say opium or wine (two of his examples), but we have to be able to identify them by their sensible properties if we are not to get accidentally intoxicated. The key task, according to Reid, was to be able to combine the information given by our senses with the scientific knowledge of the latent qualities of objects; to know, for example, what perceptible qualities of a plant (of which we otherwise have no experience) might indicate it is poisonous.

In thermodynamics, the term was introduced by Joseph Black (1728–1799) in relation to heat: the concept of latent heat assumes that heat resides already in a substance that does not feel hot to the touch but can heat up due to certain actions—by hammering, for example, or combining with other chemicals. This theory was overturned in the nineteenth century, but that it still had influence is suggested by the fact it was addressed, if only to be dismissed, in handbooks and dictionaries of photography such as Sutton and Dawson's 1867 *Dictionary of Photography*. In biology and medicine, the latent can be distinguished from the dormant, on the grounds that the latter suggests something currently not active, something sleeping, while latency suggested something which may be active but not yet manifest. So, for example, in medicine, since at least the seventeenth century, "latent" was used to describe an infection or illness that develops without immediately producing visible symptoms.

But the terms are not always that easy to distinguish. For example, Charles Darwin used concepts of dormancy and latency to describe the persistence of instinct and habit, and to explain things like reversion in evolution. According to science historian Margaret Campbell, for Darwin, hereditary units which he called "gemmules" could exist in a state of dormancy. Gemmules do not change, but their state changes if they are awakened.[9] Latency, on the other hand, is applied by Darwin to "characters," or characteristics, rather than gemmules: he wrote that "certain characters, capacities and instincts may be latent in an individual, and even in a succession of individuals."[10] Latency describes how characteristics and abilities can reappear across generations, but what makes this possible are the dormant gemmules. Campbell writes, "There is between the two concepts of dormancy and latency a clear correspondence. If the gemmules are dormant, then the character they represent is latent."[11] The close relationship between these terms is echoed elsewhere in the biological sciences.

Questioning the emphasis on agency in recent materialist philosophy, the philosopher Graham Harman draws attention to dormancy and latency. He writes: "Whether we praise objects for their agency or brashly deny that they have any, we overlook the question of what objects are when not acting. To treat objects solely as actors forgets that a thing acts because it exists rather than existing because it acts. Objects are sleeping giants holding their forces in reserve, and do not unleash their energies all at once."[12] Ideas of receptiveness, reciprocity, and sensitivity are central to my account of photographic materials. Photographic chemicals can be volatile, unstable, and reactive, but their significance is as much in their receptivity and their latent potential as their activity. How do we distinguish between something that is present and active but hidden, and something that is present but inactive, if both are imperceptible? Does the appearance of symptoms mean the latent has become manifest, or the dormant awoken, or do the symptoms merely point to the presence of something latent, just as a falling object points to the presence of gravity? Is the dormant merely sleeping or is it developing, or "incubating"? To return to Bricheleau's insect analogy—how can we know what is going on inside the cocoon?

33 MEMORY TRACES

► In the nineteenth century, the concept of latency was changing thanks to the new phenomenon of photographic latency. Not only did the latent photographic image require metaphors to explain it, but it also provided a new metaphor through which to imagine and describe other, nonphotographic processes. One well-known example is memory. L. M. Slifkin's decision to describe the silver specks as retaining a memory (see chapter 31) seems particularly appropriate, because it echoes the way that photographs are often more broadly understood as prosthetic memories or memory substitutes, and how, not long after the introduction of photography, people began to think of memory as itself photographic. Alongside other recording technologies, photography offered rich pickings for theories of memory, so much so that, according to the psychologist Douwe Draaisma, in nineteenth-century theories of visual memory, "one can trace exactly the succession of new optical processes: in 1839 the daguerreotype and the talbotype, shortly afterward stereoscopy, then ambrotypes and color photography, in 1878 'compound photography' and finally cinematography."[1]

Photography allowed writers to describe memories as traces on a surface comparable to the traces of light falling on a photographic plate, and it made possible the popular idea of a perfect "photographic" memory that can faithfully reproduce a scene from the past, a concept that draws on the faith in photographs as true records of events.[2] As Draaisma notes, many photographic metaphors emphasize a notion of memory as permanent storage and perfect image. The latency of the photographic image also shaped theories of memory, through the idea of memories as preserved sensations or deeply buried and unconscious, brought to the surface through certain events or developed in therapy.

Draaisma mentions several doctors who compared the sensory impressions underpinning memory to the invisible impressions on an exposed photographic plate that could reappear later.[3] The physiologist John William Draper, who was also one of the early pioneers of daguerreotype processes in the United States, made such a comparison in an unusually precise way. For Draper, the latent aspect of photography was key to explaining how memory worked: he argued that sensory stimuli leave permanent traces on the ganglions of the human nervous system, just as shadows cast on a photographic plate are permanently recorded. The time that passes between the moment of the stimulus being registered and its recollection is analogous to the latent image, "images ready to make their appearance as soon as the proper developers are resorted to."[4] Noting that images can be preserved on the retina of the eye, as when you look into a bright light then shut your eyes, Draper writes: "In this there is a correspondence to the duration, the emergence, the extinction, of impressions on photographic preparations. Thus I have seen landscapes and architectural views taken in Mexico, developed, as artists say, months subsequently in New York—the images coming out, after the long voyage, in all their proper forms and all their proper contrast of light and shade. The photograph had forgotten nothing. It had equally preserved the contour of the everlasting mountains and the passing smoke of a bandit-fire."[5] The photograph renders permanent the most fugitive impression and, in the process, equalizes the small and the great, the transient and intangible smoke, the solid and immovable mountains. Note that Draper is describing not just a phenomenon of time but also of space, because like memory, photographs transport us elsewhere, and take the photographer back to somewhere they have been before. The photographs are taken in Mexico and developed in New York. Smoke and mountains appear after months, but also after a "long voyage."

We can see a similar emphasis on geographic distance in other discussions of latency not necessarily shaped by photography. In medicine, the term "latent" sometimes described the existence of symptomless variations of a disease, and sometimes the appearance of symptoms far from the place where the disease's cause was thought to be located. Before the mosquito was known to spread malaria, for example, the disease was regarded as place-based, tied to specific malarial locations that were swampy, hot, and had poor sanitation. But it was also thought to be something that might not manifest immediately at the point of exposure but remain latent in bodies for many years, even as they moved away from the original malarial space. Europeans exposed to malaria in places like West Africa might start to manifest symptoms on their return to Europe. This supported narratives about the invisible dangers of tropical climates, and anxieties regarding race and the circulation of peoples in empire. This notion of the latency of malaria, together with the wide range of symptoms associated with it, added to the mystery of a disease believed to be able to drift invisibly, moving "like mist."[6]

Memories are often understood as resembling the latent images of photography in the way that they bridge place as well as time. In Marcel Proust's *Remembrance of Things Past*, the taste of a madeleine cake takes him back to another time and place, to "Sunday mornings at Combray." Involuntary memories appear out of the blue and transport us. Or rather, they are dragged up from the depths: the narrator's recollection of Combray does not come as a memory at first but merely a sensation, an "exquisite pleasure" that is "isolated, detached, with no suggestion of its origin."[7] He has to work hard, "must lean down over the abyss" to pull to the surface "this memory, this old, dead moment which the magnetism of an identical moment has travelled so far to importune, to disturb, to raise up out of the very depths of my being."[8] It's the resemblance of the present moment to the past one, the combination of taste and atmosphere they share, which makes it possible to draw it out. Though Roland Barthes invoked Proust several times in *Camera Lucida*, he also claimed there is "nothing Proustian in a photograph" since it does not "call up the past." Yet, as plenty of commentators have noted, there is much that is photographic in Proust.[9] He explicitly compares the impressions of a "beloved object" (Albertine) to the impressions received by a film, and the act of going back over those impressions at home to developing a film in an "inner darkroom," which can only be found in privacy and solitude.

For Sigmund Freud, too, the latent is a phenomenon of place, or the movement from one place to another, and, like Proust, he uses photographic metaphors—though as philosopher Sarah Kofman has noted, "whenever Freud puts a metaphor in play, he always acts with great caution: he multiplies the images, declares them crude, provisional."[10] In *Moses and Monotheism*, published in 1939, he describes psychic trauma as something that happens once one leaves the site of the event that provoked it (his example is a train collision from which a person escapes apparently unharmed). The time and distance between the event and the appearance of trauma symptoms is the period of latency. For Freud, latency is a central concept, linked to the repression or sublimation of socially unacceptable desires and drives or of traumatic experiences. Unsurprisingly, Freud's theory of latency draws on the medical concept of latency in disease: the appearance of "traumatic neurosis" arrives after a length of time that he compares to the incubation period of an infectious disease.[11] But he also explicitly makes analogies with photographic latency. For example, in *Moses and Monotheism* he writes:

> It has long since become common knowledge that the experience of the first five years of childhood exert a decisive influence on our life, one which later events oppose in vain. . . . The strongest obsessive influence derives from those experiences which the child undergoes at a time when we have reason to believe his psychical apparatus to be incompletely fitted for accepting them. The fact itself cannot be doubted, but it seems so strange that we might try to make it easier to understand by a simile; the process may be compared to a photograph, which can be developed and made into a picture after a short or long interval. . . . What a child has experienced and not understood by the time he has reached the age of two he may never again remember, except in his dreams. Only through psychoanalytic treatment will he become aware of those events.[12]

In this analogy, the latent image is visible only in the dream world, and psychoanalysis takes the role of the developer—it is the pyrogallol that brings experience to light.

Kofman critiques Freud's use of the photographic metaphor to explain trauma because she argues it reinstates "mythical and metaphysical oppositions: unconscious/conscious, dark/light, negative/positive."[13] Though I don't share her reading of the way Freud uses the metaphor, which she understands in terms of processing a negative into a positive rather than in terms of bringing a latent image into visibility, the metaphor does, as she says, imply a movement from darkness to light, and I would add from depth to surface.[14] Latency invites this: it's no coincidence that Proust's inner darkroom is also an abyss, that (certainly in the Western cultural imagination) darkness and depth go together. This is where the latent exists, lurking beneath the visible, belonging to the underground. As the literary theorist Hans Ulrich Gumbrecht remarks, latent states "involve downward movement, as in the case of something falling by the wayside and lying unnoticed until its presence is felt."[15]

34 BAD AIR

► In 1968 the Czech photographer Josef Koudelka photographed people on the streets of Prague, bravely confronting the Soviet tanks that had invaded the country to put a stop to the liberalization known as the Prague Spring. Until the invasion, Koudelka had mainly photographed performances—of Gypsy musicians and in the theater—and had not worked as a photojournalist. He has spoken about wanting to make pictures so that the violence of the invasion could not be denied by Russians keen to present it as a liberation; in this sense his photographs are intended as forms of witnessing, they testify this is happening, this happened.[1]

Despite this, Koudelka's scenes of the struggle between unarmed citizens and Soviet soldiers with guns and tanks are strangely theatrical, full of dramatic gestures and strong facial expressions. Documentary photography is frequently dramatic; its stock-in-trade is heightened emotions.[2] Yet the theatrical quality of Koudelka's images—the open mouths, the hands touching foreheads or raised in supplication—seems to take the pictures out of time, as if the scenes depicted had already been anticipated in the performative gestures of Greek tragedy.[3] Koudelka himself talked about photographing theater as if it were life, getting in among the actors, and how, through this, "I learned to see the world as theater."[4]

Koudelka's photographs of stage performances also have a strange temporality according to theater

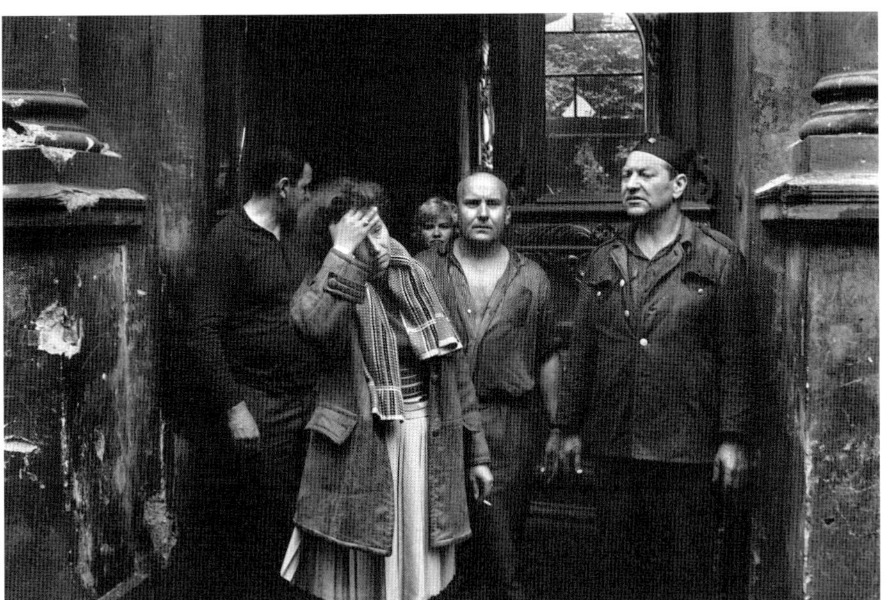

FIGURE 54. Josef Koudelka, Czechoslovakia. Prague. August 1968. Invasion by Warsaw Pact troops. © Josef Koudelka / Magnum Photos.

and photography scholar Joel Anderson. He cites the theater director Otomar Krejča, who felt a peculiar "reversal" on seeing Koudelka's images, as if the performers were acting out Koudelka's images rather than the images representing the performance. Anderson describes this effect as almost supernatural, something like a haunting, and attributes it partly to the peculiar magical quality of images that the philosopher Vilém Flusser describes.[5] Against seeing photographs (in particular) as "frozen events," Flusser argues that images translate events into states and scenes, so that there are no causes and effects, but simply elements that exist simultaneously.[6]

Famously, both Walter Benjamin and Roland Barthes were also alert to the peculiar temporality that photographs can possess. In "Little History of Photography" (1931), Benjamin contrasted early photographic portraits to painted ones, saying that the photographs have a "magical value" because they maintain the particularity of the person depicted, inviting a viewer to search for the "tiny spark of contingency . . . the inconspicuous spot where in the immediacy of that long-forgotten moment the future nests so eloquently."[7] In *Camera Lucida* Barthes picks up on Benjamin's notion, translating the "spark of contingency" into the "punctum" and into his concept of the future anterior tense of photography. Observing his own response to the handsome young Lewis Payne (also known as Lewis Thornton Powell), Barthes wrote: "I read at the same time: *This will be* and *this has been*; I observe with horror an anterior future of which death is the stake. By giving me the absolute past of the pose, the photograph tells me death in the future."[8] Yet what Anderson points to in Koudelka's photographs is captured by neither of these. These photographs suggest pasts beyond the past of the photograph. The photograph bears witness to an event, but it is also a bringing-to-the-surface of something latent—not just this image that came into being in the developing bath, but other ancient images that haunt this one, that we sense are present but can't quite see.

The photograph of the Warsaw Pact invasion that most epitomizes Koudelka's desire for proof seems to be all about time. It shows an arm wearing a wristwatch and beyond it an empty street with tramlines (Prague's iconic Wenceslas square, with the National Museum in the distance). The watch, worn on someone else's arm, acts as a time stamp on the photograph. This is the photograph as evidence, yet it is still an unreliable witness: does the watch say twenty past twelve or just past six? The exact timing of the photograph remains in doubt, despite the emphatic inclusion of the watch. Koudelka has said that the photograph records the moment, on August 22, when a protest against the invasion was to be held and the population—knowing or guessing it was a trap set up by the invading forces—abandoned it "so as not to give the Soviet occupiers a pretext for a massacre."[9] Asserting the absolute specificity of the photographed moment, the photograph suggests an indefinite waiting, or a non-event, or more precisely a manifestation (demonstration, public display) that is latent, constituted by a very present absence.

In an early example of the long-standing tradition of writing about photography in terms of death and haunting, John William Draper described the latent image as ghost: "A specter is concealed on a silver or glassy surface until, by our necromancy, we make it come

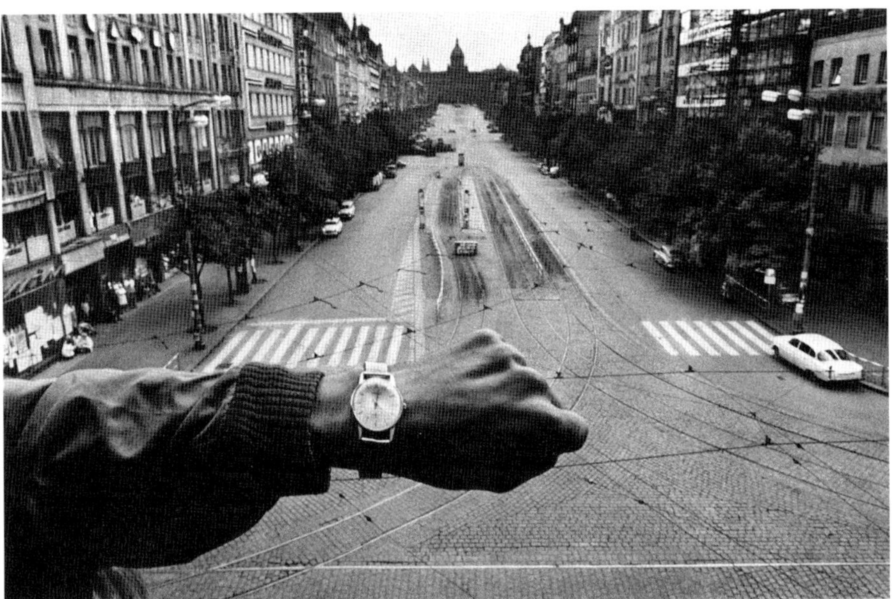

FIGURE 55. Josef Koudelka, Czechoslovakia. Prague. August 1968. Invasion by Warsaw Pact troops. © Josef Koudelka / Magnum Photos.

forth into the visible world."[10] Such metaphors are still used—in a 2014 article called "Secrets of Paper," Christopher Rovee describes the latent photographic image as a kind of "buried time" but also something haunting, "a ghost that later could be released into visibility."[11] The latent image evokes the world of the supernatural; it is occult in the mystical sense, but also in the specialist medical sense of something with no readily observable symptoms, and in the strict and old sense of something "hid from human knowledge."[12] The latent belongs to the netherworld—the creature in its cave that I used to speak about photographic sensitivity in chapter 1 could also represent latency. It hides from us cocooned in the deepest and most unreachable recesses.

Yet the latent is not easily embodied as a creature or even a ghost; it is more diffuse and atmospheric. It hovers like a threat, ambiguous, very present, and yet hard to single out or pin down. We can only recognize it retrospectively once it moves from latent to patent or manifest. The latent describes states of potentiality and ambiguity about whether that potential will ever be realized. It's no surprise the term "nuclear latency" is used to describe the purposeful ambiguity around whether a state has nuclear weapons or the capacity to make them and whether it is prepared to use them. The example of malaria shows how the concept of latency is connected to anxieties of nation and empire and fears of the unseen, to microbes, poisons, and contaminants that cannot be sensed but are nevertheless present in the atmosphere or as part of a miasma. Mal-aria: bad air. With malaria and with poison gas, latency adds to the sense of threat: while the effect of some gases is immediate, others seem to work slowly, in cases of mild exposure some symptoms manifest only later. Latency suggests that the cause is not just invisible but removed in time and space; the thing that manifests having been latent comes from elsewhere and before.

Writing in September 1930, the German Marxist philosopher Ernst Bloch observed that the triumph of National Socialism over Marxism in Germany had to do with its ability to commandeer the latent and to grasp the nonsimultaneity of time. In *Heritage of Our Times*, Bloch argued the Nazis took advantage of "an undercurrent of very old dreams," a "ghostly procession of perverted memories," the rising up of "very old pictures," and "an archaic-emotional remnant never wholly accessible or exorcizable."[13] These old dreams and pictures were latent in the culture. For Bloch, this latency undermines a linear model of history and accounts for the failure of the political left in the face of fascism. Marxism failed to make "Dionysian dreams into revolutionary ones"; it could not capitalize on what was latent. It needed a better understanding of history than the linear one it had: "History is no entity advancing along a single line, in which capitalism for instance, as the final stage, has resolved all the previous ones; but it is a *polyrhythmic and multi-spatial entity, with enough unmastered and as yet by no means revealed and resolved corners*."[14]

Writing about post-1945 Germany, in his book *After 1945*, Hans Ulrich Gumbrecht also describes something latent in German culture to do with the unshakable presence of the immediate past of the Nazi atrocities and defeat and the hope for an eventual transcendence of that past. He defines the latent as an invisible but actual presence, like a "stowaway" we can't quite locate, whose presence appears to us as a certainty even if we could not identify exactly why we are certain, and whom we might not even recognize if we met them. "Most importantly," he writes, "We have no reason to believe—at least no systematic reason—that what has entered a latent state will ever show itself or, conversely, not be forgotten one day."[15] Like Bloch, Gumbrecht speaks of the latent in atmospheric terms, as something meaningful yet impossible to straightforwardly interpret, a kind of vapor seeping through the cracks of a superficially orderly or conformist society, something that promises to fully reveal itself but never does, and which will not submit to any kind of research method or interpretative procedure. The latent manifests as mood, as ungraspable as fog, as unnarratable as the early experiments with laughing gas, though in this case much more sinister. Looking at popular German magazines of the postwar period, Gumbrecht says:

> Yet how can we be so certain that something latent is "really there" if it eludes our very perception? . . . when I peruse these emphatically peaceful and orderly postwar magazines, I am struck by the violence that often peeks through in the advertising—as occurs, for instance, in the photograph of the volcanic explosion made with a Graflex camera. In a similar vein, the quality of razor blades is advertised by showing how smoothly they pass over the soft skin of a baby's cheek. Cartoon representations of married life make jokes about husbands beating their spouses because the coffee isn't strong enough, or because the wife has forgotten to wake the breadwinner up for work.[16]

We can see this kind of latency in the *Illustrated London News* and its pages of disasters, discussed in chapter 7. Obviously, the context is different; the old dreams and pictures that haunted Britain in 1930 are culturally distinct from those that haunted postwar Germany. Like the magazines Gumbrecht examines, the *Illustrated London News* ignores some of the more immediate events—severe fogs in London coinciding with those in the Meuse Valley, the fracturing of the British Empire made apparent in the India Round Table conferences that winter. But there is an almost palpable sense of threat in the tornado images placed above the picturesque village of the Meuse, or below a ship and men who have perished in an explosion. The hiatus in time and space between the moment of photographic exposure and the moment of publication underscores the sense of threat: the street is peaceful, the ship intact, and the men posing for their collective portrait are unaware of their impending fate, which is allegorically represented by the tornado.

Like the photograph of Lewis Payne, the photographs of the men who will die on the *Artiglio* are also in the anterior future tense—they are dead and they are going to die.[17] But the arrangement of photographs in the magazine adds something else. By situating them alongside photographs of a tornado and a village that a disaster has hit or will soon hit, the newspaper diffuses the sense of horror and disturbance in a way that invokes the kind of historical latency that Gumbrecht describes. Although it seems so often to deal only with the manifest, what is present right in front of the camera, photography's ability to suspend time allows this collection of images to hint at the latent, at something that might swallow the world, ships and streets and all. The threat is diffuse, atmospheric, explosive, gaseous—present and yet unphotographable.

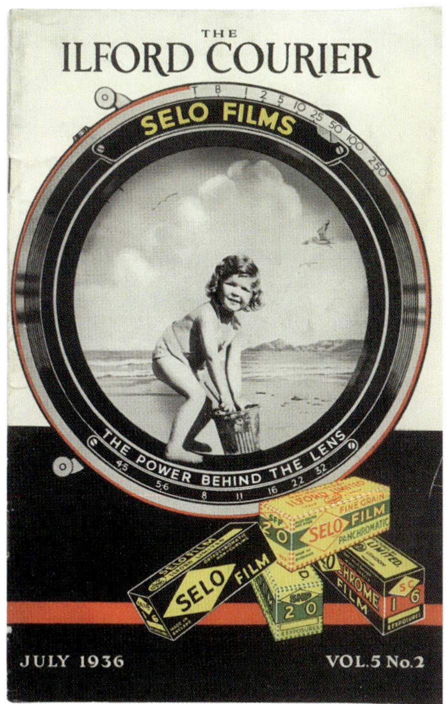

PLATE 1. *The Ilford Courier*, front cover, July 1936. Ilford Limited collections, Redbridge Museum and Heritage Centre. Courtesy of Redbridge Museum and Heritage Centre 2025.

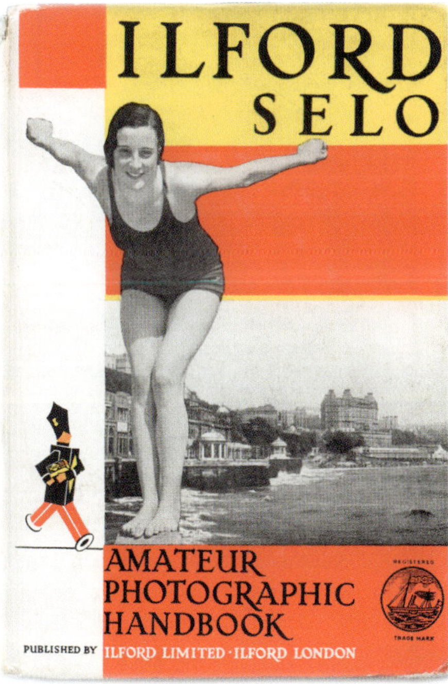

PLATE 2. *Ilford Selo Amateur Photographic Handbook*, ca. 1940, private collection.

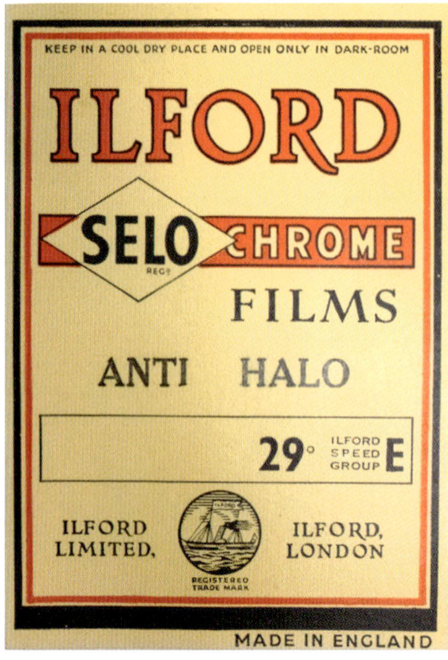

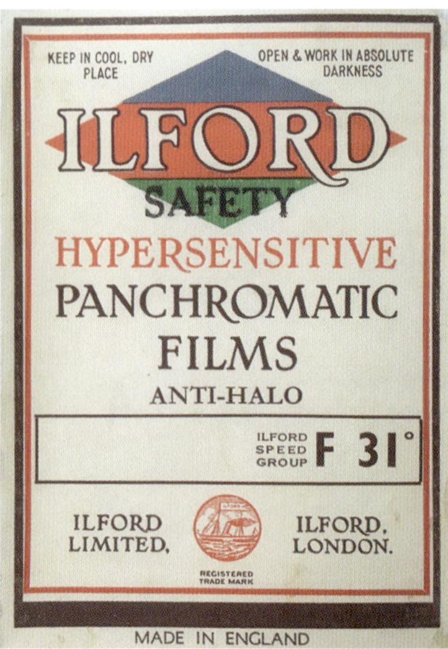

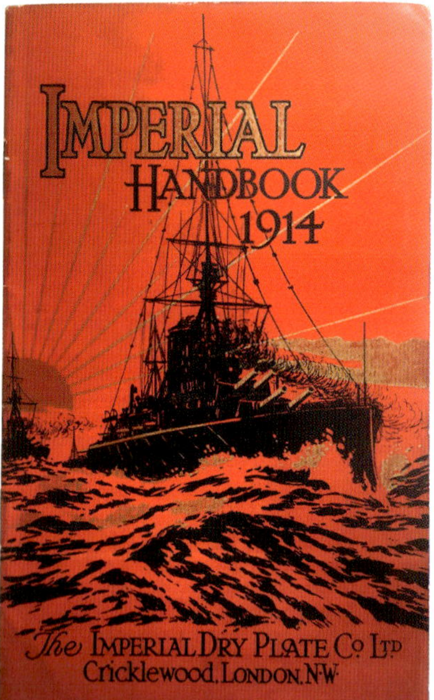

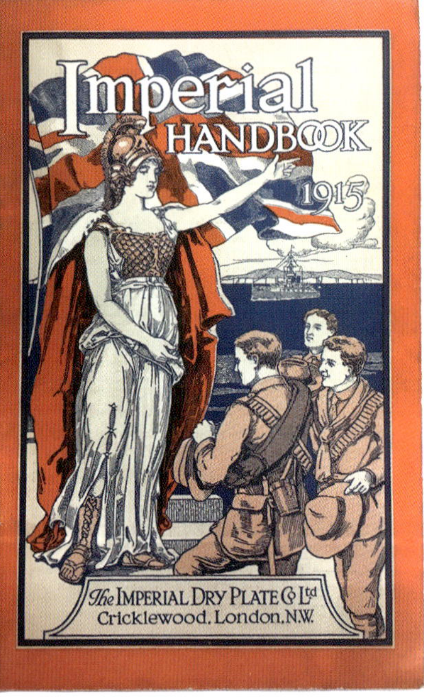

PLATE 3. Antihalo film labels and *Imperial Handbooks*. The Imperial Dry Plate Co. Ltd. amalgamated with Ilford in the 1920s, but products continued to be marketed under the Imperial name. Ilford Limited collections, Redbridge Museum and Heritage Centre. Courtesy of Redbridge Museum and Heritage Centre 2025.

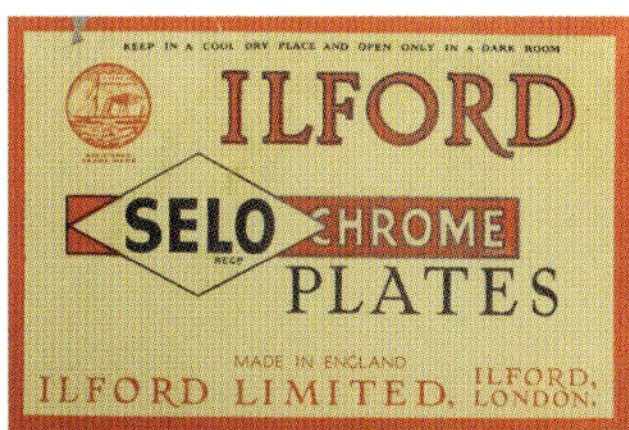

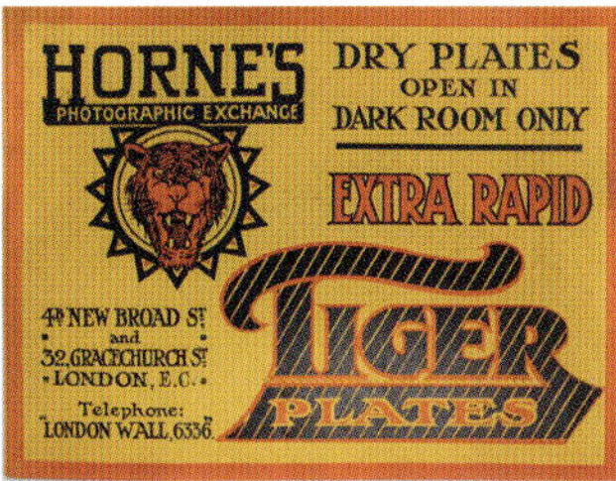

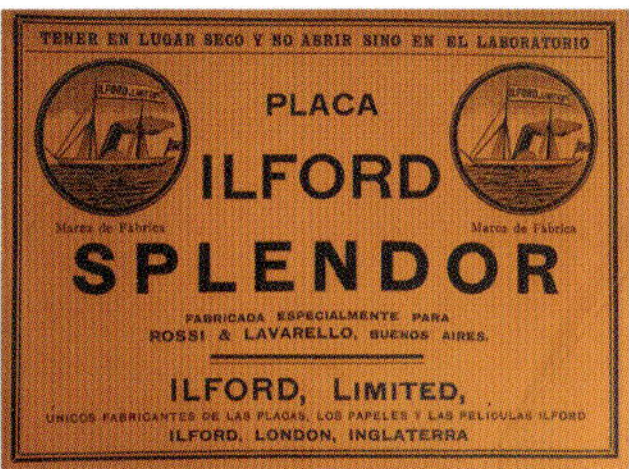

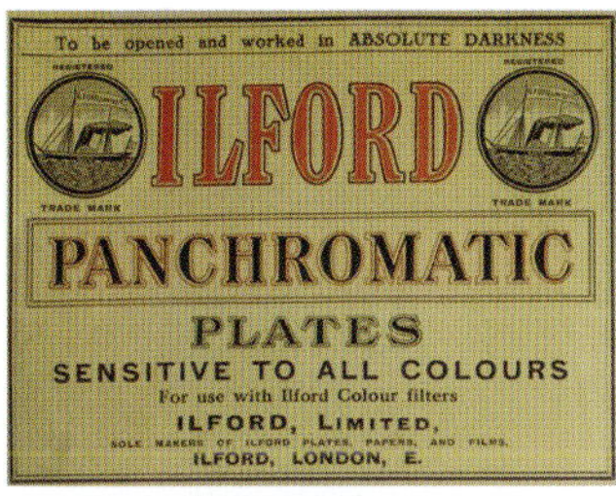

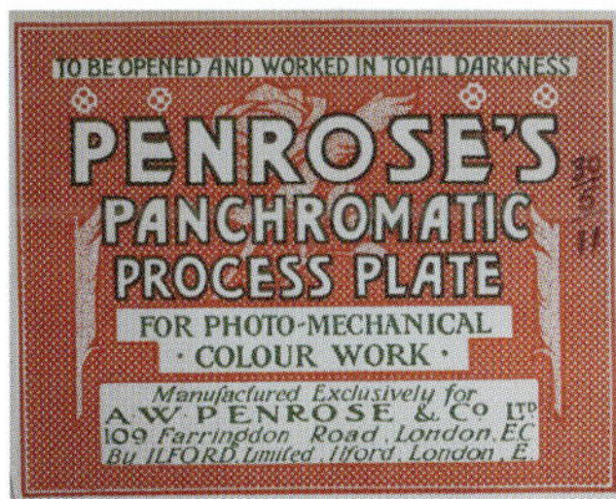

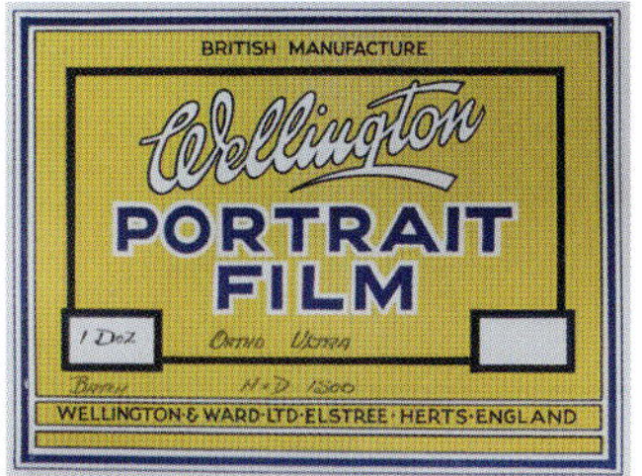

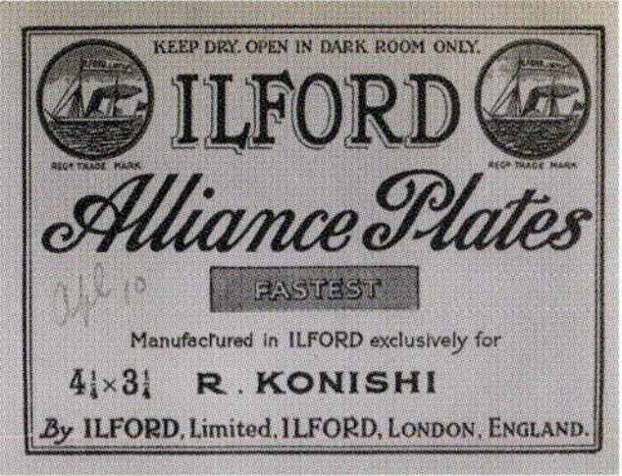

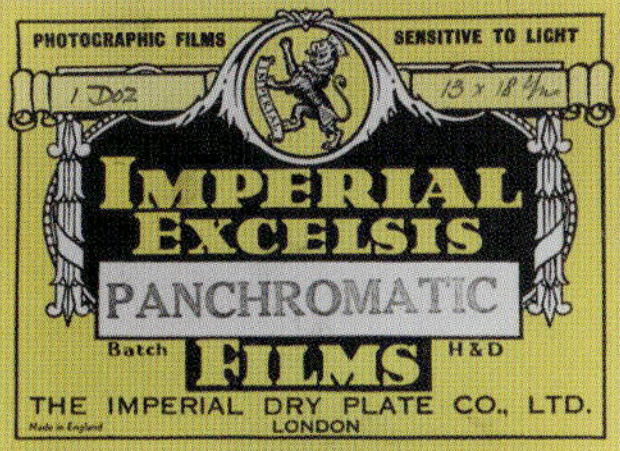

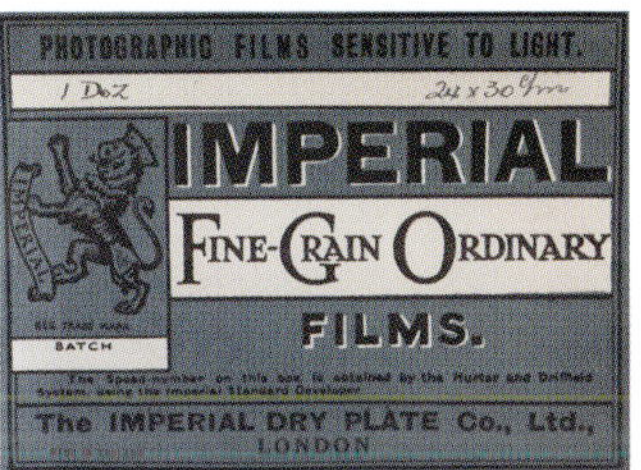

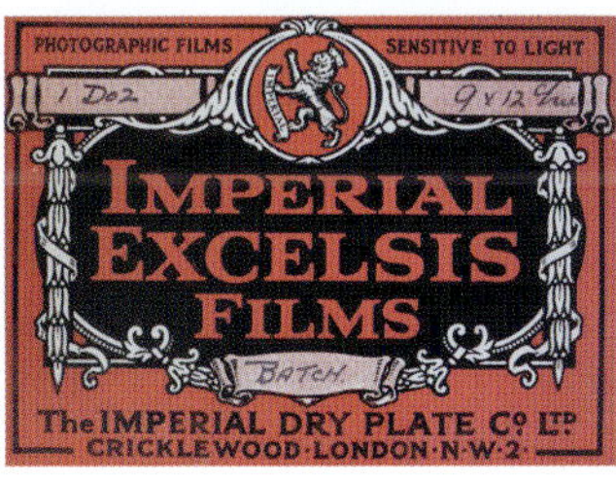

PLATE 4. Glass photographic plate labels. Ilford and the companies that it took over produced a diverse range of products, with different names and branding for different markets. Often products retained original names and branding even after the companies that manufactured them had been absorbed into Ilford Limited. Ilford Limited collections, Redbridge Museum and Heritage Centre. Courtesy of Redbridge Museum and Heritage Centre 2025.

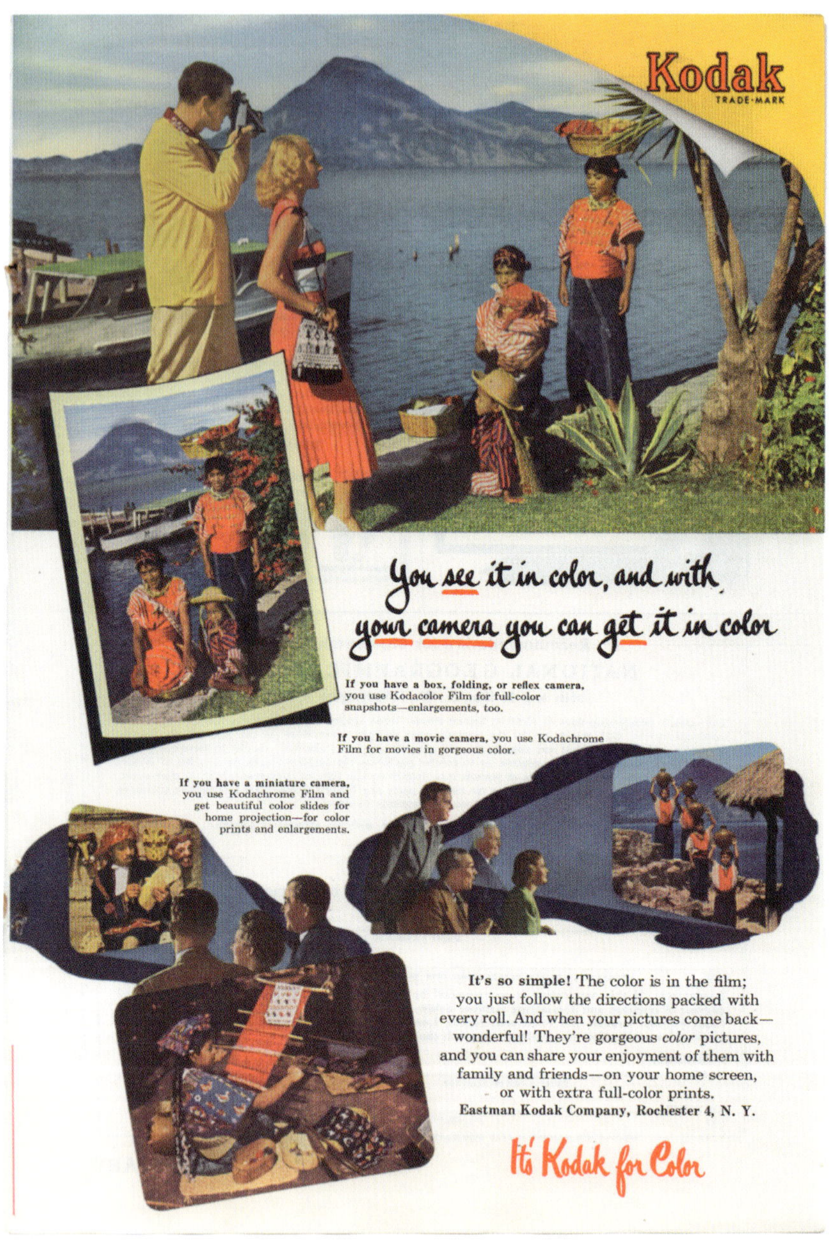

PLATE 5. Kodak advertisement, *National Geographic*, February 1952.

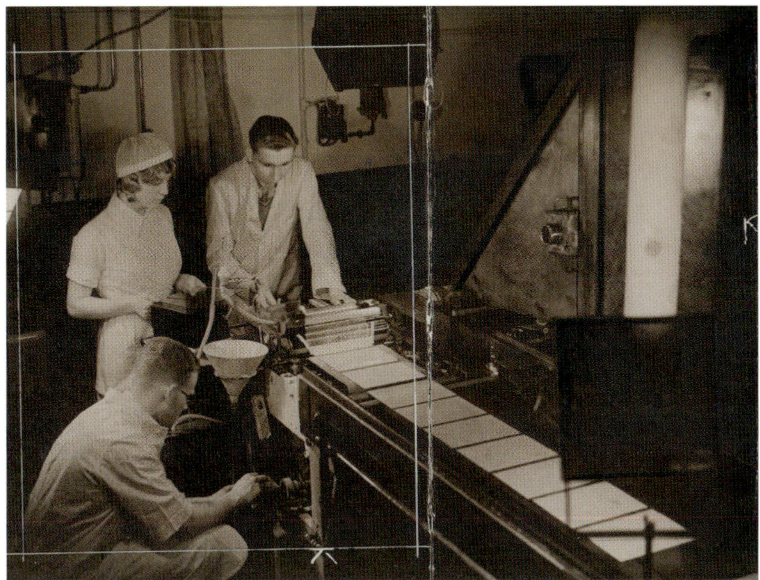

PLATE 6. James Jarché, *Seeing the Unseen*, infrared photograph of Ilford factory, showing the coating photographic plates with emulsion taking place in "almost total darkness" with infrared filters on lamps, November 1933. Daily Herald Archive, Science Museum Group © Mirrorpix/SMG Images.

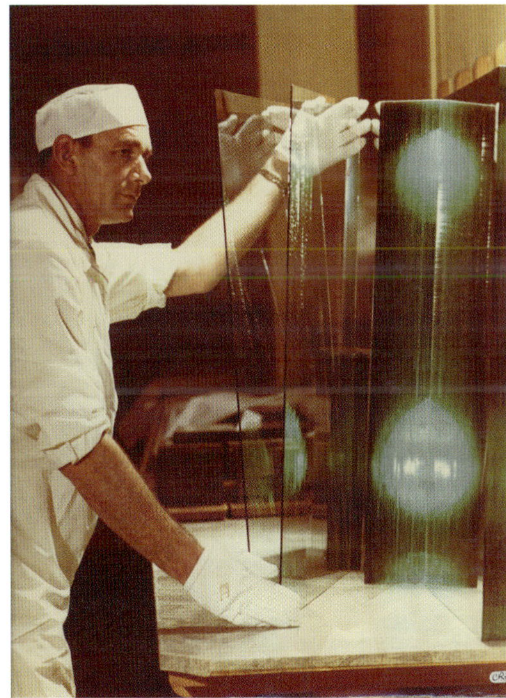

PLATE 7. Maurice Broomfield, Ilford Limited Factory, 1960–65. © The Estate of Maurice Broomfield. Courtesy of Redbridge Museum and Heritage Centre 2025.

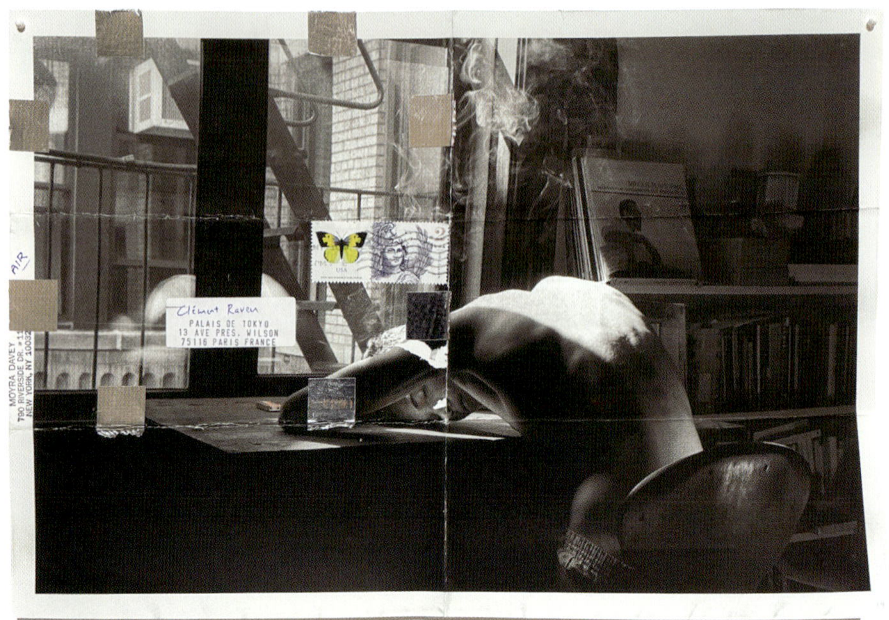

PLATE 8. Moyra Davey, *Visitor*, 2022 (one of eighteen c-prints, tape, postage, ink). Photo © Marcus Leith. Courtesy of the artist and greengrassi, London.

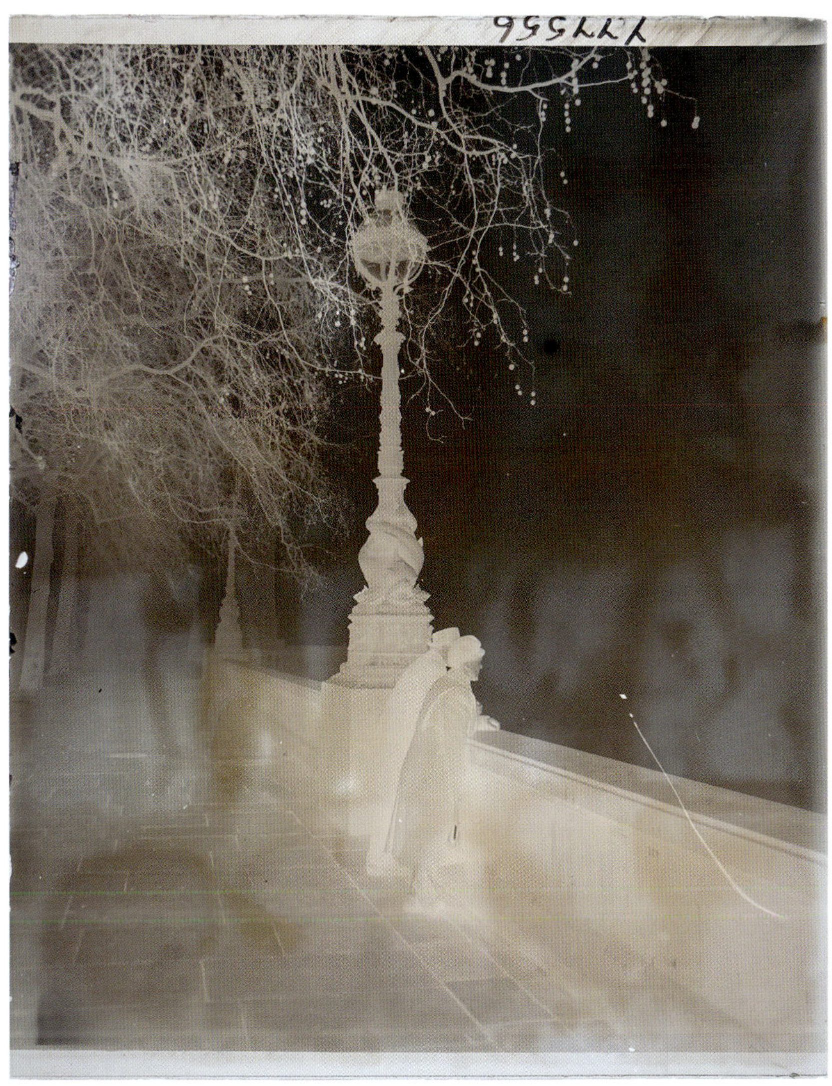

PLATE 9. Negative of Monty Fresco, *Embankment at Blackfriars*, London, December 5, 1952, rephotographed by Michelle Henning. © Getty Images, Hulton Archive.

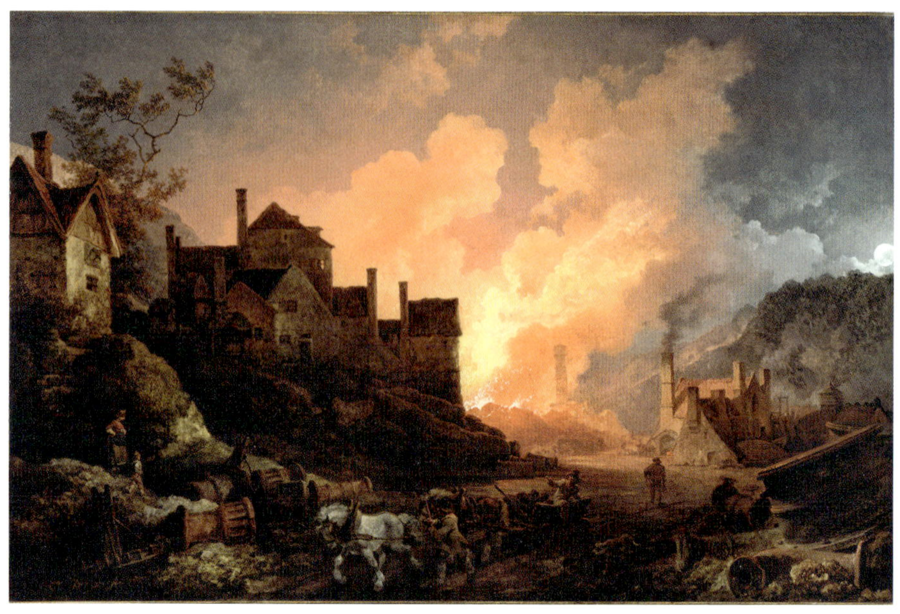

PLATE 10. Philippe Jacques de Loutherbourg, *Coalbrookdale by Night*, oil on canvas, 1801. Reproduction © Board of Trustees of the Science Museum.

35 KIDNAP GIRL MYSTERY

► There is a black-and-white photograph of me as a baby, in a packet marked "September 1968" (fig. 56). I am sat in a highchair and sticking my tongue out at my father, who is turned away from the camera but identifiable by his broken nose. He also has his mouth open, and I may be mimicking him, as it looks like he is feeding me, holding a bowl and spoon. It is hard to tell because a large stripe of white separates his side of the photograph from mine, and whatever is in his hand has been obscured. This vertical strip is a light leak, soft-edged but straight on his side, and on mine leaching and mingling with the light in the scene. The camera was not faulty, but the beginning of the film was exposed to light when the film was loaded. I know this because in the foreground, on the table, is the cardboard packaging of the film that my aunt is using to make the image. The light leak suggests that this is the first exposure on that freshly loaded film.

Despite the light leak, and even with it, the photograph could be considered a good one according to all sorts of criteria: it shows an everyday interaction between an infant and an adult, one familiar to anyone who has spent time with young children, emphasized through the exchange of gazes between us. It is beautifully lit, sunlight coming from a window out of the scene, behind and to the left, producing a catchlight in my right eye, backlighting my hair, and falling on the wall to the right. Both faces escape the light leak by a stroke of good fortune. There are newspapers on the

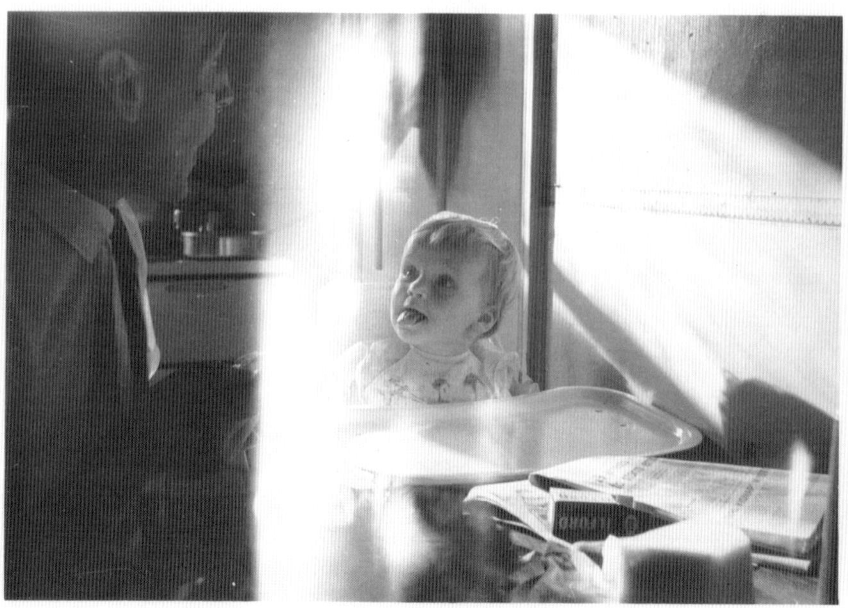

FIGURE 56. Snapshot of myself as a one-year-old with my father, September 1968.

table. The name on the film package is clear, though upside down: Ilford. The same brand name is printed diagonally in a repeat pattern across the back of the photographic print. And on the negative is written ILFORD HP3 HYPERSENSITIVE PANCHROMATIC FILM.

I first saw this photograph in March 2018, following the death of my aunt, and after I had begun researching in the archives of Ilford Limited. I am struck, of course, by the presence of that film packaging which puts Ilford there, in my early childhood. To me, Ilford was always a familiar brand. I had used Ilford films and chemicals in many darkrooms—from the makeshift one my dad made in the cupboard under the stairs, to the similarly sized one we installed at school, to the ones in several art colleges where I studied and taught. Even now, I photograph with Ilford film.

Looking at the photograph again, I am struck by something else.

It has a very specific atmosphere, caused not just by the light in the room and the grain of the film but by that light leak. When I was ten and had my own camera, my photographs often had light leaks. I thought of them as spirit visitations, ghosts that had made their way into my machine, proof of something beyond the merely visible. It was as if an angel or an alien had dropped into the scene—too bright for our eyes to make out their exact form, they dare you to look directly into the light and try to interpret its shapes.

This particular light leak I don't see as a ghost, angel, or alien but merely an effect of light-sensitive materials. I can't help looking harder at it, trying to precisely disentangle this light (the first, accidental exposure) from the light in the scene (a second exposure, this time intentional). It seems to almost coincide with the exact place in the image that the light source would be, as if it were caused by the bright light from a window rather than by light entering the open back of the camera. The image has a slightly hazy appearance, as if the light had been made opaque by smoke lingering in the air. Is there smoke? I wouldn't be surprised, practically everybody smoked then. But either way, the light leak and the light in the scene combine to appear almost tangible and atmospheric. For me, there is an emotional, nostalgic atmosphere to the picture, too. My parents tell me this is the kitchen of their house—a room I know well but don't recognize here because the wall on the right was knocked down when I was little, and the stove behind me replaced—but if it is that kitchen, I can place the light source exactly. The window faces west, and the afternoon sun and the fact that my father is wearing a tie suggests it is after 5:30 p.m. and he has just returned from work.

In my effort to date the picture more precisely, I try to read the headlines of the newspaper in the image. I can make out "19 arrested in attempt to storm embassy" and, below, "girl 'kidnap' mystery." Though I cannot locate this specific paper, I do find others from Monday, August 26, 1968, with similar headlines. In the *Daily Mirror*, two articles discuss events following the Soviet invasion of Czechoslovakia. The first, "Mob Runs Wild in Big Embassy March," describes a demonstration in London against the invasion, in which nineteen people had been arrested after breaking away from the march in Hyde Park in an attempt to storm the Russian embassy. The second, "Watch on Ports in Hunt for 'Kidnap' Girl," refers to the disappearance in England, five days previously, of seventeen year-old Nada Bilak. She was the daughter of Vasil Bilak (or Bil'ak), a Slovak communist member of the Czechoslovakian National Assembly. According to the article, she had been at an International Voluntary Service (IVS) camp in Cumberland when a Russian-speaking man and woman came to take her away.

I search for Nada Bilak and find a picture of her at Wigton in Cumberland, just north of the Lake District. She is a sturdy-looking young woman standing in water wearing a coat and wellington boots, her back to a wall, part of a line of young people who appear to be digging in the water with shovels. Other newspaper coverage mentions that she is sought by Scotland Yard and that a note which read "I'm all right" had been received, postmarked from London, at the IVS address where she was staying.[1] They also say that she had been working on "a stream widening scheme."[2]

This is turning out to be a story of fathers and daughters. Unlike my father, though, Vasil Bilak belongs to history with a capital H. He was the lead signatory of a letter inviting the Warsaw Pact countries to invade Czechoslovakia and suppress the Prague Spring. He denied this and, despite rumors, nothing was proven until 1992.[3] That year, Russian president Boris Yeltsin made public two letters from Bilak and colleagues that had been sealed in the Soviet archives and that appealed to the Communist Party of the Soviet Union "to lend support and assistance with all the means at your disposal."[4] In 1995, the memoirs of Petro Shelest, a Ukrainian member of the Soviet Politburo, revealed that it was Bilak who had handed over one of these letters on August 3, 1968, in a public toilet in Bratislava.[5]

Later newspaper reports claim that Nada was not kidnapped but sent to Moscow. Shelest said he was instrumental in her removal and that she was secretly taken to Kyiv. Bilak had been an oppressive leader whose hostility to the reforms of the Prague Spring made him unpopular among factory workers.[6] Following the invasion, he was promoted through the ranks of the Communist Party of Czechoslovakia and relentlessly pursued the participants in the Prague Spring. He was accused of multiple acts of persecution in an open letter by

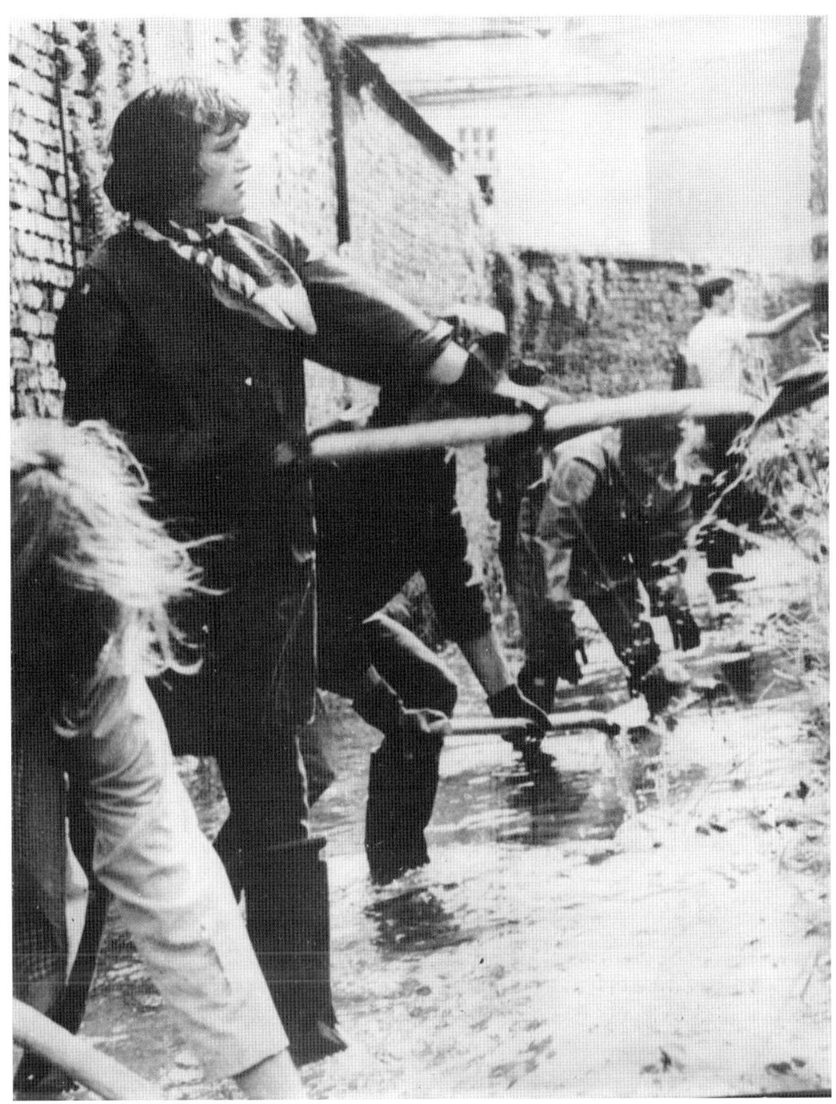

FIGURE 57. Nada Bilak 1968. Photographer unknown. Private collection.

Karel Kaplan, the erstwhile deputy director of the Historical Institute in Prague, who was forced to work as a stoker thanks to Bilak.[7]

Why is Nada in northwest England, widening a stream? Sent there to escape from the tanks her father knew would be rolling into Prague and the potential repercussions for her family? Or there already, by coincidence? There's something odd about that photograph: it is clearly intended to be "of" Nada since everyone else in the picture is bent down or far from the camera and only Nada appears, sharp, static, and in profile. Did the photographer know that she would become significant or newsworthy? There are press agency stamps on the reverse of the print. The events to which the photograph is connected but does not depict (Nada's "kidnap," the invasion of Czechoslovakia) give an otherwise uninspiring picture its significance after the fact. As I look at these photographs, time seems circular—my father sticks his tongue out and I mimic him, or I stick my tongue out and he responds. Nada stands up to shovel the water and the others around her bend down, or is it the other way round?

Photographs are hugely important to our sense of historical time even as they undermine it. These two photographs—of me as a baby at the kitchen table, and of Nada Bilak widening a stream in Cumbria—were taken chronologically close to one another, perhaps separated by only a couple of weeks. One photograph led me to another, one father-daughter relationship to another. Personal history and political history intersect, and the photographs recall others—other light leaks, other babies, the raised arm and shovel of a stoker in the gasworks or in the belly of a ship (see chapter 4).

These two images were presumably snapshots until one became a press image and both illustrations in this book. For me they evoke questions not only about the odd temporality of photographs in general, but also about what constitutes a historical event. The Prague invasion, yes, but a stream being widened? This may be history on the microscale rather than the macro, but now that we see microbes as historical actors and human history as intersecting with the deep time of geology, who is to say what the impact of a stream might be?

The stream in Wigton was being widened to prevent flooding, since Wigton is situated on the Solway plain near wetlands and salt marshes, and in summer 1968 floods were widespread across the British Isles. At the beginning of July, in a rare weather event, there were extreme hailstorms with heavy rain, colored red by dust from the Sahara desert, with hailstones large and heavy enough to break windows.[8] The British Meteorological Office identified the red dust as originating in Morocco and the Western Sahara and lifted by "unusually strong thermal currents."[9] Such storms carrying Saharan dust were rare, and meteorologists had identified only two prior occasions, one in 1903 and one in 1930, in the English Channel. Globally, dust-laden storms from the Sahara increased in frequency from the 1950s, peaking in the 1980s.[10]

The freak weather of 1968 links my photograph and that of Nada Bilak to the larger story of capitalist extraction and colonialism. Some researchers have argued that these storms are unlikely to be anthropogenic on the grounds that there are few settlements and no agricultural activity in the source areas.[11] Yet 70 percent of the world's phosphate reserves are in Morocco and Western Sahara. Like nitrate, phosphate is used as a fertilizer and can be derived from bird guano, but it can also be found in rock. French colonists mined phosphate rock in Tunisia since the 1880s, following the Nitrate War, and reflecting the rapacious appetite of Europe for soil fertilizers. The industry in Tunisia dispossessed the local nomadic populations, forcing them into laboring for the mines. From the 1940s, in Western Sahara, phosphate extraction was controlled by successive colonial forces, first Spanish, then Moroccan, again forcibly displacing nomadic peoples.[12] Today, Western Sahara is a world center of phosphate production. Researchers have found nitrates, ammonium, and sulfates in samples taken from Saharan dustfall. This is consistent with the polluting output of the large number of extractive industries in the region, which has crude oil refineries, coal power plants, phosphate rock mines, and phosphate fertilizer plants.[13] The dust is so high in phosphorus that it acts as fertilizer where it lands. The dust-filled rain can be so red that it is sometimes mistaken for blood.

In the blood-red summer of 1968, the sunlight in my parents' kitchen that lit my hair and marked the newly loaded film, and the water in which Nada Bilak stood, gamely shoveling, were not mute and eternal “nature,” outside the social and political. Just as they preserve “the everlasting mountains and the passing smoke of a bandit-fire” across time and geographical distance, photographs draw together the apparently timeless and the ephemeral, exposing all as historical actors.[14]

36 THE BLACK TRIANGLE

► Not long after Nada Bilak was widening the stream and I was being fed in my highchair (chapter 35), Josef Koudelka left Czechoslovakia, fearing for his safety. He did not reveal his authorship of the photographs of the Warsaw Pact invasion until 1984, and after the Velvet Revolution of 1989 he returned once more to his homeland. This time he did not picture violence directly aimed at people, but violence turned on the environment. He photographed the "Black Triangle," one of Europe's most environmentally devastated regions, a notoriously polluted mining area that includes the Czech Republic's northern Bohemia.[1] Under Soviet rule, the centrally planned economy had prioritized heavy industries with no regard for ecology or population health. The brown coal or lignite that was mined here and that drove the power stations produced exceptionally high levels of sulfur when burned.[2] The area was once a forest and was known at one time as the Czech paradise, but in the early 1990s it looked and felt more like hell, with a constant gray, polluted fog hovering over the region in winter. People living in these areas had the shortest life expectancy in Czechoslovakia.[3] Acid rain destroyed what forests remained after logging and open-cast mining, while soot and ash covered the landscape—in the forest, "you brushed a branch and got covered in black."[4] The area was also blighted by radioactive uranium waste, the sludge of which was dumped into lakes that supplied water to homes in the region.[5]

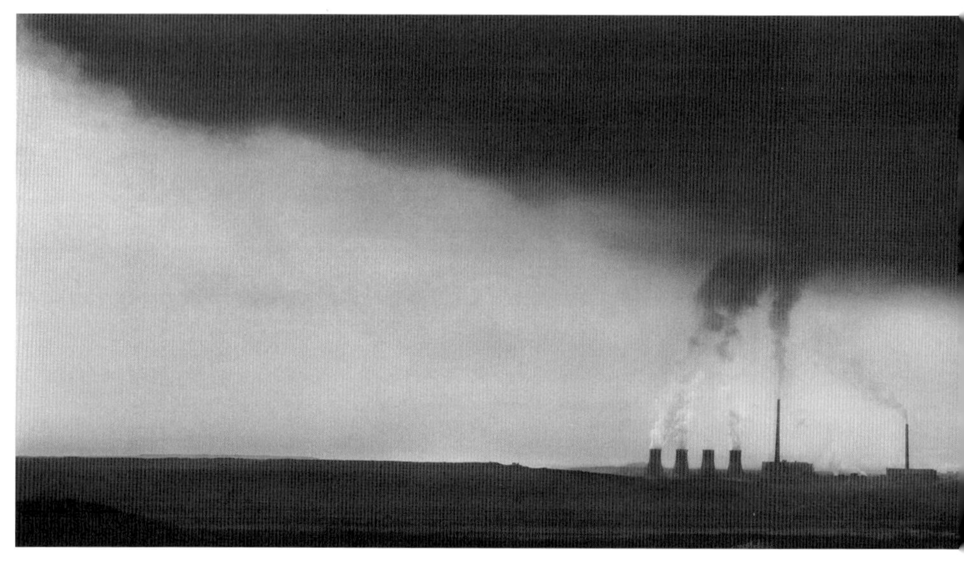

FIGURE 58. Josef Koudelka, Czechoslovakia. Region of the Black Triangle (Ore Mountains). 1991.© Josef Koudelka / Magnum Photos.

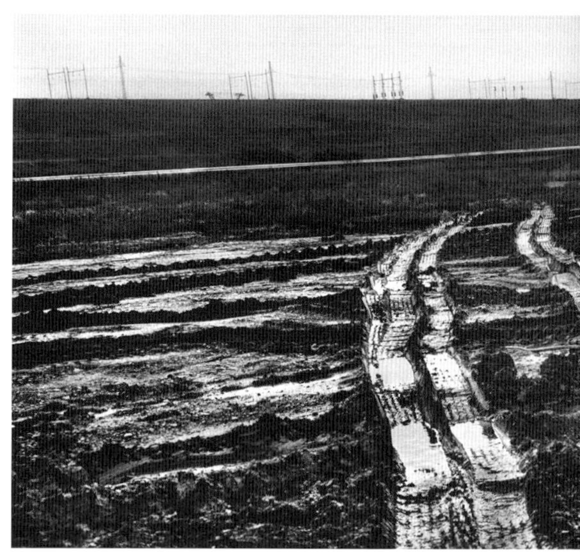

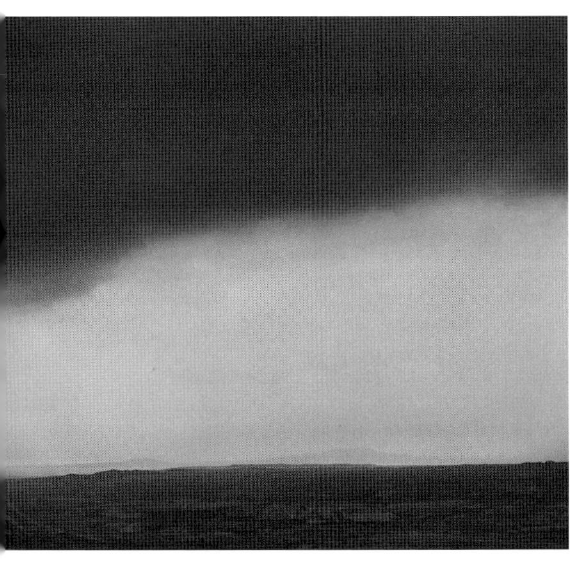

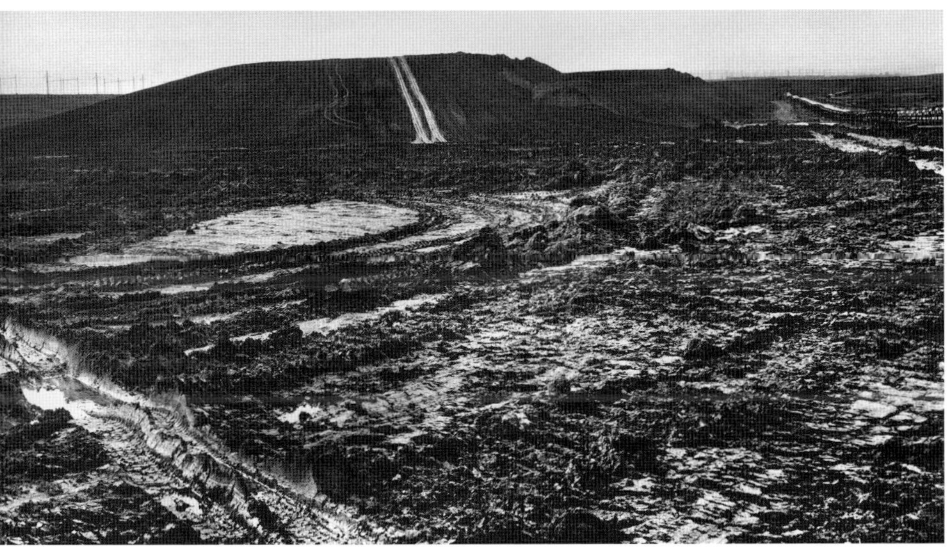

FIGURE 59. Josef Koudelka, Czechoslovakia. Region of the Black Triangle (Ore Mountains). 1992. *Recultivation*. © Josef Koudelka / Magnum Photos.

Using a panoramic camera, Koudelka photographed the former open-cast mines, printing in deep blacks and dense grays that seem to perfectly mimic the reality, as if there never was any color in this landscape or in the sky above.[6] The gray shades of these photographs also seem to echo a wider impression of the period of "normalization" between the Prague Spring of 1968 and the Velvet Revolution of 1989. Historian Miroslav Vaněk has looked at how some Czech people visualize this period as gray, arguing that "In the Czech Republic grey is, in the context of history, a synonym for flatness" but that there was also a literal grayness to the polluted landscape. As one of his interviewees says, "When you travelled around Vítkovice, everything was grey, no matter if it were a tree, a building, a road, just about everything was grey from this dust."[7]

Koudelka's Black Triangle photographs seem to depict the ecocide anticipated by Byron's 1816 poem "Darkness": "Seasonless, herbless, treeless, manless, lifeless— / A lump of death—a chaos of hard clay."[8] The strange temporality of photography is present here, too. It is unclear if Koudelka's photographs, like Byron's poem, depict a past or a future. Are they prophetic (as Jonathan Bate suggests of "Darkness") or are they records? Do they speak of futures past or a future yet to come? These images connect the latency inherent in photographic technology, which contributes to its peculiar temporality, with the historical latency of environmental destruction, something not fully visible or available to the camera but lurking, hidden in the air.

In my view, the trouble with photographs of degraded environments is not that they risk "anaesthetizing" us to environmental destruction (as many writers have argued), but that they are characterized by this uncanny relationship with time.[9] If photographs do not persuade us to act (and sometimes they do), it is not because they are beautiful and beauty is numbingly amoral and dangerous, nor that they are spectacular and spectacle turns us into mere spectators. It is partly because of a host of other factors (for example, to do with how individuals understand social and political power), but also because photographs seem to stand elsewhere, outside time, and in a different space from that of the viewer.

As Jean-Luc Nancy observes, images can be simultaneously proximate (shocking, touching, intimate) and withdrawn, far away. Images (not just photographs) are "distinct," they stand apart from the world of things even while they exist as material things. In a sentence very reminiscent of Walter Benjamin on aura, Nancy says that an image "approaches across a distance but what it brings to such proximity is distance."[10] This push-pull quality combines with the strange time of photographs—for Vilém Flusser, images are opposed to "the linear world of history" because they do not show us cause and effect or events and their consequences, but elements that simply exist together, simultaneously. Add to this general quality of images the specific qualities of photographs identified by Benjamin and by Roland Barthes: photographs, always *of* the past, nevertheless anticipate a future, and are haunted by other images that came before, events that may never happen, presences that may never appear. All of this makes them recalcitrant, untamable, and unpredictable both for activism and for history.

Some artists circumvent the debates about environmental photography's aestheticizing tendencies by shoring it up with a range of other kinds of documentary evidence. Photographer Richard Misrach and landscape architect Kate Orff did this in their 2014 book *Petrochemical America*, which combines Misrach's elegant photographs with informative text and information graphics tracing the interconnections of petrochemical pollution, plantation and slaving history, and cultural and natural history along the Mississippi River corridor between Baton Rouge and New Orleans (sometimes known as Cancer Alley).[11] Such projects use photography as one means among many to document environmental destruction, but they don't necessarily reflect on photography's own entangled histories. How can a technology so embroiled in environmental violence at the same time stand outside it, offering it back to us as if it were not itself complicit?

Other art projects directly address the role of photographic manufacturers in communities and places. For instance, Andrew Brown, a photographer and educator who works with the River Roding Trust in the London suburb of Ilford, has used the Ilford Limited archive in community workshops.[12] He combines new and old processes including infrared and composite techniques, sensitized glass plates

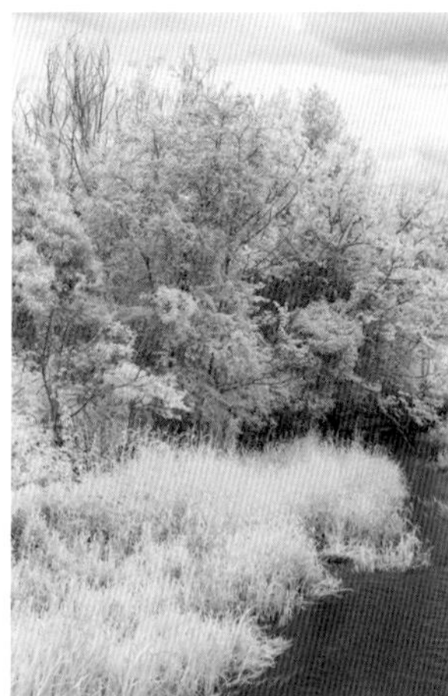
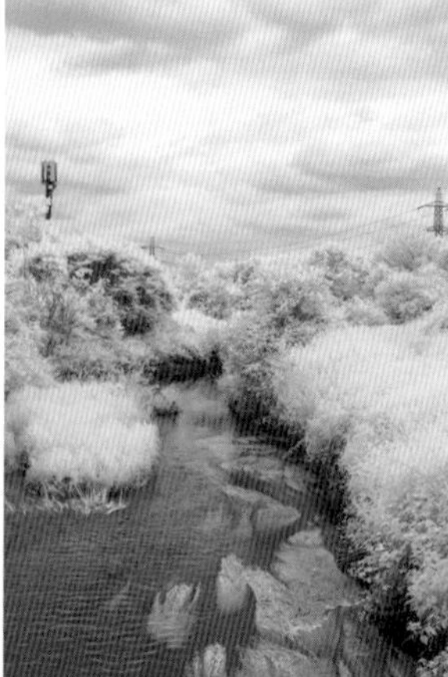
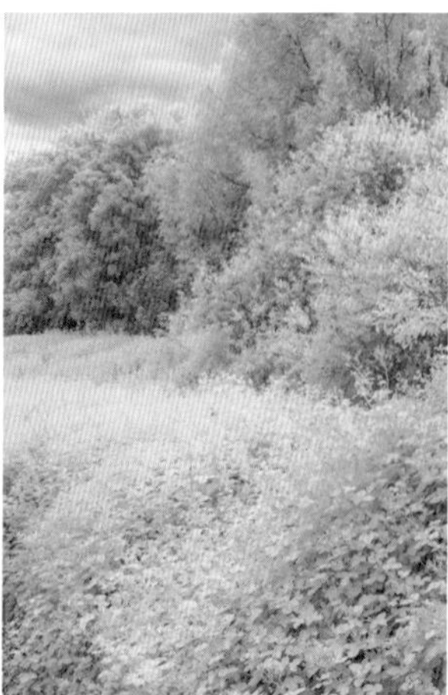

FIGURE 60. Andrew Brown, *Triptych #4*, from *River Roding/A406 series*, 2024. © Andrew Brown. Digital infrared images projected as part of an installation with soundscape. The images in this series are all made along the River Roding as it flows north to south close to the A406 (North Circular) road.

coated using emulsion formulas from the 1880s and 1890s, video, and soundscape to address the photographic history of the locality and the legacy of industrialization, particularly in and around the River Roding. In *Changing Currents* (2023), he helped local high school students use hydrophones, full-spectrum photography, and light refraction techniques to try to see what life is like for species living in the river.[13] In such community and site-specific (or location-sensitive) projects, the digital and film cameras that are used become a means to reflect on place and locality in ways that extend far beyond the photographic. But the history of photography is never far away from the history of the long-suffering river, its inhabitants, and those that live alongside it.

A growing number of contemporary artists are revisiting photography and film's contaminated legacy. As well as using photography to document environmental destruction and its origins in colonialism, empire, and extractive capitalism, these artists reflect on photography's own materiality and its involvement in larger geopolitical and environmental processes. Exhibitions such as *Mining Photography* (Hamburg, Vienna, and Winterthur, 2022–24), *Image Ecol-*

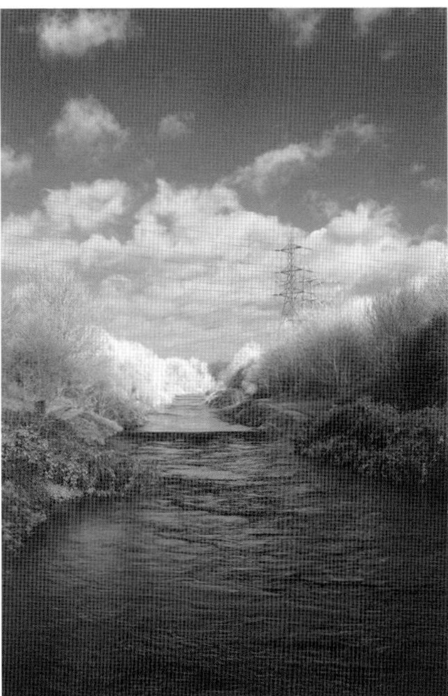

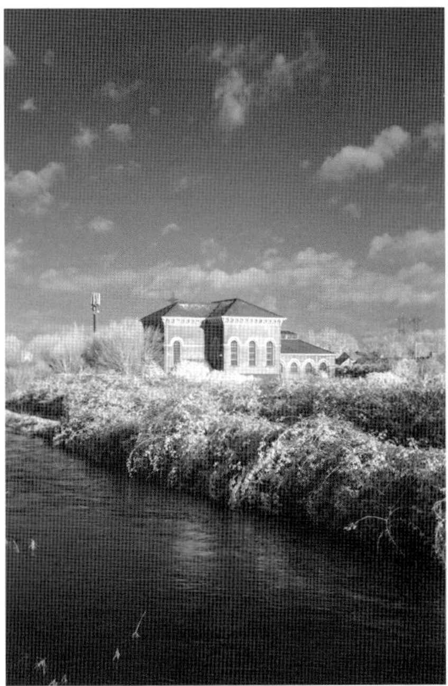

FIGURE 61. Photographs by Neeti Sayani, from *Changing Currents—can we build a better world together?* exhibition at SPACE Gallery, Ilford, London, 2023 (original in color). © Neeti Sayani. In this project students from Beal High School explored the River Roding and its pollution from the perspectives of both human and nonhuman animals, using infrared photography as well as sound recording techniques. The exhibition was curated by super/collider, supported by Andrew Brown, the River Roding Trust and Arup.

ogy (Berlin, 2023–24), *Metamorphosis* (London, 2024), *Beyond Sight* (Liverpool, 2024), and *Widening the Lens* (Pittsburgh, 2024–25), as well as artist projects like *Traces of Nitrate*, address photography's chemical and mineral constituents, its ecological impacts, its role in extraction, its connections to plant photosensitivity, and its capacity for envisaging different futures. Numerous artists including Suzanne Kriemann (Germany), Lucy Raven (US), Ackroyd and Harvey (UK), and Tomás Saraceno (Argentina/Germany) explore the wider sensing capacities of photographic materials, places, plants, animals, and ecosystems. Working in collectives like BEEF (Bristol Experimental and Expanded Film, UK) and the Sustainable Darkroom (London, UK), photographers and filmmakers use alternative plant-based and experimental processes to detoxify the darkroom.

As I understand it, these approaches are not intended to suggest viable industrial alternatives for photography but rather to resituate photography in life, while at the same time facing its industrial legacy and revisiting its contaminated sites. I started this book with an image of chemical photography as a quasi-magical creature born in Plato's cave, and I have shown (I hope) that its purification rituals came at a price. If Frankenstein's crime was to deny his Creature the bonds of community, these artists avoid that, acknowledging the full material and industrial nature of chemical photography without disavowing it or casting it out, but instead (to frame their work within my own analogy) gently returning the creature to its cave and helping it to incubate differently, reinventing its chemistry, and recultivating its atmospheric sensitivities. Like them, I refuse to cast it out. I rake through photography's dirty history, turning up its residues, and still I remain in awe of its magical and peculiar properties, thankful that companies like Harman Technology persist against the odds and that I still have Ilford films in my refrigerator.

ACKNOWLEDGMENTS

► I would like to thank the Arts and Humanities Research Council (AHRC) which funded the initial research for this project in 2018–19 under grant no. AH/R014639, and to the School of the Arts at the University of Liverpool for research leave in 2023 and for bearing some of the copyright and reproduction costs. Also the London School of Film, Media and Design at the University of West London, which supported the AHRC-funded project.

I am lucky to find myself in an international community of photography scholars which is generous and welcoming, creating remarkable spaces and opportunities for the exchange of ideas. Many of the ideas and arguments in this book have been rehearsed in papers and keynotes given at conferences by the kind invitation of colleagues. I was able to test out early ideas in a number of invited lectures and conference papers. For these opportunities I would like to thank Kelley Wilder, Gil Pasternak, and Beatriz Pichel at the Photographic History Research Centre (PHRC) at De Montfort University; Friedrich Tietjen and Maria Gourieva for including me in the *After Post-Photography 4* conference at the European University in St. Petersburg, Russia, in 2018; Patrizia DiBello and Steve Edwards at the History and Theory of Photography Research Centre, Birkbeck, London; Geoff Belknap, who gave me opportunities to speak at the National Science and Media Museum in Bradford and at the Science Museum in London; Harri Laasko, Aasko

Lehmuskallio, Maija Tammi, and all involved in *Helsinki Photomedia* (and Ben Burbridge for his company there); Estelle Blaschke and all who helped her organize the *Photo Archives VIII* conference in Basel in 2022; Sara Dominici for the conference *In the Photographic Darkroom* at the University of Westminster in 2023; Pasi Väliaho, William Nore, the Norwegian Photohistorical Society, and the University of Oslo for inviting me to talk in Oslo; and Georgina Endfield, who shared her expertise on weather history in a joint public lecture with me at Liverpool University in 2022.

Parts of chapter 3 were previously published as a journal article called "Photography's Other Sensitivities," *Media Theory* 8, no 1 (2024). Parts of chapters 12 and 14 were published in "The Worlding of Light and Air: Dufaycolor and Selochrome in the 1930s," *Visual Culture in Britain* 21, no. 2 (2020). Sections of chapters 25 to 28 appear in my article "Keeping out the Fog: Atmospheric Contamination and Control in the Industrial Darkroom," *Photoresearcher* 41 (2024). The opening chapter was inspired by my introduction, cowritten with Junko Theresa Mikuriya, to our journal special issue on Light Sensitive Material for *photographies* 14, no. 3 (2021). It was a huge pleasure and a privilege to write with Theresa, who also worked with me to organize our AHRC-funded conference *Light | Sensitive | Material* in London in 2019 (thank you, Theresa!). That conference inspired parts of this book, as did the Stuff of Photography workshops that the AHRC grant facilitated at the Photographers Gallery, London, in 2019, which were coorganized with my PhD student Rowan Lear. I would like to thank: all the participants in both events; the invited speakers Esther Leslie and Louise Purbrick; Joseph Kendra, Luisa Ulyett, and Janice MacLaren at the Photographers Gallery; and especially Rowan, an extraordinary artist and writer who also accompanied me on several archive visits. Thank you, Rowan; this book inevitably bears the marks of your influence.

I have been helped and supported by numerous archivists and their colleagues, including: Dawn-Anne Galer, Paris Snydes, Gerard Greene, Paula Wade, and Saima Qureshi at the Redbridge Museum and Heritage Centre; Kendra Bean at the National Science and Media Museum; Emma Burgham at the Museum of Science and Industry; Sophie Clapp at the Walgreens Boots Alliance Archive; and

Matthew Butson and Melanie Llewellyn at Getty Images, Hulton Archive. I would also like to thank the staff at Harman Technology Limited, who gave myself and Rowan a tour of the factory, and especially Vicky Proctor and Michelle Parr for arranging and hosting the visit. Thanks to the contemporary artists featured in the book, especially Andrew Brown for his help with images, and to Michael Pritchard for the images of the gasworks stoker from his collection. Thanks to Tim Pearse of Negative Thinking in Bristol for hand-printing the picture of me in my highchair, and to my PhD students and photography curators Hana Kaluznik and Sarah Okpokam for their help with my queries.

I would like to thank my supportive, kind, and rigorous editors at the University of Chicago Press, Karen Levine and her associate Victoria Barry; Hannah Vose for her helpful advice on permissions; the copyeditor, Trevor Perri; and the anonymous reviewers for their generous and thoughtful feedback. I am indebted to my colleagues in the Department of Communication and Media and the dean of the School of the Arts, Peter Buse, for their support and friendship and for making it such a welcoming environment in which to research and teach. Many friends and colleagues have helped, encouraged, and inspired me, among them: Peter Bennett, Gavin Butt, Garin Dowd, Becky Goddard, Alyssa Grossman, Rehan Hyder, Matilde Nardelli, and Les Roberts. Special thanks to Jordana Blejmar, Jon Cairns, Ben Highmore, Richard Hornsey, and Maria Mochnacz, who all kindly read parts of this book and gave me useful feedback. Richard and my colleague Hannah Little also helped me identify the locations in figure 2 and plate 5 (respectively). Hopey Parish gave me helpful suggestions on chapters 24 and 35, and Honor Parish was one of the first people to read and give feedback on a whole draft. Also, thank you, Honor, for the trip to Tyntesfield; our discussions there shaped chapter 22. My father, Mike Henning, helped with information about his grandfather John Devlin (thanks Dad!). The photograph of us that features in chapter 35 (fig. 56) was taken by his younger sister, my aunt Christine Henning (Auntie Tissy), who is much missed.

As always, I worry that there are others I have forgotten to mention. Something as big as a book is never (in my experience) really authored by one person, and the influences and inspirations are hard to retrace. But there are two readers I want to particularly thank: Jeremy Millar, who offered at our first and only meeting (so far!) to read the entire manuscript in draft form. In an extraordinary act of generosity, he kept his promise, read it over one weekend, and gave me the most useful feedback I have ever received. Most of all, he taught me to trust my own writing. Jeremy, the final book differs significantly from the draft you read, but I hope you still like it. The second reader is John Parish, my partner of over thirty years, who read and reread more drafts than anyone, and never once complained about my laptop accompanying me everywhere. John—even after so much practice at writing, I don't have the words to express how much I owe you.

NOTES

CHAPTER 1

01 This allusion to Plato's allegory of the cave and the incubation rituals of ancient mystics (echoing the references to Plato's cave in numerous books on photography) draws on my collaboration with Junko Theresa Mikuriya in Michelle Henning and Junko Theresa Mikuriya, "Light Sensitive Material: An Introduction," *photographies* 14, no. 3 (2021): 381–94.

02 Heike Behrend, *Contesting Visibility: Photographic Practices on the East African Coast* (Bielefeld, Germany: Transcript, 2013), 14.

03 Richard Hamblyn, *The Invention of Clouds: How an Amateur Meteorologist Forged the Language of the Skies* (London: Picador, 2001).

04 Robert J. Hercock and George A. Jones, *Silver by the Ton: The History of Ilford Limited, 1879–1979* (Maidenhead: McGraw-Hill, 1979), 47; Matthew Johnson, *Militarism and the British Left, 1902–1914* (London: Palgrave Macmillan, 2013), note 45, 202. As Lutz Alt notes, "During the 1920s, Ilford took a leaf out of the Eastman Kodak book and acquired some 15 British producers. By the time Ilford was through, it and Kodak were the British photochemical industry." "The Photochemical Industry: Historical Essays in Business Strategy and Internationalization," *Business and Economic History* 16, (1987): 187. According to company historian A. J. Catford, there were several pre-WWI attempts by Ilford to set up manufacturing plants abroad: in Germany, where they failed to agree terms with the German collaborators, in Hungary where a similar thing happened, and in Japan. A. J. Catford, "Our First 75 Years," unpublished manuscript, circa 1955, Redbridge Museum and Heritage Centre, Ilford Limited collections, Box 1361, 90/359/E1/6.

05 Simon Carter, *Rise and Shine: Sunlight, Technology and Health* (Oxford: Berg, 2007), 45. Sunlight cures were popularized, and technologies introduced to measure and control exposure (of skin to ultraviolet light) at the same time as light-meters and exposure tables began to be more widely used for photography.

06 Film speeds were rated according to the Hurter and Driffield (H&D) system that preceded the ISO system more familiar to photographers today. Available light was measured by photographers using printed tables and manual exposure meters; photoelectric light meters were introduced in the 1930s.

07 Nevertheless, film does seem to be experiencing a small revival, according to a 2024 article: "Kodak has seen demand for film roughly double in the last few years and in July, Harman, Britain's only manufacturer of 35mm film, announced a multimillion-pound investment in new equipment inspired by growing demand." (Harman is Harman Technology). Ellie Violet Bramley, "'Mistakes are Romantic': The Revival of Point-and-Shoot Cameras," *The Guardian*, August 16, 2024.

08 Boaz Levin and Esther Ruelfs, "Photography and Climate Change: The Engine of Reflection, and Its Footprint," in *Mining Photography: The Ecological Footprint of Image Production*, ed. Boaz Levin and Esther Ruelfs (Leipzig: Spector Books, 2022), 14.

09 Cited in Birgit Griesecke, "Lost Threads: Recovering the Unnarratable in the Bristol Circle Drug Experiments," *International Congress Series* 1242, *The History of Anesthesia* (Cambridge MA: Elsevier, 2002), 516.

10 Griesecke, "Lost Threads," 569.

CHAPTER 2

01 Carolyn Steedman, "Something She Called a Fever: Michelet, Derrida, and Dust," *The American Historical Review* 106, no. 4 (Oct. 2001): 1173.

02 Walter Benjamin, "On the Concept of History," in *Walter Benjamin: Selected Writings*, vol. 4, *1938–1940*, ed. Michael W. Jennings (Cambridge, MA: Harvard University Press, 2003), 401–11.

03 Robert Darnton, "How Historians Play God," *Raritan* 22, no. 1 (Summer 2002): 5.

04 Darnton, "How Historians Play God," 17.

05 Centenaries are often a "pretext for commissioned histories," according to Agnès Delahaye, Charles Booth, Peter Clark, Stephen Procter, Michael Rowlinson, "The Genre of Corporate History," *Journal of Organizational Change Management* 22, no. 1 (2009): 33.

06 Hercock and Jones, *Silver by the Ton*, 91.

07 The characteristic mode of the corporate history genre is the chronicle, which emphasizes leaders, Delahaye et al., "The Genre of Corporate History," 36.

08 Redbridge Museum and Heritage Centre itself already began this work with an exhibition about Ilford Limited, *Ilford Limited: Analogue Stories*, in 2023.

09 Cecil N. Potter, "The Works," unpublished talk, circa 1945, 6, 7. Redbridge Museum and Heritage Centre, Ilford Limited collections, Box 1362, 90/359/E1/3.

10 Potter, "The Works."

11 For an important exception see Kelley Wilder, "Photography and the Art of Science," *Visual Studies* 24, no. 2 (2009): 163–68.

CHAPTER 3

01 Benjamin makes this argument across several texts, most explicitly in "On Some Motifs in Baudelaire," in *Walter Benjamin: Selected Writings*, vol. 4, *1938–1940*.

02 Jacques Rancière, *The Politics of Aesthetics: The Distribution of the Sensible*, trans. Gabriel Rockhill (London: Continuum, 2004), 12.

03 Jane Bennett, *Vibrant Matter: A Political Ecology of Things* (Durham, NC: Duke University Press, 2010); Donna J. Haraway, *When Species Meet* (Minneapolis, MN: University of Minnesota Press, 2007) and *Staying with the Trouble: Making Kin in the Chthulucene* (Durham, NC: Duke University Press, 2016); Bruno Latour, *We Have Never Been Modern* (Cambridge, MA: Harvard University Press, 2012). The term "nonhuman" can imply that other beings can only be defined negatively in relation to humans; Haraway uses the term "critter" to avoid this. Following Latour, I use nonhuman to refer to a much wider range of entities, including concepts, technical objects, and infrastructure.

04 Amitav Ghosh, *The Nutmeg's Curse: Parables for a Planet in Crisis* (London: John Murray, 2021), 86–87. In Western culture, animism was central to the definition of primitivism.

05 Ghosh, *The Nutmeg's Curse*, 37.

06 Bruno Latour, *Facing Gaia: Eight Lectures on the New Climate Regime* (Cambridge: Polity, 2017), 141. This phrase is also in quotation marks in Latour's text, though it is unclear who he is citing.

07 Peter Geimer, *Inadvertent Images: A History of Photographic Apparitions* (Chicago: University of Chicago Press, 2018), 7.

08 Mary Shelley, *Frankenstein* (London: Collins Classics, 2010), 199, originally published in 1818.

09 See for example, Susan Sontag, *On Photography* (London: Penguin Books,

1977), and the counterarguments provided by Jane Gallop in *Living with His Camera* (Durham, NC: Duke University Press, 2003), and by Neil Evernden in "Seeing and Being Seen: A Response to Susan Sontag's Essays on Photography," *Soundings* (1985): 72–87.

10 Caroline Braunmühl describes how Western culture values a subject who can act and has power over objects, and aligns passivity with weakness. Caroline Braunmühl, "Beyond Hierarchical Oppositions: A Feminist Critique of Karen Barad's Agential Realism," *Feminist Theory* 19, no. 2 (2018): 223–40.

CHAPTER 4

01 See for example, Gisèle Freund, *La photographie en France au dix-neuvieme siècle: Essai de sociologie et d'esthétique* (Paris: A Monnier, 1936) and Lucia Moholy, *A Hundred Years of Photography* (London: Pelican Books, 1939).

02 Soraya Boudia et al., *Residues: Thinking Through Chemical Environments* (New Brunswick, NJ: Rutgers University Press, 2022), 15.

03 On Barak, *Powering Empire: How Coal Made the Middle East and Sparked Global Carbonization* (Oakland: University of California Press, 2020), 4–5.

04 François Jarrige and Thomas Le Roux, *The Contamination of the Earth: A History of Pollutions in the Industrial Age* (Cambridge, MA: MIT Press, 2020), 97.

05 Eric Hobsbawm, *Industry and Empire: From 1750 to the Present Day* (New York: The New Press, 1999), 16, originally published in 1968.

06 Barak, *Powering Empire*, 3.

07 Andreas Malm, *Fossil Capital: The Rise of Steam Power and the Roots of Global Warming* (London: Verso, 2016).

08 Malm, *Fossil Capital*, 17. The reference to the Song Dynasty is from Jussi Parikka, *A Geology of Media* (Minneapolis: University of Minnesota Press, 2015), 17.

09 Malm, *Fossil Capital*, 36.

10 Steve Edwards, *The Making of English Photography: Allegories* (University Park: Penn State University Press, 2006), 39–40.

11 On colonial relations to land, see the introduction to Max Liboiron, *Colonialism Is Pollution* (Durham: Duke University Press, 2021).

12 Ghosh, *The Nutmeg's Curse*, 58.

13 *Photograms of the Year* (London: Dawbarn & Ward Ltd., 1997), 57.

14 He was on the armored cruiser HMS *Shannon*, which fortunately was not called on to engage in the conflict. He also worked in the docks of the East End of London, and later in Bristol, as a boilermaker.

15 "Enrolment of Gas Stokers and Others in the Royal Naval Reserve," *South Metropolitan Gas Company Co-Partnership Journal* 1, no. 6 (June 1904): 90.

16 Tony Chamberlain, "'Stokers—The Lowest of the Low?' A Social History of Royal Navy Stokers 1850–1950" (PhD diss., University of Exeter, 2013), 30.

17 Chamberlain, "'Stokers,'" 91.

18 Chamberlain, "'Stokers,'" chap. 3.

19 International Energy Agency, "Global CO2 Emissions Rebounded to Their Highest Level in History in 2021," press release, March 8, 2022, https://www.iea.org/news/global-co2-emissions-rebounded-to-their-highest-level-in-history-in-2021; Hannah Ritchie, Max Roser, and Pablo Rosado, "Energy Mix" (2022), Our World in Data (website), https://ourworldindata.org/energy; "Fossil Fuels," The Environmental and Energy Study Institute, updated July 22, 2021, https://www.eesi.org/topics/fossil-fuels/description; Xiaoying You, "What Does China's Coal Push Mean for Its Climate Goals?," Carbon Brief (website), March 29, 2022, https://www.carbonbrief.org/analysis-what-does-chinas-coal-push-mean-for-its-climate-goals/; Sudarshan Varadhan, "India Expects Utilities' Annual Coal Demand to Surge About 8% after Renewables Shortfall," *Reuters*, January 10, 2023, https://www.reuters.com/world/india/india-expects-utilities-annual-coal-demand-surge-about-8-after-renewables-2023-01-10/. For a discussion of the disastrous effects of present-day coal mines and power plants in the China-Pakistan Economic Corridor on Indigenous communities, and their protests against this see Ayesha Omer, "Coal Ground," in *Cultural Studies* 35, nos. 4–5 (2021): 920–45.

20 Dipesh Chakrabarty, *The Climate of History in a Planetary Age* (Chicago: University of Chicago Press, 2021), 59.

21 In 2021 coal was the "main factor driving up global energy-related CO_2 emissions by over 2 billion tonnes, their largest ever annual rise in absolute terms." International Energy Agency, "Global CO_2 Emissions Rebounded to their Highest Level in History in 2021," press release, March 8, 2022.

22 On photography's dependence on mining and extraction, see Siobhan Angus, *Camera Geologica: An Elemental History of Photography* (Durham: Duke University Press, 2024). The phrase "daughter of coal" is taken from the Italian scientist Ciamician, discussed further in chapter 17.

CHAPTER 5

01 *The Merchandise Bulletin*, May 24, 1927, 518. In the Boots Archive, at Walgreens Boots Alliance Archive. Boots was an official Kodak dealer.

02 The term is David Edgerton's and I return to it in later chapters. David E. H. Edgerton, *Warfare State: Britain, 1920–1970* (Cambridge: Cambridge University Press, 2006).

03 Potter, "The Works," 16.

04 Boots the Chemist had over a thousand branches in Britain by 1935, most with photography departments, but from 1935 through 1940 the only color film for ordinary cameras that the company seems to have stocked (judging by price lists in the archive) was Dufaycolor. This additive color reversal (transparency) film was produced through a collaboration between the paper company Spicers Limited and Ilford Limited. Ilford introduced and then quickly discontinued a color printing service for Dufaycolor. I could not find Kodak and Agfa's subtractive color reversal films on Boots's price lists during the 1930s (held at the Walgreens Boots Alliance Archive). Although Agfa introduced their color negative film in 1939, and Kodak brought out Kodacolor in 1942, and Ektacolor in 1947, it would be decades before amateur photographers would regularly use color films.

05 *The Merchandise Bulletin*, October 30, 1928, 694.1.

06 Sales Research Services, "Report on Investigation Concerning the Sales of Photographic Roll Film with Special Reference to Selo brand," October 1934, Science and Industry Museum, Manchester, MS0232/49/9.

07 On the unruliness of the tropics, see David Arnold, "The Place of 'The Tropics' in Western Medical Ideas since 1750," *Tropical Medicine and International Health* 2, no. 4 (1997): 303–13.

08 *The Merchandise Bulletin*, no. 225, January 13, 1931, 1823.

09 The experiments were unsuccessful at identifying a way to purify the air and "failed to reveal the emulsion-fogging constituent of London fog." Edwin E. Jelley, "London Fog—A Chemical Investigation," December 22, 1931, Kodak Ltd. Report, Kodak Collection, The British Library, A2827 Research Reports, Vol. 1. H196. The study was started on October 24, 1930. For more on Edwin Jelley, see Kelley Wilder, "Science, Art, and the Business of Color," in *The Colors of Photography*, ed. Bettina Gockel (Berlin: De Gruyter, 2021), 315–16.

10 G. S. Callendar, "The Artificial Production of Carbon Dioxide and Its Influence on Temperature," *Quarterly Journal of the Royal Meteorological Society* 64, no. 275 (1938): 236.

11 See Christine L. Corton, *London Fog: The Biography* (Cambridge, MA: Harvard University Press, 2015). Though she doesn't explicitly make the connection with poison gas, she writes that "looming over" interwar representations of the London fog is the "collective memory of the First World War" (271).

CHAPTER 6

01 Benoit Nemery, Peter H. M. Hoet, and Abderrahim Nemmar, "The Meuse Valley Fog of 1930: An Air Pollution Disaster," *The Lancet* 357, no. 9257 (2001): 704–8. The original report is J. Firket, "Sur les causes des accidents survenus dans la vallée de la Meuse lors des brouillards de décembre 1930: Résultat de l'expertise judiciaire faite par MM. Dehalu, Schoofs, Mage, Batta, Bovy et Firket," *Bulletins de l'Académie royale des sciences, des lettres et des beaux-arts de Belgique* 11 (1931).

02 Jarrige and Le Roux, *The Contamination of the Earth*, chap. 4.

03 Report cited in Nemery, Hoet, and Nemmar, "The Meuse Valley Fog of 1930," 706.

04 Alexis Zimmer, "In a Fossil Fuel Economy, Air Can Become Unbreathable," Feral Atlas, accessed July 26, 2023, https://feralatlas.supdigital.org/poster/in-a-fossil-fuel-economy-air-can-become-unbreathable.

05 Zimmer, "In a Fossil Fuel Economy."

06 Nemery, Hoet, and Nemmar, "The Meuse Valley Fog of 1930," 706.

07 Respectively: *Nottingham Evening Post*, December 6, 1930; *Western Daily Press*, December 6, 1930; *Lancashire Evening Post*, December 6, 1930; *London Daily News*, December 6, 1930; *Belfast Newsletter*, December 6, 1930; "Scenes Reminiscent of the War," *London Daily News*, December 6, 1930; "Fog Horror in Belgium," *Western Morning News*, December 6, 1930 (compares to wartime gas attack); "War Time Dumps?," *Edinburgh Evening News*, December 6, 1930; "Form of Plague," *Liverpool Echo*, December 6, 1930; "Plague Theory in Belgian Valley of Terror," *Derby Daily Telegraph*, December 6, 1930; "Black Death or Poison Fumes?," *Northern Whig* (Antrim), December 6, 1930; "Deadly Germ Suspected in Stricken Valley," *The People*, December 7, 1930; "Search for Deadly Germs That Caused Fog Deaths," *Daily Mirror*, December 8, 1930.

08 "The Day That Never Came," *Evening Standard*, December 17, 1930, quoted in Stephen Legg, "'Political Atmospherics': The India Round Table Conference's Atmospheric Environments, Bodies and Representations, London 1930–1932," *Annals of the American Association of Geographers* 110, no. 3 (2020): 780–81.

09 "The Fog Disaster: Could a Similar One Occur in Britain," *London Daily News*, December 11, 1930; "False Fear of Belgian Fog Peril in England," *Birmingham Daily Gazette*, December 8, 1930; "London's Peril: Will Belgian Tragedy Be Repeated?," *Westminster and Pimlico News / Chelsea News and Advertiser*, December 12, 1930.

10 A Doctor, "Why Danger Lurks in Fog: History Repeats Itself in Belgium," *The Courier and Advertiser*, December 8, 1930.

11 Ernst Homburg and Elisabeth Vaupel, introduction to *Hazardous Chemicals: Agents of Risk and Change, 1800–2000*, ed. Ernst Homburg and Elisabeth Vaupel (New York: Berghahn, 2019), 9–10.

CHAPTER 7

01 The public rejection of tear gas meant that it was not supplied to police in Britain or the colonies until 1934. Instead, the methods that the police did use were shooting, bayoneting, and bludgeoning. From 1934, facing increasingly powerful uprisings, British colonial police forces were supplied with tear gas, and in 1939, in Burma, it was first used on a crowd of people. Erik Linstrum, "Domesticating Chemical Weapons: Tear Gas and the Militarization of Policing in the British Imperial World, 1919–1981," *The Journal of Modern History* 91, no. 3 (2019): 572–73; Simeon Shoul, "British Tear Gas Doctrine Between the World Wars," *War in History* 15, no. 2 (2008): 190.

02 "Foggy Morning," *Picture Post* 2, no. 3, January 21, 1939, 35.

03 Ernest Restell, "Foggy Morning, Ugh!," in "What Our Readers Say," *Picture Post*, February 4, 1939, 7.

CHAPTER 8

01 William Henry Fox Talbot, "Letter to Constance Talbot," June 15, 1841, *The Correspondence of William Henry Fox Talbot*, ed. Larry Schaaf, document no. 4282, accessed July 18, 2023, http://foxtalbot.dmu.ac.uk/. One of the key pioneers of photography, Talbot had invented the first negative-positive photographic process, announced to the public shortly after the daguerreotype in 1839. The calotype, introduced in 1841, involved the use of photographic developer, allowing for shorter exposure times.

02 Redbridge Museum and Heritage Centre, Ilford Limited collections, Box 1372, 90/359/B1/A13. Illingworth was one of the companies taken over by Ilford Limited.

03 National Meteorological Archive, Monthly Weather Reports, https://digital.nmla.metoffice.gov.uk/.

04 Olaf F. Bloch, "The Chemist in the Photographic Industry," *The British Journal of Photography*, February 17, 1928, 94–96.

05 Bloch, "The Chemist in the Photographic Industry," 94.

06 *Online Etymology Dictionary*, "Fog," accessed 18 July 2023, https://www.etymonline.com/word/fog.

07 On catachresis and photography, see Richard Shiff, "Phototropism (Figuring the Proper)," *Studies in the History of Art* 20 (1989): 174.

08 Collodion was invented in 1847. Geoffrey Batchen, *Negative/Positive: A History of Photography* (London: Routledge, 2021), 95.

09 Frederick Scott Archer, *The Collodion Process on Glass*, 2nd ed. (London, 1854), 27–28.

10 C. H. Bothamley, *The Ilford Manual of Photography* (London: Ilford Limited, 1921), 81. Cotton-based paper could contain metal from the metal buttons on the clothes that were recycled in its production.

11 Bloch, "The Chemist in the Photographic Industry," 94.

12 Bloch, "The Chemist in the Photographic Industry," 94.

13 See, for example, M. H. Ellis, *The Ambrotype and Photographic Instructor* (Philadelphia: Myron Shew, 1856), 59, and Charles Long, *Practical Photography on Glass and Paper: A Manual* (London: Bland & Long, 1856), 39.

14 Louis Katin, "A Fog-Bound Londoner's Lament," *Palestine Post*, February 28, 1934. The *Palestine Post* was a Zionist-Jewish English-language daily newspaper published in British-mandate Palestine from 1932 to 1948. Katin's 1934 lament was published in the context of British restrictions on Jewish immigration to Palestine in this period, even following the Nazi rise to power in Germany in 1933. Katin's idea that the best person to "see" in a fog is someone who is blind is echoed in photography companies' decisions to employ blind or visually impaired people to work in the dark spaces of the factories. See Fiona Kinsey, "Sensing in the Dark: The Role of Kodak Australasia Factory in the Making of a Photograph," paper given at the Photographic History Research Centre (PHRC) conference *The Photographer's Assistants*, June 17, 2024.

CHAPTER 9

01 Arthur Symons, *London: A Book of Aspects*, London 1909, 2.

02 "The Photographic Salon—The Eleventh Annual Exhibition" *The Amateur Photographer* 38, no. 990, (September 24, 1903) 243–45, and "Review of the Photographic Exhibition," *Country Life* 14, no. 351, (September 26, 1903) 447–48. Both cited in Robert William Crow, "Reputations Made and Lost: The Writing of Histories of Early Twentieth-Century British Photography and the Case of Walter Benington" (PhD diss., University of Gloucestershire, 2015), 117.

03 On the earlier connections between the fog and imperialism see Julia A. Alessandrini, "'London Fog' and the Symbolism of Empire: Nineteenth-

Century Artistic Representations of Atmosphere in London and Other Imperial Sites" (PhD diss., University of Western Australia 2012).

04 Brassaï cited in Richard Stamelman, "Photography: The Marvelous Precipitate of Desire," *Yale French Studies* 109 (2006): 75. Brassaï's book *Paris de nuit* (1932) influenced both Bill Brandt's *A Night in London* (1938) and Harold Burdekin's *London Night* (1934/36).

05 Stamelman, "Photography," 81. An eight-page feature of Brassaï's nighttime photos, "Paris by Night," appeared in the same issue of *Picture Post* as "Foggy Morning"—*Picture Post* 2, no. 3, January 21, 1939, 20–27.

06 Peter Geimer has written about these kinds of flaws in *Inadvertent Images*.

07 Jesse Oak Taylor, *The Sky of Our Manufacture: The London Fog in British Fiction from Dickens to Woolf* (Charlottesville: University of Virginia Press, 2016), 15.

08 Tobias Menely, *Climate and the Making of Worlds: Toward a Geohistorical Poetics* (Chicago: University of Chicago Press, 2021), 10.

09 On climate and visual culture, see Michel Serres, "Turner Translates Carnot," in *Calligram: Essays in New Art History from France*, ed. Norman Bryson (Cambridge: Cambridge University Press, 1988), 154–65, and Serres, "Science and the Humanities: The Case of Turner," *SubStance* 83 (1997): 6–21.

10 Taylor, *The Sky of Our Manufacture*, 15.

11 Corton, *London Fog: The Biography*.

12 Walter Benjamin, "Little History of Photography," in *Walter Benjamin: Selected Writings*, vol. 2, *1927–1934*, ed. Michael W. Jennings, Howard Eiland, and Gary Smith (Cambridge, MA: Harvard University Press, 1999), 518.

13 F. J. Mortimer, "Death of Pictorial Photography in Germany" (editorial), *The Amateur Photographer and Cinematographer*, January 13, 1937, 25.

14 Katerina Korola, "The Air of Objectivity: Albert Renger-Patzsch and the Photography of Industry," *Representations* 157 (2022): 93.

15 Korola, "The Air of Objectivity," 4, 102.

16 Korola, "The Air of Objectivity," 102–3.

CHAPTER 10

01 Sean O'Hagan, "Ravens by Masahisa Fukase Review—a Must for Any Serious Photobook Buff," *The Guardian*, May 30, 2017, https://www.theguardian.com/books/2017/may/30/ravens-masahisa-fukase-review-are-celebrated-photo-book.

02 Robin Kelsey, *Photography and the Art of Chance* (Cambridge, MA: Harvard University Press, 2015).

03 Walter Benington, "My Best Picture and Why I Think So." *Photographic News* 51, no. 580 (February 8, 1907), 108. Cited in Crow, "Reputations Made and Lost," 116–117.

04 Katerina Korola, "How to Photograph the Air: Photography, Cinema, and the Problem of Atmosphere in German Modernism" (PhD diss., University of Chicago, 2021), 9. Korola is talking, in particular, of the distinctive German concept of *Stimmung*, which is slightly different from the English *atmosphere*.

05 "November," *The British Journal of Photography* 45, no. 2012 (November 25, 1898): 757.

06 See Michelle Henning, *Photography: The Unfettered Image* (London: Routledge, 2018).

07 Pauline Martin, *Le Flou et la Photographie: Histoire d'une Rencontre (1676–1985)* (Rennes: Presses Universitaires de Renne, 2023), 13, my translation.

08 Martin, *Le Flou et la Photographie*, 16, my translation.

09 John Ruskin, *John Ruskin: Selected Writings*, ed. Dinah Birch (Oxford: Oxford University Press, 2004), 79.

10 Ruskin, *Selected Writings*, 267, 269, 273.

11 Neil Evernden, "Beyond Ecology: Self, Place, and the Pathetic Fallacy," *The North American Review* 263, no. 4 (1978): 19.

12 Bernard F. Dick, "Ancient Pastoral and the Pathetic Fallacy," *Comparative Literature* 20, no. 1 (1968): 27.

13 Amanda Hopkinson, "Wolfgang Suschitzky Obituary," *The Guardian*, October 7, 2016.

14 Kelsey, *Photography and the Art of Chance*, 314.

CHAPTER 11

01 Hervé Guibert, *Ghost Image*, trans. Robert Bononno (Chicago: The University of Chicago Press, 2014), e.g., 83, 89, 139.

02 Bothamley, *The Ilford Manual of Photography*, 83.

03 Catford, "Our First 75 Years," 60.

04 Bothamley, *The Ilford Manual of Photography*, 83.

05 Backings (applied again after development) were also a way to invert the negative so that it could be viewed as a positive image. See Batchen, *Negative/Positive*, 95. On antihalation backings, see Catford, "Our First 75 Years," 61; Geimer, *Inadvertent Images*, 54. The 1935 *Ilford Manual* recommends caramel, which Ilford Limited used for its earliest factory-made backed plates. George E. Brown, *The Ilford Manual of Photography* (London: Ilford Limited, 1935), 45.

06 Catford, "Our First 75 Years," 61.

07 See Michelle Henning, "The Floating Face: Garbo, Photography and Death Masks," *photographies* 10, no. 2 (2017): 157–78.

08 Benjamin, "The Artwork in the Age of Its Technological Reproducibility (2nd Version)," in *Walter Benjamin: Selected Writings*, vol. 3, *1935–1938*, ed. Howard Eiland and Michael W. Jennings (Cambridge, MA: Harvard University Press, 2002), 104–5.

09 Bothamley, *The Ilford Manual of Photography*, 111.

10 Peter Geimer argues that discussions of halation among nineteenth-century commentators imply that the photographic plate had its own "fanciful imagination." Geimer, *Inadvertent Images*, 53–55.

11 Robert Alter writes "the precise meaning of 'glory' in Wordsworth's English is 'aura.'" Robert Alter, "In the Community: On Walter Benjamin," *Commentary* 48, no 3 (1969): 86, 92.

CHAPTER 12

01 Jane Green-Pettersson, "When Did the Swallow Dive Become an Endangered Species?," *The Guardian*, February 19, 2014, https://www.theguardian.com/lifeandstyle/the-swimming-blog/2014/feb/19/swallow-dive-swimming-diving-platform.

02 Potter, "The Works," 13. Potter writes, "Before that time there was an Annual beanfeast when the factory was closed for a day and the employees went off to Yarmouth, Brighton, Margate or some such place for the day."

03 According to Michael Mackenzie, the 1928 Olympic Games in Amsterdam were the first to systematically use film and photography to measure athletes' performance. Michael Mackenzie, "The Athlete as Machine: A Figure of Modernity in Weimar Germany," in *Leibhaftige Moderne: Körper in Kunst und Massenmedien, 1918–1933*, ed. Michael Cowan and Kai Marcel Sicks (Bielefeld: transcript Verlag, 2005), 48–62.

04 Films in the interwar period had a greater exposure latitude than later films, making precise light metering less necessary. Comparing old and new film speeds is difficult, since in 1960 the rating system changed. According to Maurice Fisher's *photomemorabelia* website, Ilford HP3 roll film, introduced in 1941 and rated 125ASA, was rerated at 250ASA after 1960 without any change in the film itself (250ASA is equivalent to 250 ISO). Last modified September 21, 2024, http://www.photomemorabilia.co.uk/Ilford/Chronology.html.

05 Franz Roh and Jan Tschichold. *Foto-Auge / Oeil et Photo / Photo-Eye: 76 Photos of the Period* (Stuttgart: Akademischer verlag Dr. Fritz Wedekind, 1929);

László Moholy-Nagy, *Painting, Photography, Film* (London: Lund Humphries, 1969).

06 Michael Cowan, "Imagining the Nation through the Energetic Body: The 'Royal Jump,'" in *Leibhaftige Moderne: Körper in Kunst und Massenmedien, 1918–1933*, ed. Michael Cowan and Kai Marcel Sicks (Bielefeld: transcript Verlag, 2005), 71.

07 Patricia Vertinsky, "'Building the Body Beautiful' in The Women's League of Health and Beauty: Yoga and Female Agency in 1930s Britain," *Rethinking History* 16, no. 4 (2012): 517–42.

08 See Christina Kiaer, "Was Socialist Realism Forced Labour? The Case of Aleksandr Deineka in the 1930s," *Oxford Art Journal* 28, no. 3 (2005): 321–45; Petra Rau, "The Fascist Body Beautiful and the Imperial Crisis in 1930s British Writing," *Journal of European Studies* 39, no. 1 (2009): 5–35.

09 Carter, *Rise and Shine*; Charlotte Macdonald, "Body and Self: Learning to be Modern in 1920s–1930s Britain," *Women's History Review* 22, no. 2 (2013): 267–79.

10 Macdonald, "Body and Self," and Ina Zweiniger-Bargielowska, "The Making of a Modern Female Body: Beauty, Health and Fitness in Interwar Britain," *Women's History Review* 20, no. 2 (2011): 299–317.

11 For example, Herbert N. Casson's claims cited in Joanna Bourke, "The Great Male Renunciation: Men's Dress Reform in Inter-War Britain," *Journal of Design History* 9, no. 1 (1996): 26.

12 Bourke, "The Great Male Renunciation," 27.

13 "Dufaycolor Cine Film 9.5mm and 16mm," undated, MS0232/5/31, and "Dufaycolor Roll Film and Film Pack," 1935, MS0232/5/21, Science and Industry Museum, Manchester.

14 Mary Russo, *The Female Grotesque: Risk, Excess and Modernity* (London: Routledge, 2012), 51.

15 See Chakarabarty's reflections on the suicide note of Rohith Vemula, Dalit student activist at the University of Hyderabad, in Chakrabarty, *The Climate of History in a Planetary Age*, 114–18.

CHAPTER 13

01 *The Merchandise Bulletin*, March 15, 1927, 475, Walgreens Boots Alliance Archive.

02 *The Merchandise Bulletin*, Christmas 1937, Walgreens Boots Alliance Archive.

03 Ilford Limited, *Night Photography* (c. 1936–37).

04 *The Merchandise Bulletin* 181, February 4, 1930, 1449, Walgreens Boots Alliance Archive.

05 Ilford Limited, *Night Photography*.

06 "Advertising Department Report for the Year," Museum of Science and Industry, Manchester MS0232/49/8.

07 Ilford Limited, *Ilford Selo Amateur Photographic Handbook*, c. 1940, 1.

08 Unsuitable subjects included sexual ones but also a wide range of subjects that violated norms of intimacy and privacy. Photographing deceased family members, for example, a practice acceptable to the Victorians, was no longer acknowledged although doubtless it continued to some extent. The fact that the images would pass through the hands of workers in the developing and processing labs may have inhibited many.

09 *The Amateur Photographer and Cinematographer*, January 6, 1937, 1.

10 Jonathan Crary, *24/7: Late Capitalism and the Ends of Sleep* (London: Verso Books, 2013), 79.

11 As David Harvey explains, capitalist modernity was producing new "spatio-temporal rhythms" and overturning older ones. David Harvey, *The Condition of Postmodernity: An Enquiry into the Origins of Cultural Change* (Oxford: Blackwell, 1989), 216.

12 Ilford Limited, *Night Photography*.

13 Judith Walkowitz, *Nights Out: Life in Cosmopolitan London* (New Haven: Yale University Press, 2012), 185.

14 Anna Cottrell, *London Writing of the 1930s* (Edinburgh: Edinburgh University Press, 2017), 37–38.

15 W.L.F.W, "Day and Night Versions," *The Amateur Photographer and Cinematographer*, (February 10, 1937, 22.

16 H.V. Morton, *The Nights of London*, 5th Edition (London: Methuen, 1932), 3.

17 Mariana Valverde, "The Dialectic of the Familiar and the Unfamiliar: The Jungle in Early Slum Travel Writing," *Sociology* 30, no. 3 (1996): 493–509. This "traffic in metaphors" depends on applying familiar views (for example about class and urban parts of Britain) to unfamiliar contexts (for example Central African jungles and peoples) and then reimporting these ideas "back into the more 'familiar' arena from which it came in order to provide it with scientific authorization." In this circular exchange, a vision of social degeneration in Britain informs an interpretation of the jungle, which in turn provides a vocabulary for understanding the slums of the East End of London.

18 Morton, *The Nights of London*, 1–3.

19 Morton's writings also make use of this version of London. As Robert Crow notes, Walter Benington's *Church of England* (plate 2) was repro-

duced in two of Morton's books about London: *The Heart of London* (1925) and *London* (1940). See Crow, "Reputations Made and Lost," 122–23.

CHAPTER 14

01 Gayatri Chakravorty Spivak, "The Rani of Sirmur: An Essay in Reading the Archives," *History and Theory* 24, no. 3 (1985): 252–53.

02 Martin Heidegger, "The Origin of the Work of Art," in *Poetry, Language, Thought*, trans. Albert Hofstadter (New York: Harper Perennial Classics, 2001).

03 Spivak, "The Rani of Sirmur," 253.

04 James L. Hevia, "The Photography Complex: Exposing Boxer-Era China 1900–1901, Making Civilization," in *Photographies East: The Camera and Its Histories in East and Southeast Asia*, ed. R. C. Morris (Durham: Duke University Press, 2009), 81.

05 Paul S. Landau, "Empires of the Visual: Photography and Colonial Administration in Africa," in *Images and Empires: Visuality in Colonial and Postcolonial Africa*, ed. Paul S. Landau and Deborah Kaspin (Berkeley: University of California Press, 2002), 142 and 151.

06 Unlike the British firms, and in common with other US corporate subsidiaries in Britain, Kodak Ltd. was part of a vertically integrated concern with the advantages of control over every aspect of manufacturing and distribution. They benefitted too from the greater clout of their home economy and the research and development there.

07 Hercock and Jones, *Silver by the Ton*, 114–15.

08 Hobsbawm, *Industry and Empire*, 165.

09 Hercock and Jones, *Silver by the Ton*,114.

10 He had also tried color processes, having most success with Agfa. Report on a lecture by A. T. Schofield, "Photography in Equatorial Africa," *The Photographic Journal* 74 (April 1934): 184.

11 "Export 1911," handwritten trading reports for Imperial Dry Plate Co. Ltd., Redbridge Museum and Heritage Centre, Ilford Limited collections, Box 1352, 90/359/A1/A1. Interestingly, their sales of plates and papers in West Africa doubled by the end of the First World War.

12 Jürg Schneider, "Looking into the Past and Present: The Origins of Photography in Africa," in *Photography and Its Origins*, ed. Tanya Sheehan and Andrés Mario Zervigón (London, Routledge 2015), 174.

13 Vera Viditz-Ward, "Photography in Sierra Leone, 1850–1918," *Africa* 57, no. 4 (1987): 511.

14 Olubukola A. Gbadegesin, "'Photographer Unknown': Neils Walwin Holm and the (Ir)Retrievable Lives of African Photographers," *History of Photography* 38, no. 1 (2014): 25.

15 Charles Gore, "Intersecting Archives: Intertextuality and the Early West African Photographer," *African Arts* 48, no. 3 (Autumn 2015): 10. In a note, Gore adds that these were presumably the Ilford plates (note 9).

16 Gore, "Intersecting Archives," 10.

17 Gbadegesin, "'Photographer Unknown,'" 25.

18 Jürg Schneider, "African Photography in the Atlantic Visualscape: Moving Photographers—Circulating Images," in *Global Photographies: Memory—History—Archives*, ed. Sissy Helff and Stefanie Michels (Bielefeld: transcript Verlag, 2018), 21.

19 Gore, "Intersecting Archives," 13; Schneider, "Looking into the Past and Present," 175.

20 Gbadegesin, "'Photographer Unknown,'" 22–23.

21 Schneider, "African Photography in the Atlantic Visualscape," 33.

22 Schneider, "African Photography in the Atlantic Visualscape," 22. As Gbadegesin points out, the identities and careers of the more affluent photographers are easier to track, though much remains unknown. Gbadegesin, "'Photographer Unknown,'" 23. Gore discusses the mobility of photographers through places that are in present-day Sierra Leone, Liberia, Côte d'Ivoire, Ghana, Togo, Benin and Nigeria. Gore, "Intersecting Archives," 7.

23 Viditz-Ward, "Photography in Sierra Leone," 511 and 513. Gore writes that African photographers and traders, even those "elite professionals of African descent," were more subject to discrimination and antagonism from Europeans in some colonial centers than in others; elite West Africans responded by using European codes to assert their social and professional status. Gore, "Intersecting Archives," 15.

24 Gore, "Intersecting Archives," 7; Gbadegesin, "'Photographer Unknown,'" 25.

25 Gore, "Intersecting Archives," 9 and 10.

26 Allister Macmillan, ed., *The Red Book of West Africa: Historical and Descriptive, Commercial and Industrial Facts, Figures and Resources* (London: W.H. & L. Collingridge, 1920), 210. Both father and son are depicted in the Red Book in relation to their respective professions. Thanks to Sarah Okpokam for pointing me to this source.

27 Gore, "Intersecting Archives," 12.

28 Gore, "Intersecting Archives," 16.

29 Schneider, "Looking into the Past and Present," 174. Both Holms's studios, in Accra and Lagos, also supplied photographs to the 1920 *Red Book of West Africa*, a British commercial guide to West Africa.

30 Gbadegesin, "'Photographer Unknown,'" 30. On Decker, see Viditz-Ward, "Photography in Sierra Leone," 513.

CHAPTER 15

01 R. Dykes, "Photography in Tropical West Africa," *The British Journal of Photography* 69, no. 3238 (May 26, 1922): 310.

02 David Arnold, "The Place of 'The Tropics' in Western Medical Ideas since 1750," *Tropical Medicine and International Health* 2, no. 4 (1997): 303–6. See also Roderick S. Edmond, "Returning Fears: Tropical Disease and the Metropolis," in *Tropical Visions in an Age of Empire*, ed. Felix Driver, Luciana Martins, and Luciana de Lima Martins (Chicago: University of Chicago Press, 2005), 175.

03 Warwick Anderson, "The Natures of Culture: Environment and Race in the Colonial Tropics," in *Nature in the Global South: Environmental Projects in South and Southeast Asia*, ed. Paul R Greenough and Anna Lowenhaupt Tsing (Durham: Duke University Press, 2003), 29.

04 Arnold, "The Place of 'The Tropics' in Western Medical Ideas Since 1750," 306–9.

05 Anderson, "The Natures of Culture," 29–30.

06 Etienne S. Benson, *Surroundings: A History of Environments and Environmentalism* (Chicago: University of Chicago Press, 2020), 52–53.

07 Edmond, "Returning Fears," 183, 182.

08 J. Jackson, "The Eclipse of the Sun," *The Times*, July 20, 1932, 11.

09 Lajos Biro cited in Max Quanchi, *Photographing Papua: Representation, Colonial Encounters and Imaging in the Public Domain* (Newcastle-upon-Tyne: Cambridge Scholars Publishing, 2009), 40.

10 Dykes, "Photography in Tropical West Africa," 310.

11 Quanchi, *Photographing Papua*, 40; Dykes, "Photography in Tropical West Africa," 310.

12 Ramsay Smith cited in Quanchi, *Photographing Papua*, 40–41.

13 "News and Notes: Climate and Photography," *The British Journal of Photography* 69, no. 3256 (Sept 29, 1922): 597.

14 "Photography in the Tropics," *The British Journal of Photography* 55, no. 2506 (May 15, 1908): 374–75.

15 Quanchi, *Photographing Papua*, 41.

16 Trading results for Imperial Dry Plate Co. Ltd, Redbridge Museum and Heritage Centre, Ilford Limited collections, Box 1352, 90/359/A1/A1.

17 E. J. Steer, "Letters to the Editor: Dufaycolor in Warm Climates," *The British Journal of Photography* 84, no. 4050 (1937): 812.

18 Report on a lecture by A. Coleman, "From Suez to the Himalaya," *The Photographic Journal* 69 (March 1929): 68.

19 Elizabeth Hutchinson, "'Photographic Weather': A Posthumanist Approach to Western Survey Photography," *Panorama: Journal of the Association of Historians of American Art* 6, no. 2 (Fall 2020): 6.

20 Quanchi, *Photographing Papua*, 43–44.

21 E. Davison, "The Exposure Problem Abroad," *The Amateur Photographer and Cinematographer*, June 16, 1937, 15.

22 W. W. Crealock, "Selochrome in the Tropics," *The Ilford-Selo Record* 1, January 1935 (Palestine edition), Museum of Science and Industry Manchester, MS0232/49/5.

23 A. P. Agnew, "Development Under Tropical Conditions," *The Photographic Journal* (March 1920): 120–23.

24 Alison Feser, "Reproducing Photochemical Life in the Imaging Capital of the World" (PhD diss., University of Chicago, 2020), 236–37, drawing on Walter Clark, "Proposed Tropical Research Laboratory in Panama," 1940.

25 "Dufaycolor Processing in the Tropics," *The British Journal of Photography* 86, no. 4137 (1939): 519.

26 "The Rolleiflex in the Tropics," *The British Journal of Photography* 82, no. 3938 (1935): 684.

27 Advertised for example in *The Amateur Photographer and Cinematographer*, February 10, 1937.

28 Thanks to Ella Ravilious for drawing my attention to this, and to the advertisement for Fowke's camera in *The Photographic Journal*, March 15, 1860, 249.

29 "Apparatus and Materials," *The Photographic Journal*, June 1925, 323.

30 Anderson, "The Natures of Culture," 31.

31 W. F. A. Ermen, "Letters to the Editor: Development in the Tropics," *The British Journal of Photography* 81, no. 3844 (January 5, 1934): 11–12.

32 A. Charcois, "Correspondence: Developing in the Tropics Without Ice," *The British Journal of Photography* 67, no. 3146 (August 20, 1920): 519.

33 Jackson, "The Eclipse of the Sun"; D. M. Cuthbertson, "Photography in the Tropics." *The British Journal of Photography* 81, no. 3851 (February 23, 1934): 104.

34 Hutchinson, "'Photographic Weather,'" 6.

35 Anderson, "The Natures of Culture," 42.

36 Krista A. Thompson, *An Eye for the Tropics: Tourism, Photography, and Framing the Caribbean Picturesque* (Durham: Duke University Press, 2006), 297–98, 302.

37 Thompson, *An Eye for the Tropics*, 4–11.
38 Thompson, *An Eye for the Tropics*, 19.

CHAPTER 16

01 Patrick Zander, "Wings over Everest: High Adventure, High Technology and High Nationalism on the Roof of the World, 1932–1934," *Twentieth Century British History* 21, no. 3 (2010): 304, 315. Blacker and Clydesdale's connections with fascism are less clear, but it seems certain Clydesdale moved in far-right circles. Lady Houston ran the far-right *Saturday Review* from early 1933. Zander, "Wings over Everest," 313.
02 *Wings over Everest*, dir. Ivor Montagu and Geoffrey Barkas (British-Gaumont, 1934).
03 Joppan George, "Everest from on High: British Imperial Aerial Expedition to the Himalayan Frontiers, 1932–1934," *Verge: Studies in Global Asias* 4, no. 1 (Spring 2018): 197.
04 Zander, "Wings over Everest," 300 and 302.
05 Clydesdale cited in Lord James Douglas-Hamilton, *Roof of the World: Man's First Flight over Everest* (New York: Random House, 2013).
06 Clydesdale speech to his parliamentary constituents, Times 1932, cited in Zander, "Wings over Everest," 323, and George, "Everest from on High," 198–99.
07 Letter cited in Janette Elaine Faull, "Climbing Mount Everest: Expeditionary Film, Geographical Science and Media Culture, 1922–1953" (PhD diss., Royal Holloway, University of London, 2019), 247.
08 Zander, "Wings over Everest," 319; George, "Everest from on High," 189.
09 P. F. M. Fellowes, L. V. Stewart Blacker, P. T. Etherton, and Lord Clydesdale, *First over Everest: The Houston Mount Everest Expedition, 1933* (London: John Lane, The Bodley Head Limited, 1933), 66. This book, authored by several of the men involved in the expedition, was immediately published and rapidly went through several reprints. It was itself an elaborate production, including fold-out maps, fifty-two illustrative plates, many photographs and even a pair of 3D glasses in a specially constructed pocket at the back of the book for viewing an Anaglyph.
10 Ilford's infrared plates used a special dye to make them sensitive to the invisible end of the spectrum just beyond visible red. In daylight, many infrared plates and films also need a dark red filter to cut out the visible light.
11 Fellowes et al., *First over Everest*, 67.
12 Fellowes et al., *First over Everest*, 68–70.

13 Blacker letter, June 1932, cited in George, "Everest from on High," 208.

14 Fellowes et al., *First over Everest*, 72.

15 D. R. Crone, "Experiences in Photographic Surveying," *The Photogrammetic Record* (October 1973): 699. George, "Everest from on High," 206. Blacker had a military and engineering background, Clydesdale was an outstanding pilot but neither man was a survey photographer.

16 Blacker compared the technical difficulties with those of photography in the tropics, and concerns were raised about the possibility of emulsions being affected by cosmic rays. L. V. Stewart Blacker, "The Mount Everest Flights," *The Himalayan Journal* 6 (1934), accessed March 29, 2024, https://www.himalayanclub.org/hj/6/5/the-mount-everest-flights/.

17 Blacker, "The Mount Everest Flights."

18 Crone, "Experiences in Photographic Surveying," 704.

19 George, "Everest from on High," 206.

20 George, "Everest from on High," 206.

21 "Everest Airmen Returning," *The Times*, April 20, 1933, 10.

22 "A Conquest Complete" (Editorial), *The Times*, April 20, 1933, 11.

23 George, "Everest from on High," 209.

24 "Everest Flown Again," *The Times*, April 21, 1933, 13.

25 "The Everest Pictures," *The Times*, May 8, 1933, 15.

26 L. V. Stewart Blacker, "Everest in the Camera," *The Times*, May 8, 1933, 15.

27 "The Everest Pictures," 15.

28 Fellowes et al., *First over Everest*, 72.

29 "Infra-Red Method of Photography," *The Times*, May 18, 1933, 11.

30 "Photographs of Everest: Exhibition to Be Opened Today," *The Times*, May 17, 1933, 12.

31 According to Rebecca Sitch, Marriott pursued a "project of developing a public taste for modern art." Rebecca Sitch, "Domesticating Modern Art: Charles Marriott (1869–1957) and the Art of Middlebrow Criticism," in *Transitions in Middlebrow Writing, 1880–1930*, ed. Kate Macdonald and Christoph Singer (London: Palgrave Macmillan, 2015), 78; "Photographic Truth," *The Times*, November 30, 1933, 12.

CHAPTER 17

01 Giacomo Ciamician, "The Photochemistry of the Future," *Science* 36, no. 926 (1912): 385, 389.

02 Plutarch, "Pyrrhus," in *Plutarch's Lives*, trans. John Dryden (1906), Internet Classics Archive, accessed July 25, 2023, http://classics.mit.edu/Plutarch/pyrrhus.html.

03 Peter Sloterdijk, *Terror from the Air* (Los Angeles: Semiotext(e), 2009), 40–44.

04 Rohan Deb Roy, "Quinine, Mosquitoes and Empire: Reassembling Malaria in British India, 1890–1910," *South Asian History and Culture* 4, no. 1 (2013): 70.

05 Walter M. Jarman and Karlheinz Ballschmiter, "From Coal to DDT: The History of the Development of the Pesticide DDT from Synthetic Dyes till *Silent Spring*," *Endeavour* 36, no. 4 (2012): 136.

06 Jarman and Ballschmiter, "From Coal to DDT," 136, 139.

07 Rachel Carson, *Silent Spring* (London: Penguin, 2000), originally published 1962; Carrie Arnold, "Consequences of DDT Exposure Could Last Generations," *Scientific American*, July 1, 2021, https://www.scientificamerican.com/article/consequences-of-ddt-exposure-could-last-generations/.

08 John R. Brown and John L. Thornton, "Percivall Pott (1714–1788) and Chimney Sweepers' Cancer of the Scrotum," *British Journal of Industrial Medicine* 14, no. 1 (1957): 68.

09 On April 20, 1895, Rehn reported to the annual Congress of the German Society of Surgery (Deutsche Gesellschaft für Chirurgie) in Berlin on three cases from a group of forty-five workers. Heiko Stoff and Anthony S. Travis, "Discovering Chemical Carcinogenesis: The Case of Aromatic Amines," in *Hazardous Chemicals: Agents of Risk and Change, 1800–2000*, ed. Ernst Homburg and Elisabeth Vaupel (New York: Berghahn, 2019), 141–42.

10 Stoff and Travis, "Discovering Chemical Carcinogenesis," 145, 146.

11 Stoff and Travis, "Discovering Chemical Carcinogenesis," 141. According to Jarman and Ballschmiter, this research done in the coal tar dye industry "ultimately led to the fields of toxicology, and industrial health, which in turn, alerted Carson to the environmental dangers of chemicals, especially DDT." "From Coal to DDT," 140.

12 "The workers diagnosed ill in 1895 had been working with aniline already around 1880"—Homburg and Vaupel, introduction, 1.

13 Kirsty Sinclair Dootson, *The Rainbow's Gravity: Color, Materiality and British Modernity* (New Haven: Paul Mellon Centre / Yale University Press, 2023), 2.

14 G. S. Callendar, "The Artificial Production of Carbon Dioxide and Its Influence on Temperature," *Quarterly Journal of the Royal Meteorological Society* 64, no. 275 (1938): 223–40.

15 Jarrige and Le Roux, *The Contamination of the Earth*, 4–8.

16 See John Tyndall, "On the Absorption and Radiation of Heat by Gases and Vapors, and on the Physical Connexion of Radiation, Absorption, and Conduction," *Philosophical Transactions of the Royal Society of London*, 151

(1861), 1–36; and Eunice Foote, "Circumstances Affecting the Heat of the Sun's Rays," *The American Journal of Science and Arts* 22 (1856): 382–83.

17 W. J. Streeter, *The Silver Mania: An Exposé of the Causes of High Price Volatility of Silver* (Dordrecht: D. Reidel Publishing Co., 1984), 125–26.

18 "Fixing and Washing Photographic Prints," *Photographic News*, March 25, 1880, 150.

19 Angus, *Camera Geologica*, 112. In the UK, platinum printing was dominated by the Platinotype company of Croydon. Ilford unsuccessfully competed with their Platona paper from 1899 to 1914. Platinotype ceased production not long after due to the high costs of platinum salts. See Catford, "Our First 75 Years."

20 Angus, *Camera Geologica*, 115.

21 Francis Albert Rollo Russell, "London Fogs," (1880), in *The Dictionary of Victorian London*, ed. Lee Jackson, accessed July 25, 2023, https://victorianlondon.org/weather/londonfogs.htm.

22 S. S. Field, "Coal-Tar," *South Metropolitan Gas Company Co-Partnership Journal* 1, no. 2 (February 1904): 21.

23 S. S. Field, "Coal Tar," *South Metropolitan Gas Company Co-Partnership Journal* 1, no. 8 (August 1904): 129–32, and no. 9 (September 1904): 163–65; "Coke and Its Uses as Fuel," *South Metropolitan Gas Company Co-Partnership Journal* 6, no. 130 (October 1914): 241–47, 246, 242.

CHAPTER 18

01 Throughout, I am citing the newer translation of Benjamin's selected works: Walter Benjamin, "The Artwork in the Age of Its Technological Reproducibility (2nd Version)," in *Walter Benjamin: Selected Writings*, vol. 3, *1935–1938*.

02 Benjamin credits *La Stampa Torino*, but it also seems to have appeared, for example, in *Il Mattino Napoli* and *La Gazzetta del Popolo*. Benjamin, whose Italian was limited, may have read it in a French newspaper.

03 Ernest Ialongo, "Filippo Tommaso Marinetti: The Futurist as Fascist, 1929–37," *Journal of Modern Italian Studies* 18, no. 4 (2013): 407, 400.

04 Ialongo, "Filippo Tommaso Marinetti," 411.

05 Filippo Tommaso Marinetti, "Estetica futurista della guerra," *Il Mattino Napoli*, October 27, 1935.

06 The extract from "The Futurist Aesthetic of War" reproduced in the English translation of Benjamin's essay may have passed through more than one language, since Benjamin probably read it first in French. Marinetti himself reworked and recycled the text later, first in issue no. 13–14 of the

Turin review *Stile Futurista*, and in *Il Poema African della Divisione '28 Ottobre* (Milan: A. Mondadori, 1937).

07 Filippo Tommaso Marinetti, *Teoria e invenzione futurista*, ed. Luciano De Maria (Milan: A. Mondadori, 1983), cited in Jeffrey T. Schnapp, "The Fabric of Modern Times," *Critical Inquiry* 24, no. 1 (Autumn 1997).

08 Schnapp, "The Fabric of Modern Times," 209.

09 This conclusion made his friend Theodor Adorno bristle. Adorno viewed art-for-art's-sake as *both* a symptom of capitalist alienation *and* one of the few means by which a space of freedom might be preserved, so he read Benjamin's attack on art-for-art's-sake as undialectical. Adorno ignored Benjamin's direct engagement with Marinetti and the Ethiopian war, which was happening when Benjamin first composed the essay in the autumn of 1935. See Theodor W. Adorno and Walter Benjamin, *The Complete Correspondence, 1928–1940*, ed. Henri Lonitz, trans. Nicholas Walker (Cambridge: Polity Press, 2003), 128.

10 Alexander De Grand, "Mussolini's Follies: Fascism in Its Imperial and Racist Phase, 1935–1940," *Contemporary European History* 13, no. 2 (2004): 127. On Marinetti's fascination with Africa, see Giovanna Trento, "From Marinetti to Pasolini: Massawa, the Red Sea, and the Construction of Mediterranean Africa in Italian Literature and Cinema," *Northeast African Studies* 12, no. 1 (2012): 283.

11 De Grand, "Mussolini's Follies," 136.

12 De Grand, "Mussolini's Follies," 140. Mussolini's declaration of victory in 1936 (although resistance persisted for years afterwards) also reasserted the racial inferiority of Ethiopians. Italo Brandimarte, "Breathless War: Martial Bodies, Aerial Experiences and the Atmospheres of Empire," *European Journal of International Relations* (2023): 7, drawing on A. Sbacchi, *Legacy of Bitterness: Ethiopia and Fascist Italy, 1935–1941* (Lawrenceville, NJ: Red Sea Press, 1997).

13 "I shall do everything in my power to prevent a colonial conflict from taking on the aspect and weight of a European war." Mussolini, radio speech, October 2, 1935. Available online in Italian at the Imperial War Museum, accessed August 26, 2023, https://www.iwm.org.uk/collections/item/object/80033686, and in English translation at History Central, accessed August 26, 2023, https://www.historycentral.com/HistoricalDocuments/Mussolini%27sSpeech.html.

14 Benjamin, "The Artwork in the Age of Its Technological Reproducibility," 105.

15 Walter Benjamin, "One-Way Street," in *Walter Benjamin: Selected Writings*, vol. 3, *1935–1938*, 486–87.

16 Benjamin, "One-Way Street," 487.

17 Benjamin understands war as diverting the potential and power of technology and of the mobilized working classes. Benjamin, "The Artwork in the Age of Its Technological Reproducibility (2nd Version)," 121.

18 Benjamin "One-Way Street," 487.

CHAPTER 19

01 Sloterdijk, *Terror from the Air*, 23.

02 Filippo Tommaso Marinetti, "La Guerra Elettrica (Visione-Ipotesi Futurista)," in *Teoria e Invenzione Futurista*, ed. L. De Maria (Milan: Mondadori, 1968), 273–78; Brandimarte, "Breathless War," 10.

03 Sloterdijk, *Terror from the Air*, 28. What Sloterdijk terms terrorism connects "domicide" (the deliberate destruction of home) with "ecocide" (the destructions of ecosystems). Writing about the seventeenth-century genocide of the Bandanese in Indonesia, for the sake of the nutmeg trade, Amitav Ghosh says, "the elimination of the Bandanas was brought about not just by targeted killings of humans, but by destroying the entire web of nonhuman connections that sustained a certain way of life." Ghosh, *The Nutmeg's Curse*, 41.

04 "Discours prononcé par Sa Majesté Haylé Sélassié Ier, empereur d'Éthiopie, à l'Assemblée de la Société des Nations, à la session de juin-juillet 1936" (Speech by His Majesty Haile Selassie I, Emperor of Ethiopia, at the Assembly of the League of Nations, at the Session of June-July 1936). The speech, in Amharic and French, is available at the Library of Congress, accessed July 5, 2023, https://www.loc.gov/item/2021667904.

05 Brandimarte, "Breathless War," 15.

06 Imperial War Museum, "Voices of the First World War: Gas Attack at Ypres," accessed July 17, 2023, https://www.iwm.org.uk/history/voices-of-the-first-world-war-gas-attack-at-ypres.

07 Sloterdijk, *Terror from the Air*, 23.

08 Brandimarte, "Breathless War," 15–16.

09 Speech by His Majesty Haile Selassie I.

10 Brandimarte, "Breathless War," 15. On the racialization of breath and breathing, see Christina E. Sharpe, *In the Wake: On Blackness and Being* (Durham, NC: Duke University Press, 2016) and Achille Mbembe, "The Universal Right to Breathe," *Critical Inquiry* 47 (2021): S58–S62. The phrase "I Can't Breathe" was mobilized as part of the Black Lives Matter movement and in response to the police killings by asphyxiation of several Black American men, including Eric Garner in 2014 and George Floyd in 2020.

11 Imperial War Museum, "Voices of the First World War."
12 Brandimarte, "Breathless War," 20.
13 Benjamin, "The Artwork in the Age of Its Technological Reproducibility," 122.
14 Benjamin, "The Artwork in the Age of Its Technological Reproducibility," 105.
15 Benjamin, "The Artwork in the Age of Its Technological Reproducibility," 105.
16 Benjamin, "The Artwork in the Age of Its Technological Reproducibility," 104–5.
17 Benjamin, "One-Way Street," 486.
18 Benjamin, "The Artwork in the Age of Its Technological Reproducibility," 105, 106, 122.
19 Karl Marx and Friedrich Engels, "Manifesto of the Communist Party," in *Karl Marx and Friedrich Engels: Selected Works* (London: Lawrence and Wishart, 1968), 38.
20 Benjamin, "One-Way Street," 487. Benjamin is referring to particular epileptic seizures known as "ecstatic seizures," but the image also evokes the surrealist André Breton's notion of "convulsive beauty." On convulsive beauty and surrealist photography see Hal Foster, *Compulsive Beauty* (Cambridge, MA: MIT Press, 1995), 27–29.
21 Schnapp, "The Fabric of Modern Times," 192.
22 Benjamin, "The Artwork in the Age of Its Technological Reproducibility," 122.

CHAPTER 20

01 Originally published in *Punch* 95 (July 7, 1888), 123; this satirical poem of eleven verses was reprinted in the *South Metropolitan Gas Company Co-Partnership Journal*, February 1904, 22–23, accompanied by a text by S. S. Field describing the company's contribution to the coal tar industry and calling the poem "very clever" and "very true."
02 Jarman and Ballschmiter, "From Coal to DDT," 131.
03 Anthony S. Travis, *The Rainbow Makers: The Origins of the Synthetic Dyestuffs Industry in Western Europe*, (London: Associated University Presses, 1993), 34. Aniline belongs to a class of organic chemicals known as "aromatic amines," though their smell is usually considered unpleasant. See Stoff and Travis, "Discovering Chemical Carcinogenesis," 137.
04 Travis, 34–41.
05 Desmond Reilly, "Salts, Acids and Alkalis in the 19th Century: A Compar-

ison between Advances in France, England and Germany," *Isis* 42, no. 4 (1951), 294. Nineteenth-century supplies of saltpeter initially flowed from India. On its use for photographic lighting, see Niharika Dinkar, "Pyrotechnics and Photography: Saltpeter and the Colonial History of Photographic Lighting," *photographies* 14, no. 3 (2021): 395–420.

06 Andrew Wybrant-Penrose, "On Some Probable Causes of Failure in Photo-Engraving," *The Photographic Journal*, January 30, 1897, 107–8.

07 Photo-engravers used ammonia liquid from the gasworks, too. Wybrant-Penrose, "On Some Probable Causes of Failure," 105.

08 Travis, *The Rainbow Makers*, 53–55.

09 Travis, *The Rainbow Makers*, 80.

10 David E. H. Edgerton, "Industrial Research in the British Photographic Industry, 1879–1939," in *The Challenge of New Technology: Innovation in British Business since 1850*, ed. Jonathan Liebenau, (Aldershot: Gower, 1988), 107.

11 Jarman and Ballschmiter, "From Coal to DDT," 134.

12 In the 1880s the largest German firm had 25,000 employees; the largest British firm had about 100–140. R. D. Welham, "The Early History of the Synthetic Dye Industry," *Journal of the Society of Dyers and Colorists* 79, no. 4 (1963): 151.

13 Welham, "The Early History of the Synthetic Dye Industry," 152.

14 Gem Dry Plate Co. Ltd. export report, Redbridge Museum and Heritage Centre, Ilford Limited collections, Box 1352, 90/359/A1/A1.

15 Trade union membership in the UK more than tripled between 1910 and 1920. When the revolution began in Russia, strikes across Britain fueled anxieties about factory labor and class conflict. See Vicky Long, "Industrial Homes, Domestic Factories: The Convergence of Public and Private Space in Interwar Britain," *Journal of British Studies* 50, no. 2 (2011): 462.

16 APM was a group of seven small photographic firms: Marion and Co. of Soho Square; the Paget Prize Plate Company of Watford, manufacturers of plates and papers; Marion and Foulger Limited, of Bedford, specialist in manufacturing mounts for plates; Rajar of Mobberley in Cheshire, which manufactured sensitized papers; Kershaw and Sons of Leeds, which made "optical goods"; and the Rotary Photographic Company of West Drayton, a photographic publishing company.

17 Hercock and Jones, *Silver by the Ton*, 37–38; Michael Pritchard, "The Development and Growth of British Photographic Manufacturing and Retailing, 1839–1914" (PhD diss., De Montfort University, 2010), 211. In 1918 there was an "agreement for mutual co-operation" between Ilford Limited and a new company called British Photographic Plates and Paper

Limited, which had acquired a controlling interest in Imperial and Gem. In 1919, Ilford bought all the issued share capital of the photographic paper company Thomas Illingworth & Co. Limited (based in Park Royal), setting up another Limited company to hold these shares. By 1927, Ilford had all the share capital of Imperial and Gem. See *Confidential Report*, Redbridge Museum and Heritage Centre, Ilford Limited collections, Box 1362, 90/359/E1.

18 Catford, "Our First 75 Years." Catford had transferred to Ilford when it took over APM, which had already absorbed his original employers Wellington and Ward. Hercock and Jones, *Silver by the Ton*, 61.

19 "The Trust Movement in Great Britain—Further Illustrations—IV," *Economist*, January 12, 1924, 47–48. See also "Amalgamated Photographic Manufacturers, Limited," *Economist*, February 12, 1921, 313.

20 For the details of this, see Hercock and Jones, *Silver by the Ton*, 54–58 and 99–105.

21 Esther Leslie, *Synthetic Worlds: Nature, Art and the Chemical Industry* (London: Reaktion, 2006), 153–54.

22 Leslie, *Synthetic Worlds*, 154.

23 E.g., Ulrich Marsch, "Strategies for Success: Research Organization in German Chemical Companies and IG Farben until 1936," *History and Technology* 12 (1994): 23–77.

24 Leslie, *Synthetic Worlds*, 169.

25 Hobsbawm, *Industry and Empire*, 200; see also Edgerton, *Warfare State*.

CHAPTER 21

01 Wilder, "Photography and the Art of Science," 167.

02 Geoffrey B. Harrison, "The Laboratories of Ilford Limited," *Proceedings of the Royal Society of London* 142 (1954): 13.

03 Quinine is fluorescent, as noted by Sir John Herschel in 1845.

04 William Jackson Pope, "Presidential Address: The Future of Pure and Applied Chemistry," *Journal of the Chemical Society* 113 (1918): 296.

05 Professor W. H. Mills was also important in this. See Harrison, "The Laboratories of Ilford Limited." Harmer would go on to work at Kodak.

06 Pope, "Presidential Address," 296.

07 W. F. Meggers and F. J. Stimson, "Dyes for Photographic Sensitizing," *Journal of the Optical Society of America* 4 (1920): 102.

08 Harrison, "The Laboratories of Ilford Limited," 13.

09 I discuss these in the article "Photography's Other Sensitivities," which

formed the basis of chapter 3. Michelle Henning, "Photography's Other Sensitivities," *Media Theory* 8, no. 21 (2024), https://journalcontent.mediatheoryjournal.org/index.php/mt/article/view/1069.

10 Book of labels, Redbridge Museum and Heritage Centre, Ilford Limited collections, Box 1379, 90/359/C1/B.

11 Pope, "Presidential Address," 293.

12 Catford, "Our First 75 Years," 73–75. Catford attributes the increased profits to growing demand from the Royal Flying Corps, demand for X-ray plates, and the YMCA "snapshots from home" scheme, in which amateur photographers photographed soldiers' homes and relatives (as well as "pets and hollyhocks") and sent them to the front. He also talks of the difficulty in obtaining materials during World War I: British paper was contaminated with iron particles, glass was becoming scarce, and recycled plates were not satisfactory.

13 Leslie, *Synthetic Worlds*, 16.

14 Edgerton, *Warfare State*.

15 Edgerton, *Warfare State*, chap. 1.

16 Edgerton, *Warfare State*, 10.

17 Pope, "Presidential Address," 291, 293.

18 Pope, "Presidential Address," 293–94.

19 Edgerton, *Warfare State*, 12 and 277.

20 Jarman and Ballschmiter, "From Coal to DDT," 134.

21 C. E. Kenneth Mees, "Sensitizing Dyes and their Use in Scientific Photography," *Nature* 137, no. 3470 (1936): 726–30.

22 C. E. Kenneth Mees, "The Kodak Research Laboratories," *Proceedings of the Royal Society of London*, Series B—Biological Sciences, 135:879 (1948): 140.

23 By October 2013, Wratten & Wainwright had relocated from their factory in West Croydon to the Kodak plant, and from then on, all their goods and services were invoiced via Kodak Ltd. Letter in Kodak Ltd. Archive, British Library.

24 *Imperial Handbook*, 1933, Redbridge Museum and Heritage Centre, Ilford Limited collections, 90/359/C2/A2.

25 In the late 1920s, the movie industry adopted panchromatic film at the same time as it introduced sound. Noisy arc lights had been replaced with quieter but dimmer incandescent lights, so a faster film was needed. However, the warmer cast of the new lights combined with the Panchromatic film made the actors' faces appear "muddy." Max Factor devised a range of special reflective "panchromatic" makeup to get around the problem, which worked on film but looked strange in daylight. This makeup was also used in still studio photography.

26 Potter, "The Works," 15. Fiona Kinsey's research on Kodak Australasia factory workers at Museums Victoria finds that Kodak Australasia began bringing in blind or visually impaired staff in the 1940s, and indeed had many of these staff, and some continued to have full careers there. Kinsey, "Sensing in the Dark."

27 John Lehmann, *The Whispering Gallery: Autobiography*, vol. 1 (London: Longmans, Green and Co., 1955), 289.

28 Lehmann, *The Whispering Gallery*, 290. The film Lehmann was using, Kodak's Panatomic, was brought to market around 1934 on nitrate stock and 1937 on safety film—no wonder the Georgian photographer was unfamiliar with it.

CHAPTER 22

01 Ben Quinn, "Insurgents to Bring War on 'Wokeness' to National Trust AGM," *The Guardian*, October 1, 2021, https://www.theguardian.com/uk-news/2021/oct/01/insurgents-bring-war-wokeness-national-trust-agm.

02 The National Trust, "Tyntesfield's History," accessed July 14, 2023, https://www.nationaltrust.org.uk/visit/bath-bristol/tyntesfield/history-of-tyntesfield.

03 "A. Gibbs & Sons guano business operated from 1842 to 1861. During this time, workers' conditions, including pay and medical care, improved under Gibbs's control, coupled with pressure from external observers." The National Trust, "Tyntesfield's History."

04 *The Times*, 1882, cited in Lola Loustaunau, Mauricio Betancourt, Brett Clark, and John Bellamy Foster, "Chinese Contract Labor, the Corporeal Rift, and Ecological Imperialism in Peru's Nineteenth-Century Guano Boom," *The Journal of Peasant Studies* 49, no. 3 (2022): 512–13.

05 On the uses of saltpeter for artificial lighting for photography and the other purposes, see also Niharika Dinkar, "'Our Best Machines Are Made of Sunlight': Photography and Technologies of Light," in *Ubiquity: Photography's Multitudes*, ed. Jacob W. Lewis and Kyle Parry (Leuven: Leuven University Press, 2021), 102.

06 Jarrige and Le Roux, *The Contamination of the Earth*, chap. 3.

07 Jill Kinahan and John Kinahan, "'A Thousand Fine Vessels are Ploughing the Main . . .': Archaeological Traces of the Nineteenth-Century 'Guano Rage' on the South-Western Coast of Africa," *Australasian Historical Archaeology* 27 (2009): 45.

08 Lesley Kinsley, "The Significance of Peruvian Guano in British Fertilizer History (c. 1840–1880)," *Agricultural History Review* 70, no. 2 (2022): 223.

09 Jarrige and Le Roux, *The Contamination of the Earth*, chap. 4. They give a figure of nearly 13 million metric tons (14.33 million US tons).

10 Purbrick, "Nitrate Traffic." See also Ignacio Acosta, Louise Purbrick, and Xavier Ribas, *Traces of Nitrate*, accessed August 2, 2023, https://tracesofnitrate.org/.

11 Joseph Robert Brown, "The Chilean Nitrate Industry in the Nineteenth Century" (PhD diss., Louisiana State University, 1954), 6.

12 Purbrick, "Nitrate Traffic."

13 Brown, "The Chilean Nitrate Industry," 5.

14 Benson, *Surroundings*, 26.

15 Homburg and Vaupel, introduction, 2. Their figure is given in metric—145 million metric tons.

16 John Bellamy Foster, Brett Clark, and Richard York, *The Ecological Rift* (New York: Monthly Review Press, 2010), 15.

17 Referring to a list of nine "planetary boundaries" proposed by scientists as measures or guides for a global environment in which humanity can exist, Foster, Clark, and York write: "With respect to the nitrogen cycle, the boundary is concerned with the amount of nitrogen removed from the atmosphere for human use in millions of tons per year. Before the rise of industrial capitalism (more specifically before the discovery of the Haber-Bosch process early in the twentieth century), the amount of nitrogen removed from the atmosphere was 0 tons. The proposed boundary, to avoid irreversible degradation of the earth system, is 35 million tons per year. The current status is 121 million tons per year." Foster et al., *The Ecological Rift*, 15.

18 Pap A. Ndiaye, *Nylon and Bombs: DuPont and the March of Modern America*, trans. Elborg Forster (Baltimore: Johns Hopkins University Press, 2007), 102, 8–9, 76–80. Ndiaye argues that nylon helped shift the company's reputation from an association with the military to a household name in consumer products for a civilian market (104–5). The original gunpowder was black powder made from charcoal, sulfur, and saltpeter (later, sodium nitrate) (8). DuPont had diversified into manufacturing dynamite in the 1880s. Dynamite, invented by Swedish chemist Alfred Nobel, is nitrate based—using nitroglycerin. Though explosives remained their core business, the new products included paint, celluloid, artificial leather, cellophane, and rayon—all nitrocellulose based (64).

19 LV Chilton, "Frank Forster Renwick: His Life and Work," Renwick Memorial Lecture, *Photographic Journal*, Section A (November 1945): 228.

20 Alice Lovejoy, "Celluloid Geopolitics: Film Stock and the War Economy, 1939–47," *Screen* 60, no. 2 (Summer 2019): 231.

21 Leslie, *Synthetic Worlds*, 17.

22 Ndiaye, *Nylon and Bombs*, 90 and 102.

CHAPTER 23

01 Serres, *Malfeasance: Appropriation Through Pollution* (Stanford: Stanford University Press, 2011), 3; 40–41, 43.

02 Liboiron, *Pollution Is Colonialism*, 6 and note 19, 6–7.

03 Liboiron, *Pollution Is Colonialism*, 13–15, 39–47; 51 and 57.

04 Angus, *Camera Geologica*, 91–92.

05 Lawrence K. Wang, Yung-Tse Hung, Howard H. Lo, and Constantine Yapijakis, *Handbook of Industrial and Hazardous Waste Treatment* (New York: Marcel Dekker, 2006), 309.

06 See Arun B. Mukherjee and Abhoy P. Mukherjee, "An Overview of Silver in India," *International Journal of Surface Mining, Reclamation and Environment* 11, no. 4 (1997): 195.

07 G. W. Bryan and W. J. Langston, "Bioavailability Accumulation and Effects of Heavy Metals in Sediments with Special Reference to United Kingdom Estuaries: A Review," *Environmental Pollution* 76 (1992): 89–131.

08 Samuel N. Luoma, Y. B. Ho, and G. W. Bryan, "Fate, Bioavailability and Toxicity of Silver in Estuarine Environments," *Marine Pollution Bulletin* 31, nos. 1–3 (1995): 44; Natalia Tsepina, Sergey Kolesnikov, Tatiana Minnikova, and Alena Timoshenko, "Soil Contamination by Silver and Assessment of its Ecotoxicity," *Reviews in Agricultural Science* 10 (2022): 186–205.

09 See Harman Technology (website), accessed September 20, 2024, https://www.harmantechnology.com/antimicrobial-technology/.

10 Romina Juncos, Linda Campbell, Marina Arcagni, Romina Daga, Andrea Rizzo, María Arribére, and Sergio Ribeiro Guevara, "Variations in Anthropogenic Silver in a Large Patagonian Lake Correlate with Global Shifts in Photographic Processing Technology," *Environmental Pollution* 223 (2017): 689. See also Feser, "Reproducing Photochemical Life," 42.

11 Carolyn Roberts, "Cleaning Up the Thames: Success or Failure?," Gresham College, September 28, 2017, https://www.gresham.ac.uk/sites/default/files/2017-03-22_CarolynRoberts_CleaningUpThamesT.pdf; H. McCormick, T. Cox, J. Pecorelli, and A. J. Debney, eds., *The State of the Thames 2021: Environmental Trends of the Tidal Thames* (London: ZSL, 2021). The declaration of the Thames as dead was from A. Wheeler, "The Fishes of the London Area," *London Naturalist* 37 (1958): 80–101.

12 Bryan and Langston, "Bioavailability Accumulation and Effects of Heavy Metals," 89; 91, 95.

13 Clarence A. Mills, "Urban Air Pollution and Respiratory Diseases," *American Journal of Epidemiology* 37, no. 2 (March 1943): 139.

14 Rowan Lear, "A Photographing Body" (PhD diss., University of West London, 2021), 313. Michelle Murphy describes the tendency to treat polluting chemicals individually as a "technoscientific epistemic habit" which makes it very difficult to see "the fullness of our chemical relations." Murphy, "Alterlife and Decolonial Chemical Relations," 496.

15 Bennett, *Vibrant Matter*, 23–24.

16 Boudia et al., *Residues*, 8.

17 Christopher H. Vane, Grenville H. Turner, Simon R. Chenery, Martin Richardson, Mark C. Cave, Ricky Terrington, Charles J. B. Gowing, and Vicky Moss-Hayes, "Trends in Heavy Metals, Polychlorinated Biphenyls and Toxicity from Sediment Cores of the Inner River Thames Estuary, London, UK," *Environmental Science: Processes and Impacts* 22, (2020): 364.

18 Jarrige and Le Roux, *The Contamination of the Earth*; Travis, *The Rainbow Makers*, 196–98, 119, and 74. During the 1860s dye companies dumped arsenic waste into major rivers across Europe. Workers were employed in shifts to avoid continuous exposure to toxic fumes. Only late in the decade did the firms succeed in developing less toxic processes to avoid the use of arsenic and the resulting conflict with city authorities. Travis, *The Rainbow Makers*, 96–101.

19 Hercock and Jones, *Silver by the Ton*, 29.

20 Richard Maxwell and Toby Miller, *Greening the Media* (Oxford: Oxford University Press, 2012), 72.

21 On River Roding pollution, see Adam Vaughan, "Toxic Timebomb of 'Forever Chemicals' in England's Rivers," *The Times* (London), February 28, 2023, https://www.thetimes.co.uk/article/toxic-timebomb-forever-chemicals-england-s-rivers-clean-it-up-jwxccc02j; and Bill Latto (dir.), *Life after Laporte* (c. 1987), video about the clean-up at the Laporte Works, Ilford, formerly known as Howards & Sons Chemicals, accessed September 14, 2024, https://www.londonsscreenarchives.org.uk/title/2749/.

22 Vaughan, "Toxic Timebomb." See also Patrick Barkham, "'The Roding Is Sacred and Has Rights': The Hammer-Wielding Barrister Fighting for London's Forgotten River," *The Guardian*, December 5, 2022, https://www.theguardian.com/environment/2022/dec/05/river-roding-barrister-paul-powlesland-london-polluters-footpaths. In 2024, Paul Powlesland of the River Roding Trust became the first juror in British history to swear his oath in court on river water rather than a holy book. See Patrick Barkham, "Environmentalist Becomes First Juror to Swear Oath

on River Water," *The Guardian*, August 2, 2024, https://www.theguardian.com/environment/article/2024/aug/02/environmentalist-becomes-first-juror-to-swear-oath-on-river-water.

23 Angus, *Camera Geologica*, 94.

24 Feser, "Reproducing Photochemical Life," 19.

25 Lear, *A Photographing Body*, 314.

26 Maxwell and Miller, *Greening the Media*, 73; Elizabeth A. Povinelli, "Fires, Fogs, Winds," *Cultural Anthropology* 32, no. 4 (2017), 507.

27 Richard Maxwell, "Green Accounting for a Creative Economy," in *Cultural Industries and the Environmental Crisis: New Approaches for Policy* 29, eds. Kate Oakley and Mark Banks (Berlin: Springer, 2020).

28 Feser, "Reproducing Photochemical Life."

29 Jennifer Tucker, "Chemical Affinities: Photography, Extraction, and Industrial Heritage in Nineteenth-Century Northern England," *Nineteenth-Century Contexts* 44, no. 5 (2022): 528.

30 On the London fog as symbolizing imperial power, see Alessandrini, "'London Fog' and the Symbols of Empire."

31 Boudia et al., *Residues*, 10.

32 Povinelli, "Fires, Fogs, Winds," 505.

33 Murphy, "Alterlife and Decolonial Chemical Relations," 497.

34 Boudia et al., *Residues*, 119.

CHAPTER 24

01 Geoffrey Batchen, *Burning with Desire: The Conception of Photography* (Cambridge, MA: MIT Press, 1999); Stephen Connor, *The Matter of Air: Science and Art of the Ethereal* (London: Reaktion Books, 2010), 25.

02 Davy's "Essay on Heat and Light" was published by Beddoes in 1799. Davy thought oxygen was a compound of light as oxygenation could give off light. David Knight, *Humphry Davy: Science and Power* (Cambridge: Cambridge University Press, 1996), 24.

03 Wedgwood and Davy first met in Davy's native Cornwall. Joseph Priestley had recorded nitrous oxide in 1772, but Davy was the first to systematically assess its effects on people.

04 The text credits the invention to Wedgwood "with observations by H. Davy."

05 Mike Jay, *The Atmosphere of Heaven* (New Haven: Yale University Press, 2009), 82–83. He had experimented with quicksilver and nitrate of silver at his father's ceramics factory, and had invented "silvered ware," a style of black earthenware decorated with silver, in 1791. Batchen suggests Wedg-

wood may have got the instructions for silver nitrate from James Watt, Joseph Priestley, or Scheele. Geoffrey Batchen, "Tom Wedgwood and Humphry Davy: 'An Account of a Method,'" *History of Photography* 17, no. 2 (1993): 172, 174.

06 Santiago Colás, "Aesthetics vs Anaesthetic: How Laughing Gas Got Serious," *Science as Culture* 7, no. 3 (1998): 339. Colás points out that Davy's book *Researches Chemical and Philosophical, Chiefly Concerning Nitrous Oxide* (1800) emphasizes the pleasurable intensification of physical sensation, rather than numbing.

07 Cited in Connor, *The Matter of Air*, 68.

08 Jordan Bear, "Self-Reflections: The Nature of Sir Humphry Davy's Photographic 'Failures,'" in *Photography and Its Origins*, ed. Tanya Sheehan and Andres Zervigón (London: Routledge, 2014), 188.

09 Bear, "Self-Reflections," 188–92.

10 Davy was based at the Pneumatic Institution from October 1799 until February 1801.

11 See Colás, "Aesthetics vs Anaesthetic," 347; Jan Golinski, *The Experimental Self: Humphry Davy and the Making of a Man of Science* (Chicago: University of Chicago Press), 173–74; Tim Fulford, "Science and Poetry in 1790s Somerset: The Self-Experiment Narrative, the Aeriform Effusion, and the Greater Romantic Lyric," *ELH* 85, no. 1 (2018): 88–89.

12 Jan Golinski, "Humphry Davy: The Experimental Self," *Eighteenth-Century Studies* (2011): 21.

13 Fulford, "Science and Poetry in 1790s Somerset," 89. See also Lisa Ann Robertson, "'Swallowed Up in Impression': Humphry Davy's Materialist Theory of Embodied Transcendence and William Wordsworth's 'Tintern Abbey,'" *European Romantic Review* 26, no. 5 (2015): 591–614.

14 "Dr. Beddoes' sanatorium turned into a reckless experimental site, both in terms of chemicals and language." Griesecke, "Lost Threads," 567.

15 Cited in L. M Griffiths, "The Reputation of the Hotwells (Bristol) as a Health-Resort," *Bristol Medical Chirurgical Journal* 75 (March 1902): 1–26; contd. 76 (June 1902): 142–52 and 77 (September 1902): 193–208.

16 Griesecke, "Lost Threads," 569.

17 Griffiths, "The Reputation of the Hotwells," 12, 24, 14.

18 Carrick and Sutherland cited in Griffiths, "The Reputation of the Hotwells," 24, 19.

19 David Hussey, "'From the Temple of Hygeia to the Sordid Devotees of Pluto': The Hotwell and Bristol; Resort and Port in the Eighteenth Century," in *Resorts and Ports: European Seaside Towns since 1700*, ed. Peter Borsay and John K. Walton (Bristol: Channel View Publications, 2011), 51.

20 Hussey, "'From the Temple of Hygeia,'" 56.

21 Cited in Hussey, "'From the Temple of Hygeia,'" 51. As early as 1739, Alexander Pope, who visited the city for a Hotwells cure, described the glass works as "twenty odd pyramids smoking over the town," and Horace Walpole described the city as "the dirtiest great shop I ever saw." Peter T. Marcy, *Eighteenth Century Views of Bristol and Bristolians* (Bristol: Bristol University, 1966), 7, 8.

22 Jay, *The Atmosphere of Heaven*, 79.

23 Griffiths, "The Reputation of the Hotwells," 201 and 203; Hussey, "'From the Temple of Hygeia,'" 60.

24 Carrick cited in Griffiths, "The Reputation of the Hotwells," 203; Latimer cited in Hussey, "'From the Temple of Hygeia,'" 60.

25 Mills, "Urban Air Pollution and Respiratory Diseases," 138. From the 1930s, studies of air pollution could be extraordinarily thorough. One survey of air pollution in New York City under Roosevelt's Works Progress Administration analyzed all the fuel movement in and out of the city to measure its consumption of oil and solid fuel, mapped the smoke across the city and measured the soot-fall. Sol Pincus and Arthur C. Stern, "A Study of Air Pollution in New York City," *American Journal of Public Health and the Nation's Health* 27, no. 4 (1937): 321–33.

26 As of 2023, possession of nitrous oxide for recreational use became a criminal offence in England and Wales.

CHAPTER 25

01 Hercock and Jones, *Silver by the Ton*, 15–16.

02 Peter Haff, "Technology as a Geological Phenomenon: Implications for Human Well-Being," *Geological Society of London Special Publications* 395 (2014): 301–9.

03 Chris Otter, "Encapsulation: Inner Worlds and their Discontents," *Journal of Literature and Science* 10, no. 2 (2017): 56.

04 Bill Jay, "Death in the Darkroom: Poisonings of Nineteenth Century Photographers," *Phoebus* 3 (1981): 85–99.

05 C. H. Bothamley, *The Ilford Manual of Photography* (London: Ilford Limited, 1904 edition), 48–49. See also Jennifer Tucker, "'Diseases of the Darkroom' in the Long Nineteenth Century," in "The Darkroom," ed. Sara Dominici, special issue, *Photoresearcher* 41 (2024): 16–31.

06 Hercock and Jones, *Silver by the Ton*, 16.

07 Catford, "Our First 75 Years."

08 Hercock and Jones, *Silver by the Ton*, 24.

09 Hercock and Jones, *Silver by the Ton*, 17–19, 28–29.

10 Patented by A. S. Haslam, an engineer who had pioneered refrigeration for the transporting of food on ships. Catford, "Our First 75 Years."

11 Colin Runeckles, "Looking Back at Roden Street," *Ilford Historical Society Newsletter* 130 (August 2019), 4–7.

12 Long, "Industrial Homes, Domestic Factories," 461.

13 Potter, "The Works," 11.

14 Hercock and Jones, *Silver by the Ton*, 131, 138–39.

15 Thomas Illingworth experiment book, Redbridge Museum and Heritage Centre, Ilford Limited collections, Box 1372, 90/359/B1/A18.

CHAPTER 26

01 Cramer cited in Reyner Banham, *The Architecture of the Well-Tempered Environment* (London: Architectural Press, 1969), 81.

02 Dustin Valen, "On the Horticultural Origins of Victorian Glasshouse Culture," *Journal of the Society of Architectural Historians* 75, no. 4 (December 2016): 403.

03 Banham, *The Architecture of the Well-Tempered Environment*, 174.

04 Marsha Ackermann, *Cool Comfort: America's Romance with Air-Conditioning* (Washington, DC: Smithsonian Institution Press, 2002).

05 "London Fog," *The British Journal of Photography* 17, no. 544 (October 7, 1870): 476.

06 "Dust and Fog," *The British Journal of Photography* 28, no. 1079 (January 7, 1881): 2–3.

07 "November," *The British Journal of Photography* 45, no. 2012 (November 25, 1898): 757.

08 Hercock and Jones, *Silver by the Ton*, 47.

09 Although over the years new processes reduced smoke and gas emissions in some plants, and new works were built to process the ammoniacal liquors rather than just releasing them into the environment, the coal-fired gasworks remained highly polluting. A. O. Thomas and J. N. Lester. "Gaswork Sites as Sources of Pollution and Land Contamination: An Assessment of Past and Present Public Perceptions of Their Physical Impact on the Surrounding Environment," *Environmental Technology* 14, no. 9 (1993): 803.

10 Thorsheim, *Inventing Pollution*, 137.

11 Cited in Thorsheim *Inventing Pollution*, 140–41.

12 By the 1910s, an unusually high number of the population commuted to London for work in professional and clerical positions. W. R. Powell, ed., *A History of the County of Essex*, vol. 5 (London: Victoria County History, 1966), 9–21. British History Online, accessed August 1, 2023, http://www.british-history.ac.uk/vch/essex/vol5/, 9–21.

13 Richard Farmer, "Meteorology and British Film Studios: An Article of the London Fog," *Historical Journal of Film, Radio and Television* (2021): 4.

14 Typewritten description of the Selo Factory at Brentwood, Redbridge Museum and Heritage Centre, Ilford Limited collections, Box 1362, 90/359/E1/2.

15 Powell, *A History of the County of Essex*, 37–47.

16 Catford, see also Hercock and Jones, *Silver by the Ton*, 47.

17 When the factory at Brentwood reopened in 1920 as the Selo factory, it used gauze filters treated with silver nitrate to remove sulfur compounds from the air. In other words, the sensitivity of photographic materials to the contaminants, is also used as a means of filtering, much like the early gas masks of the First World War, reportedly coated with hypo (fixer). See Hercock and Jones, *Silver by the Ton*, 54, 58, 135.

18 Hercock and Jones, *Silver by the Ton*, 48.

CHAPTER 27

01 "How was the air to be picked out of its surroundings, when air was ambiance itself?" Connor, *The Matter of Air*, 17, 26, 30.

02 Banham describes Carrier as the "father" of air-conditioning, a title Ackermann says Carrier actively pursued. Ackermann, *Cool Comfort*, 21; Banham, *The Architecture of the Well-Tempered Environment*, 81.

03 Catford, "Our First 75 Years."

04 Banham, *The Architecture of the Well-Tempered Environment*, 172.

05 Hercock and Jones, *Silver by the Ton*, 132.

06 Hercock and Jones, *Silver by the Ton*, 131. Emulsion was passed through a hydraulic shredding machine, and the original teapot was replaced by coating machines. These machines were silver-plated to prevent contamination, and "a few were constructed from solid silver." The hand-cranked coating machines were later driven by steam then electricity. Hercock and Jones, *Silver by the Ton*, 132.

07 Hercock and Jones, *Silver by the Ton*, 132.

08 Otter, "Encapsulation," 58. For a fascinating analysis of the emergence and development of these climate bubbles in the Pacific area, see Yuriko

Furuhata, *Climatic Media: Transpacific Experiments in Atmospheric Control* (Durham: Duke University Press, 2022).

09 By 1895, the company installed electric lighting at the plant. Before that (and probably after to some extent), they used candles and oil lamps, so the plates were at risk of contamination not just from the outside but also from soot and smoke from these lighting sources inside the factory. See Hercock and Jones, *Silver by the Ton*, 29.

10 Hercock and Jones, *Silver by the Ton*, 35.

11 Geoffrey B. Harrison, "Automatic Control of Mains Voltage," Museum of Science and Industry, Manchester, MS0232/49/19.

12 Hercock and Jones, *Silver by the Ton*, 132.

13 Otter, "Encapsulation," 59.

14 Feser notes that at Kodak, "In order to keep dust, lint, and other debris from sticking to the surface of the film, the machinery cabinets were positively pressured. They pushed the air, that is the solvents, away from the film, into the room, and out into the open air of the Park"—resulting in solvent vapors across the neighborhood. Feser, "Reproducing Photochemical Life," 114–15.

15 Chakrabarty, *The Climate of History in a Planetary Age*, 96–99.

16 Cited in Chakrabarty, *The Climate of History in a Planetary Age*, 99.

17 Roberto Eposito, *Immunitas: The Protection and Negation of Life* (Cambridge: Polity, 2011), 141.

18 Chakrabarty, *The Climate of History in a Planetary Age*, 113.

19 Rem Koolhaas, "Junkspace," *October* 100 (2002): 175.

20 Walter Benjamin, *The Arcades Project*, trans. Howard Eiland and Kevin McLaughlin, ed. Rolf Tiedemann (Cambridge, MA: Harvard University Press, 1999), 217 (I.2.6), and Walter Benjamin, "Main Features of My Second Impression of Hashish" (January 15, 1928), in *Walter Benjamin: Selected Writings*, vol. 2, *1927–1934*.

21 The concept of "thermal comfort" is discussed in Ackermann, *Cool Comfort*, 23–25.

22 Valen, "On the Horticultural Origins of Victorian Glasshouse Culture," 403.

23 Valen, "On the Horticultural Origins of Victorian Glasshouse Culture," 406–7.

24 Peter Sloterdijk, "Atmospheric Politics," in *Atmospheres of Democracy*, ed. Bruno Latour and Peter Wiebel (Cambridge, MA: MIT Press, 2005), 945.

25 Barak, *Powering Empire*, 16.

26 Dustin Valen, "Imperial Atmospheres: Race and Climate Control on the Niger," *ABE Journal: Architecture beyond Europe* 17 (2020).

CHAPTER 28

01 Colonial plantations contributed to the dangers that white settlers identified with the tropics, for example, through deforestation and stagnant water around plantations, bringing mosquitoes carrying yellow fever and malaria. Benson, *Surroundings*, 53.

02 Ackermann, *Cool Comfort*, 23–25.

03 Vladimir Jancović, *Confronting the Climate: British Airs and the Making of Environmental Medicine* (London: Palgrave Macmillan, 2010), 90 and 91.

04 Jancović, *Confronting the Climate*, 9.

05 "Fog and Photography," *Scientific American* 76, no. 23 (June 5, 1897): 355. Steve Edwards mentions the use of a steam engine in Oliver Sarony's Scarborough studio in the 1860s, though he notes it is the only example he has found from the period. Edwards, *The Making of English Photography*, 84.

06 Potter cited in "Behind the Scenes at the Ilford Factory," *The Ilford-Selo Record*, March 1935, 4. Museum of Science and Industry Manchester, MS0232/49/5.

07 *The Ilford Courier* 8, no. 2 (1939), 5.

08 See Jancović, *Confronting the Climate*, 39.

09 Michael Osman defines regulation as "a practical mode of living, working, and thinking; it refers to an assemblage of techniques—mechanical, legal, administrative, and scientific—that defined a range of deviations from normal in which modern life could retain both the appearance of order and the functionality of organization." Michael Osman, *Modernism's Visible Hand: Architecture and Regulation in America* (Minneapolis: University of Minnesota Press, 2018), viii.

10 Bernhard Siegert, *Cultural Techniques: Grids, Filters, Doors and Other Articulations of the Real* (New York: Fordham University Press 2015), 13.

11 Siegert, *Cultural Techniques*, 23.

12 An example is the use of unit operations in photographic materials manufacture. These are sets of repeatable, step-by-step procedures associated with modern, efficient industrial practices of standardization that had emerged in American systems of scientific management. Ndiaye, *Nylon and Bombs*, 35.

13 "Behind the Scenes at the Ilford Factory," *The Ilford-Selo Record*, March 1935, 4. Museum of Science and Industry Manchester, MS0232/49/5.

14 Potter, "The Works," 11.

15 These examples are taken from the indexes of the Thomas Illingworth experiment books contained in Box 1372, 90/359/B1/, at the Redbridge Museum and Heritage Centre, Ilford Limited collections.

16 Osman, *Modernism's Visible Hand*, 48.

17 Jarrige and Le Roux, *The Contamination of the Earth*, 77.

18 Jarrige and Le Roux, *The Contamination of the Earth*, 77–78.

19 Arthur Pereira, "The Exhibition Reviewed: 'The Life of a Film,'" *The Photographic Journal* 75 (December 1935): 637.

20 Typewritten description of the Selo Factory at Brentwood, Redbridge Museum and Heritage Centre, Ilford Limited collections, Box 1362, 90/359/E1/2.

21 Potter, "The Works," 10.

CHAPTER 29

1 Dinkar, "'Our Best Machines Are Made of Sunlight,'" 96–97.

2 Edwards, *The Making of English Photography*, 43–44.

3 See for example Junko Theresa Mikuriya, *A History of Light: The Idea of Photography* (London: Bloomsbury, 2016).

4 Joanna Zylinska, "Photography after Extinction," in *Extinction*, ed. Richard Grusin (Minneapolis: University of Minnesota Press, 2018), 54, her emphasis.

5 Dinkar, "'Our Best Machines are Made of Sunlight.'"

6 Jacques Derrida, *Writing and Difference*, trans. Alan Bass (Chicago: University of Chicago Press, 1978), 27. See also Mikuriya, *A History of Light*.

7 Lynall, *Imagining Solar Energy: The Power of the Sun in Literature, Science and Culture* (London, Bloomsbury 2020), 83.

8 This is a reference to the argument made by Oliver Wendell Holmes in his essay "The Stereoscope and the Stereograph," *Atlantic Monthly* 3, no. 1 (1859).

9 This painting sprung to mind because I grew up with it: a reproduction hangs in the hallway of my parents' house, only a few steps away from where I was sat in my highchair in figure 56 of this book.

10 "Disentangling the cultural impacts of the possible weather event associated with Tambora from this context of disruption and hardship is challenging." Lucy Veale and Georgina H. Endfield, "Situating 1816, the 'Year Without Summer,' in the UK." *The Geographical Journal* 182, no. 4 (2016): 319.

11 John Clubbe, "The Tempest-toss'd Summer of 1816: Mary Shelley's Frankenstein," *Byron Journal* 19 (1991): 27.

12 Clubbe, "The Tempest-toss'd Summer of 1816," 28.

13 Clubbe dates the volcano thesis to 1913. Clubbe, "The Tempest-toss'd Summer of 1816," 30.

14 Wood, *Tambora*, 9.

15 Mary Shelley, *The Letters of Mary Wollstonecraft Shelley*, ed. Betty T. Bennett (Baltimore: Johns Hopkins University Press, 1980–89), 17.

16 Shelley, *The Letters*, 20.

17 Niépce is credited with producing the earliest surviving photograph, *View from the Window at Le Gras*, from 1826–27.

18 On how *View from the Window at Le Gras* became known as the "first," see Jessica McDonald, "Helmut Gernsheim and the 'World's First Photograph,'" in *Photography and Its Origins*, 15–28.

19 Niépce, letter to Claude Niépce, June 16, 1816, from Chalon-sur-Saône, letter 254 in Nicéphore Niépce, *Niépce correspondence et papiers*, ed. Manuel Bonnet and Jean-Lous Marignier (Saint-Loup-de-Varennes: Maison Nicéphore Niépce, 2003), 407 (my translation), https://niepce-letters-and-documents.com/ (accessed September 14, 2023).

20 Niépce, letter to Claude Niépce, May 9, 1816, from Saint-Loup-de-Varennes, letter 248 in *Niépce correspondence et papiers*, 390 (my translation).

21 Victor Foque, *La verité sur l'invention de la photographie* (The truth concerning the invention of photography) (Paris: Librairie des Auteurs, 1867), accessed June 27, 2024, https://gallica.bnf.fr/, my translation. Foque was an early advocate of Niépce's role in the invention of photography.

22 Niépce, letter to Claude Niépce, June 8, 1816, from Chalon-sur-Saône, letter 257 in *Niépce correspondence et papiers*, 420, my translation.

23 Niépce, letter to Claude Niépce, June 16, 1816, from Chalon-sur-Saône, letter 254 in *Niépce correspondence et papiers*, 408, my translation.

24 Niépce, letter to Claude Niépce, June 16, 1816, from Chalon-sur-Saône, letter 254 in *Niépce correspondence et papiers*, 412, 410.

25 Emily Doucet et al., "Editors' Introduction: The Aerial Image," *Grey Room* 83 (2021): 8–9.

26 Jonathan Bate, *The Song of the Earth* (London: Picador, 2000), 98.

27 Bate, *The Song of the Earth*, 52.

28 Bate, *The Song of the Earth*, 54.

29 Geoffrey Batchen, *Burning with Desire*.

30 Douglas R. Nickel, "Notes Towards New Accounts of Photography's Invention," in *Photography and its Origins*, ed. Tanya Sheehan and Andres Zervigón (London: Routledge, 2014), 82–93, 87–90.

31 Serres, "Science and the Humanities," 14.

CHAPTER 30

01 Crace Calvert, "Cantor Lectures. Ox Dyes and Dye-Stuffs other than Aniline by Dr. Crace Calvert, F.R.S." Lecture IV. —Delivered Tuesday, February 28, *The Journal of the Society of Arts* 19, no. 988 (1871): 845.

02 T. Frederick Hardwich, *A Manual of Photographic Chemistry: Including the Practice of the Collodion Process* (London: J. Churchill, 1857), 27; Calvert, "Cantor Lectures," 843.

03 Calvert, "Cantor Lectures," 843.

04 On March 30, 1839, Talbot purchased one dram of gallic acid in a bottle, at a cost of one shilling and six pence, from Alexander Garden's chemist shop at 372 Oxford Street. R. Derek Wood, "Latent Developments from Gallic Acid, 1839," *The Journal of Photographic Science* 28, no. 1 (1980): 36–42.

05 "The ingenious scheme of developing a latent image was founded on the observation that AgCl [silver chloride] blackened faster if it was formed on leather. . . . A piece of fine detective work later established that the essential factor in the effect of leather was the gallic acid used to tan it." L. M. Slifkin, "The Photographic Latent Image," *Science Progress* (1972): 166.

06 Wood, "Latent Developments," 36.

07 Tannin from galls was also used as a preservative in a version of the dry plate process called the tannin process. This is described in nineteenth-century handbooks such as Désiré van Monckhoven, *A Popular Treatise on Photography*, trans. W. H. Thornthwaite (London, 1863); or John Towler, *Dry Plate Photography: Or, The Tannin Process, Made Simple and Practical for Operators and Amateurs* (New York: Joseph H. Ladd, 1865).

08 Details of the carriage of goods by camels, and the weight of the sacks of galls, are in "The Commerce of Syria," *Hunt's Merchants Magazine*, June 1832, 489–511.

09 Hannah Drayson, "To the Bitter End: Affect, Experience and Chemical Ecology," *Curare: Journal of Medical Anthropology* 42, nos. 3–4 (2019).

10 The monarch butterfly is Drayson's example, the cinnabar moth my own local equivalent. See Drayson, "To the Bitter End."

11 Sianne Ngai, *Ugly Feelings* (Cambridge, MA: Harvard University Press, 2007), 6.

12 Drayson, "To the Bitter End." In the nineteenth century, galls were one of the principal exports to Britain from the Levant—a return trade otherwise very limited according to D. C. M. Platt. He claims "dye-stuffs were Smyrna's main export to the United Kingdom, and galls the only article sent in quantity from Aleppo." Platt, "Further Objections to an 'Imperialism of Free Trade', 1830-60," *The Economic History Review* 26, no. 1 (1973): 80.

CHAPTER 31

01 The 1935 edition of *The Ilford Manual of Photography* admits that in the case of the latent image, "it cannot be said that its exact chemical or phys-

ical constitution has been determined." George E. Brown, *The Ilford Manual of Photography* (London: Ilford Limited, 1935), 131.

02 Slifkin, "The Photographic Latent Image," 153.

03 Douwe Draaisma makes this point about much scientific use of metaphor—often "no literal alternative is available." Douwe Draaisma, *Metaphors of Memory: A History of Ideas about the Mind* (Cambridge: Cambridge University Press, 2000), 11.

04 See, to name just a few, Thomas Sutton and George Dawson's *Dictionary of Photography* (London: Sampson, Low, Son & Marston, 1867), Elbert Anderson's *The Skylight and the Darkroom* (Philadelphia: Benerman and Wilson, 1872), William de Wiveleslie Abney's *A Treatise on Photography* (London: Longmans, Green & Co., 1878). The term "invisible image" was also used repeatedly in writings by influential figures such as Josef Maria Eder and Herman Vogel.

05 James Mitchell, *The Ilford Manual of Photography* (London: Ilford Limited, 1945), 58.

06 William Henry Fox Talbot, "Fine Arts: Calotype (Photogenic) Drawing," *Literary Gazette and Journal of Belles Lettres, Science and Art* 1256 (February 13, 1841): 108.

07 A. L. M. Sowerby, *Dictionary of Photography*, 18th ed. (New York: Philosophical Library, 1956).

08 *The Imperial Handbook*, 1914, Redbridge Museum and Heritage Centre, Ilford Limited collections, Box 1357, 90/359/C2/A2.

09 Hito Steyerl, "'Politics of Post-Representation': In Conversation with Marvin Jordan," *DIS Magazine*, June 2014, https://dismagazine.com/disillusioned-2/62143/hito-steyerl-politics-of-post-representation/.

10 Kate Palmer Albers, "Parafiction and the New Latent Image," in *Ubiquity: Photography's Multitudes*, ed. Jacob W. Lewis and Kyle Parry (Leuven: Leuven University Press, 2021), 205.

11 Yanai Toister, "Latent Digital," *Journal of Visual Art Practice* 19, no. 2 (2020): 140.

12 Toister, "Latent Digital," 133.

13 Moyra Davey, *Index Cards* (London: Fitzcarraldo Editions, 2020), 62–63.

CHAPTER 32

01 R. Derek Wood, "The Daguerreotype and Development of the Latent Image: 'Une Analogie Remarquable,'" *The Journal of Photographic Science* 44, no. 5 (1996): 165.

02 Wood, "The Daguerreotype and Development of the Latent Image," 165.
03 Bricheleau cited in Wood, "The Daguerreotype and Development of the Latent Image," 165.
04 Bricheleau cited in Wood, "The Daguerreotype and Development of the Latent Image," 165–66.
05 John Towler, *The Silver Sunbeam* (New York: Joseph H. Ladd, 1864), 94.
06 Thomas Reid, *Essays on the Intellectual Powers of the Human Mind: A Critical Edition*, ed. Derek Brookes and Knud Haakonssen (Edinburgh: Edinburgh University Press, 2002), 240. The book is based on lectures given in the 1770s.
07 Reid, *Essays on the Intellectual Powers of the Human Mind*, 216.
08 Reid, *Essays on the Intellectual Powers of the Human Mind*, 240.
09 M. Campbell, "The Concepts of Dormancy, Latency, and Dominance in Nineteenth-Century Biology," *Journal of the History of Biology* 16, 409–31 (1983): 418–19.
10 Darwin cited in Campbell, "The Concepts of Dormancy, Latency, and Dominance," 420.
11 Campbell, "The Concepts of Dormancy, Latency, and Dominance," 420.
12 Graham Harman, *Immaterialism: Objects and Social Theory* (London: Polity, 2016), 7. Harman is referring to "objects" in the philosophical sense.

CHAPTER 33

01 Draaisma, *Metaphors of Memory*, 3.
02 Kate Flint has written about the concept of photographic memory, especially in relation to flash, and Geoffrey Batchen and others have written about the mnemonic functions of photography. See Kate Flint, *Flash! Photography, Writing and Surprising Illumination* (Oxford: Oxford University Press, 2017), and Geoffrey Batchen, *Forget Me Not: Photography and Remembrance* (New York: Princeton Architectural Press, 2004).
03 Draaisma, *Metaphors of Memory*, 122.
04 John William Draper, *History of the Conflict Between Religion and Science*, 3rd ed. (New York: D. Appleton and Company, 1875), 133.
05 Draper, *History of the Conflict Between Religion and Science*, 133–34.
06 Joseph Fayrer, "First Troonian Lecture," cited in Rohan Deb Roy, *Malarial Subjects: Empire, Medicine and Nonhumans in British India, 1820–1909* (Cambridge: Cambridge University Press, 2017), 149.
07 Marcel Proust, *Remembrance of Things Past*, vol. 1, trans. C. K. Scott Moncrieff and Terence Kilmartin (London: Chatto & Windus, 1981), 48. First

published in French between 1913 and 1927. More recent English translations have it as *In Search of Lost Time*.

08 Proust, *Remembrance of Things Past*, 50.

09 Roland Barthes, *Camera Lucida: Reflections on Photography*, trans. Richard Howard (London: Vintage Books, 2000), 82. Originally published in 1980/1981.

10 Sarah Kofman, *Camera Obscura of Ideology*, trans. Will Straw (Ithaca: Cornell University Press, 1999), 77. Originally published in French in 1973.

11 Sigmund Freud, *Moses and Monotheism*, trans. Katherine Jones (London: Hogarth Press and the Institute of Psycho-Analysis, 1939), 84.

12 Freud, *Moses and Monotheism*, 198–99.

13 Kofman, *Camera Obscura of Ideology*, 26.

14 There are two moments of latency in a negative-positive process: the first when the film is exposed but undeveloped; the second between when the paper is exposed by the light of the enlarger and when it is placed in the developing bath. If Kofman's reading of the photographic analogy is confused, her larger point is still valid, which is that the analogy does not work because it is linear (exposure–latency–development), when in fact Freud's work argues against such a linear reading.

15 Hans Ulrich Gumbrecht, *After 1945: Latency as Origin of the Present* (Stanford: Stanford University Press, 2013), 31.

CHAPTER 34

01 See Melissa Harris, "Koudelka's Prague, Fifty Years Later," *Aperture*, August 20, 2018, https://aperture.org/editorial/josef-koudelka-68/; and Josef Koudelka, "Josef Koudelka: The 1968 Prague Invasion," Magnum, accessed August 26, 2023, https://www.magnumphotos.com/newsroom/josef-koudelka-invasion-prague-68/. There was a risk of his film being confiscated and that his anonymously published photographs could be used against the protesters—that the Soviet authorities would use them to identify people, as happens to the photographer in Milan Kundera's novel *The Unbearable Lightness of Being (London: Faber and Faber, 1984)*.

02 John Tagg, *The Burden of Representation: Essays on Photographies and Histories* (Minneapolis: University of Minnesota Press, 1993), 12. Tagg describes 1930s documentary in terms of an "emotionalized drama of experience" and an "ethnographic theatre."

03 Oscar Wilde expresses this idea of life imitating art in Wilde, "The Decay of Lying" (1891), in *The Decay of Lying* (London: Penguin Books, 2020).

04 Harris, "Koudelka's Prague."

05 Joel Anderson, "Theatre Photography between Theatre and Performance," *Focales* 3 (2019), accessed May 25, 2024, https://journals.openedition.org/focales/803.

06 Vilém Flusser, *Towards A Philosophy of Photography* (London: Reaktion, 2000), 9.

07 Benjamin, "Little History of Photography," 510.

08 Barthes, *Camera Lucida*, 96.

09 "I wanted to take a photograph of the Soviet tanks and soldiers alone in Wenceslas Square after the people of Prague had decided not to demonstrate so as not to give the Soviet occupiers a pretext for a massacre—the Czechs realized they were being set up. In my photograph of the hand with the watch, you don't see the Soviets." Koudelka in Harris, "Koudelka's Prague."

10 Draper, *History of the Conflict Between Religion and Science*, 133.

11 Christopher Rovee, "Secrets of Paper," *Word & Image* 30, no. 4 (2014): 395. Hervé Guibert's *Ghost Image* also refers to latency: his book "speaks only of ghost images, images that have not yet issued, or rather of latent images, images that are so intimate that they become invisible," 114.

12 Reid, *Essays on the Intellectual Powers of the Human Mind*, 216.

13 Ernst Bloch, *Heritage of Our Times* (Berkeley: University of California Press, 1991), 57–60. Originally compiled in 1935, first published in German in 1962.

14 Bloch's own emphasis, *Heritage of Our Times*, 60, 62.

15 Gumbrecht, *After 1945*, 23.

16 Gumbrecht, *After 1945*, 24.

17 I am referencing Barthes's caption to Alexander Gardner's 1865 portrait of Lewis Payne in *Camera Lucida*: "He is dead and he is going to die . . . ," *Camera Lucida*, 95.

CHAPTER 35

01 "Yard Hunt Czech girl," *The Observer*, August 25, 1968.

02 "No Trace of Czech Girl," *The Guardian*, August 26, 1968; David Jenkins, "Why Nada (the Missing Czech Girl) was Working in Britain," *The Newcastle Evening Chronicle*, August 30, 1968, 12.

03 Harold Gordon Skilling, *Czechoslovakia's Interrupted Revolution* (Princeton: Princeton University Press, 1984), 717–18.

04 Translated by Mark Kramer. In Mark Kramer, "A Letter to Brezhnev: The

Czech Hardliners' 'Request' for Soviet Intervention, August 1968," *Cold War International History Project Bulletin* 2 (1992): 16.

05 Luboš Veselý, "The Ukrainian Factor of the Prague Spring? Petro Shelest and the Czechoslovak Year 1968 in the Light of Documents of the Ukrainian Security Service," *Czech Journal of Contemporary History* 8, no. 8 (2020): 99.

06 Pete Dolack, "Workers' Councils in the Prague Spring," *Socialism and Democracy* 32, no. 2 (2018): 32–55.

07 "The persecution for which you are co-responsible . . . has, of course, not stopped at matters of everyday life, at losing one's job or at what amounts to forced labour. You have limited civil rights, you organise house searches, you confiscate notes, books and manuscripts from scholars and writers, you have thrown dozens of people in jail where you still hold many of them. And you torture people by persecuting even their children, e.g., by not admitting them to schools, no matter how talented they are. I know exactly what I am talking about, Mr Secretary. I have experienced it all myself." Karel Kaplan, "An Open Letter from Karel Kaplan to Vasil Bilak, Secretary and Presidium Member of the Central Committee of the Communist Party of Czechoslovakia," *Critique: Journal of Socialist Theory* 6, no. 1 (1976): 102.

08 "Giant Hailstones Shattered Windows," *Pontypridd Observer*, July 4, 1968.

09 A. F. Pitty, "Particle Size of the Saharan Dust Which Fell in Britain in July 1968," *Nature* 220 (October 26, 1968): 364.

10 See C. M. Stevenson, "The Dust Fall and Severe Storms of 1968," *Weather* (1969): 126–32; A. S. Goudie and N. J. Middleton, "Saharan Dust Storms: Nature and Consequences," *Earth-Science Reviews* 56 (2001): 195–96.

11 Goudie and Middleton, "Saharan Dust Storms," 197.

12 The conference *Extractivism/Activism* (Autograph / Paul Mellon Centre, March 13, 2024), included two papers that I have drawn on here to challenge the view of these storms as purely natural meteorological events: FRAUD, "Undergrounding the Critical Mineral," and Nancy Demerdash, "Fuelling Foment: (Counter)colonial Histories of Phosphate Extraction in Tunisia." I am indebted to both. Phosphate mines had first opened in Morocco in the 1920s under the French colonial administration; see Simon Jackson, "The Phosphate Archipelago: Imperial Mining and Global Agriculture in French North Africa," *Jahrbuch für Wirtschaftsgeschichte* 57, no. 1 (2016): 187.

13 S. Rodríguez, A. Alastuey, S. Alonso-Pérez, X. Querol, E. Cuevas, J. Abreu-Afonso, et al., "Transport of Desert Dust Mixed with North African In-

dustrial Pollutants in the Subtropical Saharan Air Layer," *Atmospheric Chemistry and Physics* 11 (2011): 6663–85.

14 Draper, *History of the Conflict Between Religion and Science*, 133–34.

CHAPTER 36

01 Josef Koudelka, *The Black Triangle: The Foothills of the Ore Mountains; Photographs 1990–1994* (Prague: Podkrusnohri / Magnum Photos, 1994). The Black Triangle includes parts of two other Eastern European countries then under Communist rule—in Germany's southern Saxony (then in East Germany) and Poland's Lower Silesia.

02 Vaněk writes, "There was no escape from the ecological consequences of the socialist experiment with its mania for heavy industry and 'philosophy of iron.'" Miroslav Vaněk, "Twenty Years in Shades of Grey? Everyday Life During Normalisation Based on Oral History Research," in Kevin McDermott and Matthew Stibbe, *Czechoslovakia and Eastern Europe in the Era of Normalisation, 1969–1989* (London: Palgrave Macmillan, 2022), 162.

03 David Lu, "Air Pollution Regulation in the Czech Republic: Environmental Protection in the Context of Political and Economic Transition," *Wisconsin International Law Journal* 13 (1994): 565 and 570.

04 R. Tichý and V. Mejstřík, "Heavy Metal Contamination from Open-Pit Coal Mining in Europe's Black Triangle and Possible Remediation," *Environmental Reviews* 4, no. 4 (1996): 323, quote from interviewee cited in Vaněk, "Twenty Years in Shades of Grey?," 161.

05 The sixty-minute film *Black Triangle* (dir. Nick Davidson, 1990) depicts children playing on radioactive slag heaps, and radioactive sludge being piped into the lakes that supply water to homes in the region.

06 Koudelka described his early method of underexposing and overdeveloping in hot developer to push the film, creating very dense negatives, and prints with heavy black shadows. Koudelka in James Estrin, "Josef Koudelka: A Restless Eye," *The New York Times*, Lens Blog, November 20, 2013, https://archive.nytimes.com/lens.blogs.nytimes.com/2013/11/20/josef-koudelka-a-restless-eye/.

07 Vaněk, "Twenty Years in Shades of Grey?," 148.

08 Lord Byron (George Gordon), "Darkness," In *Lord Byron: The Major Works*, ed. Jerome J. McGann (Oxford: Oxford University Press, 1986).

09 Zylinska, "Photography after Extinction," 66.

10 Jean-Luc Nancy, *The Ground of the Image* (New York: Fordham University Press, 2005), 4.

11 Richard Misrach and Kate Orff, *Petrochemical America* (New York: Aperture, 2014).

12 These were the result of a collaboration with UP Projects, a public art commissioning organization, and SPACE, a London artists' studio provider.

13 As part of an exhibition at SPACE Ilford called *Changing Currents: Can We Build a Better World?* See Andrew Brown (personal website), accessed January 27, 2025, https://www.andrewjohnbrown.com/.

SELECTED BIBLIOGRAPHY

Angus, Siobhan. *Camera Geologica: An Elemental History of Photography*. Durham: Duke University Press, 2024.

Barthes, Roland. *Camera Lucida: Reflections on Photography*. Translated by Richard Howard. London: Vintage Books, 2000.

Batchen, Geoffrey. *Burning with Desire: The Conception of Photography*. Cambridge, MA: MIT Press, 1999.

Batchen, Geoffrey. *Negative/Positive: A History of Photography*. London: Routledge, 2021.

Bate, Jonathan. *The Song of the Earth*. London: Picador, 2000.

Benjamin, Walter. *Walter Benjamin: Selected Writings, Vols. 2–4*, edited by Michael W. Jennings, Howard Eiland, and Gary Smith. Cambridge, MA: Harvard University Press, 1999–2003.

Bennett, Jane. *Vibrant Matter: A Political Ecology of Things*. Durham, NC: Duke University Press, 2010.

Benson, Etienne S. *Surroundings: A History of Environments and Environmentalism*. Chicago: University of Chicago Press, 2020.

Boudia, Soraya, Angela N. H. Creager, Scott Frickel, Emmanuel Henry, Nathalie Carsten Reinhardt, and Jody A. Roberts. *Residues: Thinking through Chemical Environments*. New Brunswick, NJ: Rutgers University Press, 2022.

Chakrabarty, Dipesh. *The Climate of History in a Planetary Age*. Chicago: University of Chicago Press, 2021.

Connor, Stephen. *The Matter of Air: Science and Art of the Ethereal*. London: Reaktion Books, 2010.

Dinkar, Niharika. "'Our Best Machines Are Made of Sunlight': Photography and Technologies of Light." In *Ubiquity: Photography's Multitudes*, edited by Jacob W. Lewis and Kyle Parry. Leuven: Leuven University Press, 2021.

Doucet, Emily, Matthew C. Hunter, and Nicholas Robbins. "Editors' Introduction: The Aerial Image." *Grey Room* 83 (2021): 6–23.

Dootson, Kirsty Sinclair. *The Rainbow's Gravity: Color, Materiality and Brit-*

ish Modernity. New Haven: Paul Mellon Centre / Yale University Press, 2023.

Draaisma, Douwe. *Metaphors of Memory: A History of Ideas about the Mind*. Cambridge: Cambridge University Press, 2000.

Edgerton, David E. H. *Warfare State: Britain, 1920–1970*. Cambridge: Cambridge University Press, 2006.

Edwards, Steve. *The Making of English Photography: Allegories*. University Park, PA: Penn State University Press, 2006.

Flusser, Vilém. *Towards A Philosophy of Photography*. London: Reaktion, 2000.

Geimer, Peter. *Inadvertent Images: A History of Photographic Apparitions*. Chicago: University of Chicago Press, 2018.

Ghosh, Amitav. *The Nutmeg's Curse: Parables for a Planet in Crisis*. London: John Murray, 2021.

Guibert, Hervé. *Ghost Image*. Chicago: University of Chicago Press, 2014.

Gumbrecht, Hans Ulrich. *After 1945: Latency as Origin of the Present*. Stanford: Stanford University Press, 2013.

Haraway, Donna J. *When Species Meet*. Minneapolis: University of Minnesota Press, 2007.

Hercock, Robert J., and George A. Jones. *Silver by the Ton: The History of Ilford Limited, 1879–1979*. Maidenhead: McGraw-Hill, 1979.

Hevia, James L. "The Photography Complex: Exposing Boxer-Era China 1900–1901, Making Civilization." In *Photographies East: The Camera and Its Histories in East and Southeast Asia*, edited by R. C. Morris. Durham: Duke University Press, 2009.

Homburg, Ernst, and Elisabeth Vaupel, eds. *Hazardous Chemicals: Agents of Risk and Change, 1800–2000*. New York: Berghahn, 2019.

Jarrige, François, and Thomas Le Roux. *The Contamination of the Earth: A History of Pollutions in the Industrial Age*. Cambridge, MA: MIT Press, 2020.

Jay, Mike. *The Atmosphere of Heaven*. New Haven: Yale University Press, 2009.

Kelsey, Robin. *Photography and the Art of Chance*. Cambridge, MA: Harvard University Press, 2015.

Korola, Katerina. "The Air of Objectivity: Albert Renger-Patzsch and the Photography of Industry," *Representations* 157 (2022): 90–114.

Landau, Paul S., and Deborah Kaspin, eds. *Images and Empires: Visuality in Colonial and Postcolonial Africa*. Berkeley: University of California Press, 2002.

Latour, Bruno. *Facing Gaia: Eight Lectures on the New Climate Regime*. Cambridge: Polity, 2017.

Latour, Bruno. *We Have Never Been Modern*. Cambridge, MA: Harvard University Press, 2012.

Leslie, Esther. *Synthetic Worlds: Nature, Art and the Chemical Industry*. London: Reaktion, 2006.

Levin, Boaz, and Esther Ruelfs, eds. *Mining Photography: The Ecological Footprint of Image Production*, Leipzig: Spector Books, 2022.

Liboiron, Max. *Colonialism Is Pollution*. Durham: Duke University Press, 2021.

Malm, Andreas. *Fossil Capital: The Rise of Steam Power and the Roots of Global Warming*. London: Verso, 2016.

Martin, Pauline. *Le Flou et La Photographie: Histoire d'une Rencontre (1676–1985)*. Rennes: Presses Universitaires de Renne, 2023.

Maxwell, Richard, and Toby Miller. *Greening the Media*. Oxford: Oxford University Press, 2012.

Menely, Tobias. *Climate and the Making of Worlds: Towards a Geohistorical Poetics*. Chicago: University of Chicago Press, 2021.

Murphy, Michelle. "Alterlife and Decolonial Chemical Relations." *Cultural Anthropology* 32, no. 4 (2017): 494–503.

Ndiaye, Pap A. *Nylon and Bombs: DuPont and the March of Modern America*. Translated by Elborg Forster. Baltimore: Johns Hopkins University Press, 2007.

Purbrick, Louise. "Nitrate Traffic." In *Nitrate*. Barcelona: Museu d'Art Contemporani de Barcelona, 2014.

Quanchi, Max. *Photographing Papua: Representation, Colonial Encounters and Imaging in the Public Domain*. Newcastle-upon-Tyne: Cambridge Scholars Publishing, 2009.

Rancière, Jacques. *The Politics of Aesthetics: The Distribution of the Sensible*. Translated by Gabriel Rockhill. London: Continuum, 2004.

Sheehan, Tanya, and Andres Zervigón, eds. *Photography and Its Origins*. London: Routledge, 2014.

Sloterdijk, Peter. *Terror from the Air*. Semiotext(e), 2009.

Taylor, Jesse Oak. *The Sky of Our Manufacture. The London Fog in British Fiction from Dickens to Woolf*. Charlottesville: University of Virginia Press, 2016.

Thorsheim, Peter. *Inventing Pollution: Coal, Smoke, and Culture in Britain since 1800*. Athens, OH: Ohio University Press, 2006.

Travis, Anthony S. *The Rainbow Makers: The Origins of the Synthetic Dyestuffs Industry in Western Europe*. London: Associated University Presses, 1993.

Wilder, Kelley. "Photography and the Art of Science." *Visual Studies* 24, no. 2 (2009): 163–68.

Wilder, Kelley. "Science, Art, and the Business of Color." In *The Colors of Photography*, edited by Bettina Gockel. Berlin: De Gruyter, 2021.

INDEX

Page numbers in italics refer to figures.